Analysis of Lipid Oxidation

Editors

Afaf Kamal-Eldin
Department of Food Science
Swedish Institute of Agricultural Sciences
Uppsala, Sweden

Jan Pokorný
Department of Food Chemistry and Analysis
Faculty of Food and Biochemical Technology
Institute of Chemical Technology
Prague, Czech Republic

AOCS PRESS
Urbana, Illinois

AOCS Mission Statement

To be a global forum to promote the exchange of ideas, information, and experience, to enhance personal excellence, and to provide high standards of quality among those with a professional interest in the science and technology of fats, oils, surfactants, and related materials.

The paper used in this book is acid-free and falls within the guidelines established to ensure permanence and durability.

Library of Congress Cataloging-in-Publication Data

Analysis of lipid oxidation / editors Afaf Kamal-Eldin, Jan Pokorný.
p. cm.
Includes bibliographical references and index.
ISBN 1-893997-86-3 (alk. paper)
1. Lipids--Oxidation--Research--Methodology. I. Kamal-Eldin, Afaf.
II. Pokorný, Jan, 1928-

QP751.A456 2005
612′.01577--dc22 2005005229

Printed in the United States of America.
12 11 10 6 5 4 3

Preface

Lipid oxidation, though researched since the beginning of the 20th century, still gives no complete and satisfactory information on the composition of oxidized lipids. One important factor contributing to these gaps in our knowledge about lipid oxidation relates to the shortages in analytical methodology. Traditional analytical methods have been increasingly replaced by modern sophisticated instrumental methods, but lipid oxidation still presents a challenge in regard to its detailed mechanism, as well as its implications in the stability of biological tissues/compartments and *inter alias* human health. These shortages are very much connected to the complexity of parallel and consecutive, but overlapping, free radical-driven reactions and to the instability of a wide range of products.

Analytical methods suitable for oxidized lipids were often reviewed in the last decade, but mostly from the aspect of determination of individual oxidized lipid classes, such as peroxides, aldehydes, polar lipids or polymers. In this book, they are treated from the standpoint of types of analytical methods used. In modern lipid laboratories, analytical chemists are usually specialized to a single instrumental equipment so that the approach used in this book will be more useful than the traditional presentation. It will show, what could be achieved using the particular instrumental technique. On the contrary, for those, who are not familiar with the respective technique, and will thus be obliged to ask for help of a specialist, the book will give a basic information, what they can ask, and what they can expect from the technique.

The eleven chapters of *Analysis of Lipid Oxidation* aim to review the state-of-the-art of some of the methods currently used in studying lipid oxidation. Chapter 1 provides a short review of the primary and secondary products of lipid oxidation, as well as the problems associated with sample preparation and chemical and instrumental methods of analysis. Chapter 2 presents different volumetric methods used for the analysis of lipid hydroperoxides, free fatty acids, carbonyl oxidation products, epoxides, and residual double bonds following lipid oxidation. Chapter 3 reviews different UV-visible spectrometric methods used for the analysis of lipid radicals, hydroperoxides, and carbonyl compounds formed during the reaction. Analysis of non-volatile lipid oxidation products in different lipid matrices by high performance size-exclusion chromatography (HPSEC) is discussed in Chapter 4. Chapter 5 provides a review of the use of nuclear magnetic resonance spectroscopy (NMR) in the structural characterization of different compounds formed as a result of lipid oxidation. The analysis of intermediate radical species by electron spin resonance spectroscopy (ESR) is reviewed in Chapter 6. The use of differential scanning calorimetry (DSC) in the analysis of lipid oxidation is covered in Chapter 7, the use of chemiluminescence in Chapter 8, and the use of accelerated stability tests in Chapter 9. Different approaches used for the evaluation of the kinetics of lipid oxidation are discussed in Chapter 10. The last chapter of the book reviews the analysis of interaction products of oxidized lipids with amino acids, proteins, and carbohydrates. This book is essential for further developments in analytical methodology and hyphenated techniques, with which more understanding of the reaction kinetics, mechanism, and implications will take place.

The editors are, indeed, grateful to the authors of the different chapters for making this publication possible. We also acknowledge, with great gratitude, the professional work of the AOCS staff that put this book into this shape.

Jan Pokorny
Afaf Kamal-Eldin
February 15, 2005

Contents

Chapter 1

Lipid Oxidation Products and Methods Used for Their Analysis

Afaf Kamal-Eldin[a] and Jan Pokorný[b]

[a]Department of Food Science, Swedish Institute of Agricultural Sciences, Uppsala, Sweden; and [b]Department of Food Chemistry and Analysis, Faculty of Food and Biochemical Technology, Institute of Chemical Technology, Prague, Czech Republic

Introduction

All natural food materials contain lipid oxidation products, at least in minute amounts. They are produced by the catalytic action of enzymes or by the action of singlet oxygen in living organisms, e.g., oilseeds and animal tissues used for fats and oils processing. During the isolation from the raw material, some oxidation can occur. Lipids are further oxidized either when stored or during heating in the course of food preparation. Lipid oxidation also proceeds *in vivo* after lipid ingestion or due to the leak of endogenous or exogenous free radicals and is implicated in a number of physiologic malfunctions that might lead to disease.

The analysis of lipid oxidation products is an important task, one often encountered by lipid analytical chemists. This task is difficult because the lipid oxidation reactions are consecutive but at the same type overlapping (see below). Therefore, the analytical methods used should be selected and/or adapted to the composition and amount of lipid oxidation products. The different chapters in this book provide different methods that can be used for the analysis of different oxidation products or stages. Knowledge of these alternative methods will enable the analyst to choose those appropriate for the question at hand.

Mechanism of Lipid Oxidation

To be able to choose an appropriate analytical method, it is important to understand the complexity of the lipid oxidation reaction and the products thereof. The lipid oxidation reaction consists of the following steps: (i) formation of free lipid radicals, initiating the oxidation process; (ii) formation of hydroperoxides as primary reaction products; (iii) formation of secondary oxidation products; and (iv) formation of tertiary oxidation products. The reaction mechanism is very complex, and only a short review will be given here for information. More details can be found elsewhere (e.g., Chan 1987, Kamal-Eldin 2003).

Formation of Lipid Free Radicals and Primary Lipid Oxidation Products

Lipid oxidation is initiated by the formation of free radicals, even in the case of enzyme-catalyzed lipid oxidation. A relatively high activation energy, necessary for the formation of the first lipid free radicals, may be supplied by heat energy, natural radioactivity, singlet oxygen or other sources. Unsaturated fatty acids, especially di- and triunsaturated acids, are more easily converted into free radicals than saturated fatty acids because an atom of hydrogen is more easily abstracted from the molecule if a double bond is located on the adjacent carbon atom. Substantially lower energy is sufficient to produce free radicals from traces of lipid oxidation products, especially in the presence of transient valency metals, such as copper, iron, or manganese. Free radicals are also produced in the tissues of living organisms. The determination of the level of free radicals is important both for food scientists and in biological or medical research.

Oxygen is essentially a biradical so that it reacts with a free carbon-centered lipid radical with formation of a peroxy radical. Free radicals can be detected and investigated by electron spin resonance. Peroxyl radicals possess high energy so that they can abstract an atom of hydrogen from the lipid molecule. The peroxyl radical is thus converted into a molecule of lipid hydroperoxide, and another lipid radical is produced. The process can be repeated many times. When present in abundance, free radicals are very reactive and readily react with other free radicals forming dimers polymers. The general, often cited, scheme of lipid oxidation is presented below:

Initiation $$X^{\bullet} + LH \rightarrow L^{\bullet} + H^{\bullet} \quad [1]$$

Propagation $$L^{\bullet} + O_2 \rightarrow LOO^{\bullet} \quad [2]$$

$$LOO^{\bullet} + LH \rightarrow LOOH + L^{\bullet} \quad [3]$$

Termination $$2\,LOO^{\bullet} \rightarrow \text{nonradical products} \quad [4]$$

$$LOO^{\bullet} + L^{\bullet} \rightarrow \text{nonradical products} \quad [5]$$

$$2\,L^{\bullet} \rightarrow \text{nonradical products} \quad [6]$$

The structure of lipid hydroperoxides depends on the structure of the original fatty acid, so that a mixture of isomeric hydroperoxides is produced. Fats and oils contain many different triacylglycerols, and each fatty acid bound in a triacylglycerol is oxidized with formation of a few or several isomeric hydroperoxides. To judge the progress of the overall oxidation reaction, it is sometimes sufficient to determine the total content of hydroperoxides (see Chapter 2), but for mechanistic understanding, it is important to determine the individual hydroperoxide species in the mixture. To date, no satisfactory method is available to accomplish the later task.

Hydroperoxides present in food lipids are unsaturated, and can be oxidized by similar mechanisms into dihydroperoxides or hydroperoxides containing an additional cyclic peroxy group. Bicyclic diperoxides may also be produced from polyunsaturated fatty acids. Special methods exist for the determination of hydroperoxides in the presence of cyclic peroxides.

The rate of lipid oxidation depends on several factors. As mentioned above, polyunsaturated lipids are more rapidly oxidized than monounsaturated lipids, and saturated lipids are almost stable. The oxidation rate increases with increasing temperature, oxygen pressure, and irradiation. The oxidation is catalyzed by heavy metals, and inhibited by antioxidants. Water and various nonlipidic food components also can greatly affect the process. Lipid analysts should be aware that oxidation proceeds slowly in the case of refrigerated storage and under reduced oxygen pressure or in inert gas and that antioxidants can only reduce the rate of oxidation, not stop it completely.

Hydroperoxide Decomposition into Secondary Oxidation Products

Hydroperoxides, especially those of polyunsaturated fatty acids, are very unstable. The rate of their degradation is catalyzed by heavy metal traces, metal ions, as well as some complexes and undissociated salts. Three types of degradation products are formed:

1. Monomeric degradation products of hydroperoxides are formed by various reactions of hydroperoxides. Epoxides (oxirane derivatives) are produced by the interaction of hydroperoxides with a double bond; hydroperoxides are transformed by the reduction of the hydroperoxyl group into hydroxyl derivatives, or by dehydration into ketones. Cyclic monomeric derivatives may also occur.
2. Low-molecular-weight compounds result from the cleavage of the hydroperoxide chain, most often at a carbon atom adjacent to the hydroperoxyl group. Aldehydes, ketones, alcohols, and hydrocarbons are formed in these reactions. The rancid off-flavor of oxidized lipids is due to these volatile compounds.
3. High-molecular-weight combination products are formed in the course of polymerization of free radical degradation products of hydroperoxides, or of copolymerization of free radical decomposition products with other food components. Dimers or trimers may be aliphatic, monocyclic, or bicyclic.

Oxidized lipids contain a mixture of all of the above-mentioned secondary products. For the analysis of such complicated samples, prefractionation into simpler mixtures may be necessary before analysis by a technique such as high performance size exclusion chromatography (HPSEC). Monomeric and polymeric secondary oxidation products, which are still unsaturated, are further oxidized similarly to the original fatty acids. The oxidation also proceeds at carbon atoms adjacent to the oxygen-containing functional groups in the chain.

Further Reactions of Secondary Lipid Oxidation Products (Formation of Tertiary Oxidation Products)

The secondary lipid oxidation products are also unstable; aldehydes in particular are very reactive. They are easily oxidized into peroxoacids, which are unstable, and decompose into a mixture of other products. Unsaturated aldehydes, alcohols, or ketones are oxidized into hydroperoxides, and compounds with a shorter chain length are formed by their cleavage. Formic acid and other low-molecular-weight fatty acids are the end products that are measured by techniques such as the Rancimat or the Oxidative Stability Index. The formation of volatile fatty acids is used as an indicator in some accelerated methods for the determination of lipid stability. Aldehydes change even in the absence of oxygen by various aldolization and retroaldolization reactions.

Interactions of Lipid Oxidation Products with Other Food Components

Foods and biological tissues contain not only lipids, but also many other components that can react with lipid free radicals, hydroperoxides, aldehydes, epoxides, and other reactive oxidation products. Natural minor components of fats and oils, such as sterols, tocopherols, or other phenolic derivatives, readily react with lipid free radicals or with hydroperoxides so that oxysterols or tocopherol oxidation products are always found in oxidized lipids. Polyphenols, such as flavonoids and anthocyanins, are frequently found in foods of plant origin. They react with lipid free radicals or with hydroperoxides similarly to antioxidants. Phospholipids are present only in traces in refined oils, but their content is much higher in many foods of plant or animal origin. Because they are rich in polyunsaturated fatty acids, they are oxidized with the formation of products analogous to those of triacylglycerols. Free radicals, hydroperoxides, and aldehydes react with the nitrogen functional groups of the phospholipid molecules, forming brown-colored products. Amino acids, peptides, and proteins present in foods react with lipid hydroperoxides, epoxides, hydroxyketones, and aldehydes. Their reaction products cannot be determined by common analytical methods and require special procedures.

Problems with the Analysis of Lipid Oxidation Products

Before the analysis of oxidized lipids, it is necessary to decide whether the determination of individual species is required or whether the determination of classes of compounds is sufficient. Often, the individual species cannot be determined without previous fractionation and risk of artifact formation, whereas the determination of classes such as total peroxides or total aldehydes is much simpler. In isolated oxidized lipids, various oxidized minor components such as oxidized sterols are also present. They increase the fraction of oxidized lipids and may complicate the analysis. Usually, it is less difficult to analyze the whole mixture than to separate oxidized lipids from the oxidized minor substances and analyze each fraction separately. To achieve this goal, selective methods must be used.

Polar lipids such as monoacylglycerols, free fatty acids, or phospholipids have polarities similar to those of oxidized lipids and may interfere with their determination. Different classes of oxidation products give similar reactions, e.g., aldehydes and ketones, so that the use of specific analytical procedures is necessary. The polarities of oxidized monomers and dimers are often very similar, so that HPLC should be replaced by HPSEC.

The preliminary isolation of oxidized lipids often consists of several steps; because it is time consuming, further oxidation may take place. Use of inert gas or antioxidants will reduce the oxidation, but not eliminate it entirely. The isolation of oxidized lipids from common foods, which are mixtures of lipids with many nonlipidic components, is particularly difficult, especially in the case of those oxidized lipid fractions that are bound to proteins and similar compounds with covalent bonds. Different markers are often used instead of a complete analysis.

Instead of the determination of the oxidized products, it is possible to determine the loss of precursor substrates. The fatty acid composition is altered during the oxidation because of the different oxidation rates of individual fatty acids. The relative content of polyunsaturated acids decreases, whereas that of saturated fatty acids increases. This approach has the disadvantage that the fatty acid composition of the original, fresh sample should be known. The method is not very sensitive, because the relative increase of a saturated acid is much lower that the amount of oxidized polyunsaturated acid. The method is therefore suitable only for samples oxidized to a high degree.

No single analytical method gives reliable results. Therefore a few or several methods, based on different principles, are necessary. In the case of sophisticated instrumental methods, it is sometimes difficult to build an experienced team of operators. In the case of biological samples, such as blood plasma, the amount of sample is usually very small so that suitable microanalytical methods should be developed or modified. Some of these methods are available in the literature, but they should be adapted to the material undergoing analysis and the level of sensitivity required in most cases.

Preparation of Oxidized Lipids for the Analysis of Oxidation Products

The preparation of the sample for analysis is usually complicated, and it is better to test the procedure first using model samples, and then adapt it to the particular material. Extraction is frequently the first step. Nonpolar solvents, such as hexane or cyclohexane, used for the extraction of fresh lipids may not be suitable for oxidized lipids, and more polar solvents give better results. Polar groups of oxidized lipids are bound to proteins and other polar food components by hydrogen bonds, for example, so that mixtures containing methanol or ethanol may be necessary. In such a case, the extract is contaminated with water, and during evaporation of the solvent, some lipid components may distill off with water vapor. Most organic solvents are toxic; therefore, the selection of a suitable solvent or solvent mixture is difficult because harmless solvents are often less efficient. Liquid carbon dioxide may be considered in future developments.

The solvent is removed from the lipid extract at high temperature, low pressure, or a combination of both principles. Lipids are not very volatile, but some oxidation products formed by the cleavage of hydroperoxides could be lost during the evaporation. If the evaporation is carried out in air, the extracted lipids could be oxidized. It may be preferable to use the extract without solvent removal for the analyses and to determine the weight of extracted lipids separately in an aliquot volume of the extract. If absolutely required, removal of solvents should be performed in an atmosphere of an inert gas, i.e., nitrogen or argon.

A mixture of original and oxidized lipids is obtained by extraction, so that the extract is purified mainly on prepacked silica gel or alumina. The unoxidized lipids are eluted with a nonpolar solvent, and the oxidized fraction is then eluted using a more polar solvent mixture. A risk always exists that some very polar or polymeric oxidation products will remain in the column. after the removal of nonpolar lipids. It is possible to fractionate the oxidized lipid fraction by column chromatography on various solid phases. It is also possible to use selective membranes or molecular sieves, but these methods are seldom used for the separation of oxidized lipids.

Chemical and Instrumental Methods of Analysis of Lipid Oxidation Products

After isolation of the fraction of oxidized lipids, they are analyzed by chemical, physical, or instrumental methods. The chemical methods are the oldest; they are very simple and relatively selective. Their disadvantage is the use of large amounts of organic solvents. Volumetric and colorimetric methods are commonly used, whereas gravimetric and electrometric methods are applied only rarely for the analysis of oxidized lipids.

Physical methods, such as the determination of viscosity, specific density, refractometry, dielectrical constant, optical rotation, for example, are suitable only for orientation or monitoring of a process. The conductivity measurements are used in some accelerated tests of lipid stability against oxidation.

Instrumental methods are progressive methods, used increasingly in recent years. Instruments necessary for the analysis are expensive, but the analysis itself is short and relatively inexpensive. The selectivity can be improved by derivatization. Spectral, calorimetric, and chromatographic methods will be discussed in detail in different chapters of this book.

For the analysis of oxidized lipids, it is better to combine several methods, e.g., chemical, chromatographic, and spectrometric methods. Because most operators are experts in only a single instrumental method, it is better to form a team of analysts. The work should be coordinated so that the analyses are conducted at the same time or within short time intervals. It is important that a specialist in the analysis of oxidized lipids be present.

Concluding Remarks

The analysis of oxidized lipids is a difficult task because the material, which is a very complex mixture of different compounds, is unstable during storage and analytical operations. The best procedure is to use at least three analytical methods based on different principles. The interpretation of results requires a scientist with long experience in lipid analysis. Different techniques that can be used for the analysis of different products are discussed in the ensuing chapters of this book.

References

Chan, H.W.-S., ed. (1987) *Autoxidation of Unsaturated Lipids*, Academic Press Inc., London.

Kamal-Eldin, A., ed. (2003) *Lipid Oxidation Pathways*, AOCS Press, Champaign, IL.

Chapter 2

Volumetric Analysis of Oxidized Lipids

Jan Pokorný

Department of Food Chemistry and Analysis, Faculty of Food and Biochemical Technology, Institute of Chemical Technology, Prague, Czech Republic

Introduction

Chemical methods were the first analytical methods used for the estimation of lipid oxidation products. Volumetric methods, i.e., methods based on titration, were proposed more than 50 years ago because they are very simple, rapid methods that require no specific equipment. Their disadvantage is that they require the use of organic solvents and other toxic chemicals. Most volumetric methods have been replaced by instrumental methods, but some of them are widely used even now.

Volumetric methods were developed and standardized several decades ago, particularly for fresh fats and oils; they have changed little since that time. Therefore, most of the references cited in this chapter are very old. Nevertheless, their discussion is useful even now because most lipid scientists and technologists, who use them for the analysis of oxidized lipids, are not familiar with their limitations, the effect of various factors on the results, or the environmental aspects.

Iodometric Determination of Lipid Hydroperoxides

Principle of the Reaction

Hydroperoxides belong to the most important primary oxidation products. The hydroperoxide group is located on the carbon atom adjacent to a double bond or to a system of two conjugated double bonds. Hydroperoxides can be still further oxidized because they are unsaturated compounds. Dihydroperoxides or cyclic peroxides are formed in the later oxidation stages. All of these peroxidic compounds react with iodide ions. They are reduced to hydroxy derivatives, whereas iodide is oxidized into free iodine (Fig. 2.1A). In the presence of excess iodide, a complex ion that reacts in the same way as free iodine is formed (Fig. 2.1B). Iodine is then titrated, usually with a solution of sodium thiosulfate, which is oxidized into a tetrathionate (Fig. 2.1C). Reducing agents other than sodium thiosulfate may be used for the titration, e.g., sodium arsenite, which would be preferable but is unfortunately much more toxic than thiosulfate. A starch solution is added as an indicator because it forms a deep violet product with iodine. At the end of titration, the reaction mixture is decolorized.

(A) reaction of hydroperoxides with iodide ions
$$R\text{-}OOH + 2\,I^- + 2\,H^+ \rightarrow R\text{-}OH + I_2 + H_2O$$

(B) complex formation of iodine with iodide ions
$$I_2 + I^- \rightarrow I_3^-$$

(C) reduction of free iodine with thiosulfate
$$I_2 + 2\,S_2O_3^{2-} \rightarrow S_4O_6^{2-} + 2\,I^-$$

Fig. 2.1. Mechanism of iodometric determination of the peroxide value.

Analytical Procedures and the Effect of Air Oxygen

The result of the analysis is called the peroxide value (PV) for historical reasons, but the hydroperoxide value would be a more appropriate term because only lipid hydroperoxides react quantitatively, and the other peroxides only partially because of their low reactivity. Fortunately, hydroperoxides are present almost exclusively in fats and oils at a low degree of oxidation.

The sample is dissolved in a mixture of chloroform and acetic acid (or another solvent mixture of a similar polarity), and a saturated aqueous potassium iodide solution is added. The reaction takes place in the dark or at least under diffuse daylight. After the reaction, water is added, and the liberated iodine is titrated with a solution of sodium thiosulfate. Near the end point, when the colour of dissolved iodine becomes faint, the starch solution is added, and the titration is finished. The solution should be shaken only gently; otherwise, the reaction mixture would be contaminated with air oxygen. The resulting PV would then be higher than the actual value.

Several slightly diferent procedures are available for the determination of peroxides; those proposed by Wheeler (1932), Lea (1952), and Sully (1954) are the basis of further development of procedures that are now standardized. Amer *et al.* (1961) compared three methods of iodometric PV determination, and observed great differences among the results. These methods differed in the reaction time, temperature, and the presence of inert gas. The factors affecting the results are reviewed in Table 2.1. If the peroxide content in a sample is low, the reaction time has no great effect, and determination at ambient temperature is satisfactory. A reaction time of 5 min at ambient temperature was selected as the optimum (Yanishlieva and Popov 1972), but only 1 min is generally sufficient in the analysis of fresh fats and oils.

The presence of oxygen leads to an overestimation of the PV and is an important factor, especially in the case of low PV. Thus, it is preferable to remove oxygen from the reaction vessel by a stream of nitrogen or carbon dioxide before the analysis. The gas exchange takes rather a long time, e.g., a level of 5% oxygen in the atmosphere is attained after >7 min in a 300-mL flask with a flow rate of nitrogen >100 mL/min (Kubota *et al.* 1974). The gases should be free of oxygen

TABLE 2.1
Factors Affecting the Peroxide Value (PV)

Factor	Effect on the PV
Time	Increasing
Temperature	Increasing
Shaking	Increasing
Oxygen	Increasing
Inert gas (nitrogen, carbon dioxide)	Decreasing
Increasing pH value	Decreasing
Water	Increasing
Light	Increasing
UV radiation	Increasing
Polyunsaturation	Decreasing
Cupric and ferric ions	Increasing

because traces of oxygen could affect the results. Iodide ions are stabilized against oxidation by air oxygen by the addition of cadmium salts (Takagi et al. 1978), but cadmium is a toxic metal. If the blank is carried out in the same way and the difference between the sample and the blank is recorded, the inert gas is often omitted. In the case of a 1-min reaction time, the introduction of an inert gas is unnecessary. Water added after the end of the reaction also must be free of oxygen and trace metal ions. In samples with a high PV, complete elimination of oxygen is not so crucial as in fresh samples. The PV rises by the action of other oxidants, such as ferric ions (Gutfinger *et al.* 1976). Wheeler's procedure was found suitable for the analysis of dry soap (Popov and Yanishlieva 1968), in which 0.1–1.0 g of sample was dissolved directly into the solvent mixture.

According to the IUPAC standard procedure (Paquot and Hautfenne 1987), the reaction time is 5 min in diffuse daylight at ambient temperature, and the inert gas is not required. According to the AOCS standard procedure (Firestone 1996), the reaction time is 1 min at ambient temperature and under diffuse daylight, and the inert gas is also omitted. Isooctane can be used instead of chloroform. The standard method as proposed by Wheeler was modified to a micromethod that requires <0.1 g sample and a 2-min reaction time (Yanishlieva *et al.* 1978).

Lipid hydroperoxides may partially polymerize during the reaction, leading to the formation of less reactive products. This side reaction becomes important at high sample PV. Hydroperoxides may be stabilized by the addition of boric acid so that the PV is higher and corresponds better to the real concentration of the hydroperoxides than if the standard method is used (Amer *et al.* 1961). The authors suspected that a part of the iodine, formed by the reaction of hydroperoxides, was reabsorbed on double bonds of the analyzed sample, thus reducing the PV (Amer *et al.* 1960).

Chloroform and acetic acid are usually used as solvents. Chloroform is a good lipid solvent, and the addition of acetic acid is important as a medium suitable for the interaction of the reactants. For the analysis of biological samples, chloroform can be replaced by the Folch reagent (chloroform and methanol, 1:1, vol/vol) so that the extract can be used immediately as the reaction medium. The use of hydrocarbon solvents that are less toxic than halogen solvents, such as isooctane, was proposed for PV determination. The disadvantage of isooctane is that a fine emulsion is obtained during the titration, making determination of the end point very difficult and less accurate.

The end of the titration can be determined visually in the presence of starch only with difficulty, especially in darker samples. The end of titration is more precisely measured using the potentiometric indication (Szumilak and Gudaszewski 1985). The electrometric determination gave different results from those of the standard titration procedure (Bogs 1975). The potentiometric determination was modified to allow the use of very small samples (Hara *et al.* 1982), such as 3 mL of human serum (Hara *et al.* 1985). The titration can be replaced by coulometric reduction of free iodine (Fiedler 1974), and the results are in good agreement with those of the titrimetric determination.

Effect of the Reaction Time and the Reaction Temperature

For the analysis of samples oxidized to high PV at ambient temperature, the reaction time should be longer. If samples of oil heated to a high temperature are analyzed, e.g., frying oils, it is necessary to increase the reaction temperature during the analysis; otherwise the reaction will be incomplete. For example, in the case of sunflower oil oxidized at cooking or frying temperatures, 30 min of reaction time at 70°C was necessary to reach a constant PV (Pokorný and Čmolík, 1961). The peroxidic fraction in samples at an advanced stage of oxidation comprises a mixture of peroxides of different reactivities. The more stable peroxides do not react with potassium iodide after the standard procedure, but react with hydriodic acid, determined after refluxing the reaction mixture for 7–10 min (Said *et al.* 1964). Very small amounts of stable peroxides are also present in fresh fats.

Other Methods for Hydroperoxide Determination

Several other methods exist for the determination of hydroperoxides. The thiocyanate method and other spectrophotometric methods are discussed in Chapter 3. Hydroperoxides can be determined by many other instrumental procedures, either directly or after derivatization.

How to Express the Peroxide Value

The results (termed the peroxide value, in spite of the presence of hydroperoxides as major components) can be expressed in three ways, which is confusing. They are most often expressed in milliequivalents of active oxygen per kg lipids (mEq/kg or μEq/g). However, according to the IUPAC/IUB recommendations, this unit should not be used; the unit millimole of active oxygen per kg lipids (mmol/kg) was proposed. The PV expressed in this unit = 50% of the previous PV. After another proposal, supported by IUPAC Fat and Oil Standardization Committee, the content of active oxygen was expressed in mg per kg lipids (or μg/g); the PV expressed in this unit is 8 times higher than the PV in mEq/kg, and 16 times higher than the PV expressed in mmol/kg. Therefore, in the determination of PV, the unit should always be given, even if mEq/kg is now used almost exclusively. It should also be remembered that in the case of low PV, the coefficient of variation (the confidence coefficient) is ~5% of the result. The error is high in deeply oxidized oils or in the case of oil heated to a high temperature.

Determination of Free Fatty Acids

Mechanism of Free Fatty Acid Production

During lipid oxidation, short-chain fatty acids are produced by secondary oxidation of unsaturated aldehydes and other products originating from the cleavage of lipid hydroperoxides. Formic acid is present in particularly high amounts. More fatty acids may also be formed by the hydrolytical cleavage of triacylglycerols, e.g., under conditions of deep fat frying. Therefore, the free fatty acid content is an important marker of frying oil degradation.

Procedure of Fatty Acid Determination

The procedure to determine free fatty acids in oxidized fats and oils is the same as in the case of fresh samples. The sample is dissolved in a nonpolar or a medium polar solvent, and neutralized ethanol or methanol is added to enable rapid neutralization of the fatty acids. They are determined by titration with a solution of potassium hydroxide. For environmental reasons, nonpolar solvents are often omitted. Phenolphthalein is often used as an indicator, but in the case of deep brown samples, e.g., frying oils, thymolphthalein or another suitable indicator may be used. The result is expressed in mg KOH necessary for the neutralization of free fatty acids in 1 g of sample.

The presence of phospholipids and some synergists would interfere with the reaction because they also contain acidic groups. The acid value, found in oxidized lipids, involves both the free fatty acids formed by hydrolysis and the volatile acids formed in the later stages of autoxidation. Gas chromatographic analysis of the respective methyl esters is necessary to distinguish between the two groups of fatty acids.

Determination of Carbonyl Oxidation Products Using an Acidimetric Titration

Aldehydes and ketones, formed as secondary oxidation products, contain carbonyl groups. Hydroxylamine hydrochloride solution in 90% aqueous ethanol is used as a reactant. Its solution should be neutralized by titration with potassium hydroxide,

$R_2C{=}O + H_2NO{\cdot}HClH \rightarrow R_2C{=}N\text{-}OH + HCl$
ketones

$R\text{-}C(H){=}O + H_2NO{\cdot}HClH \rightarrow R\text{-}C(H){=}N\text{-}OH + HCl$
aldehydes

Fig. 2.2. Determination of carbonylic products with hydroxylamine hydrochloride.

using bromophenol blue or bromocresol green as indicators. The hydroxylamine hydrochloride solution is then added to a solution of oxidized lipids, and the reaction mixture is shaken for 30 min. The reaction takes place (Fig. 2.2), liberating a molecule of hydrochloric acid in an amount equivalent to a carbonyl group. The reaction mixture is then titrated with a solution of potassium hydroxide to the same hue as that of the reactant solution before the analysis. The results may be expressed in mmol/kg or any other unit suitable for the aim of research. The method is rapid and relatively easy, but the titration is not very sensitive and is suitable only for samples oxidized to a high degree. Moreover, the volumetric methods require large amounts of organic solvents. Spectrophotometric methods (Chapter 3) would be preferred in most samples because they are much more sensitive.

Determination of Epoxides Using an Acidimetric Titration

Epoxy (oxirane) derivatives are formed by reaction of a hydroperoxide group with a double bond or via several other pathways. Epoxides are very often neglected among the oxidation products by lipid scientists, but they are very important because of their high reactivity. Epoxides can be determined on the basis of their reaction with hydrochloric acid (Fig. 2.3). An exact amount of a solution of anhydrous hydrochloric acid in anhydrous diethyl ether is titrated with a standard potassium hydroxide solution using phenolphthalein as an indicator. The same amount of the hydrochloric acid solution is added to the sample solution. After a 2-h reaction, the reaction mixture is titrated again, and the loss of acidity is equivalent to the content of hydrochloric acid bound to epoxides (Swern *et al.* 1947; Knight, Swern 1949). Lipid hydroperoxides do not react under the reaction conditions. The content of epoxides is expressed in mmol epoxide groups per kg of the sample.

R^1-CH-CH-R^2 + HCl → R^1-CH-CH-R^2
\ / | |
O OH Cl

R^1-CH-CH-R^2 + HBr → R^1-CH-CH-R^2
\ / | |
O OH Br

Fig. 2.3. Reaction of an epoxide (oxirane) group with hydrochloric or hydrobromic acid.

Because the use of anhydrous hydrochloric acid in diethyl ether is somewhat incovenient, a modification was proposed (King 1949) using a solution of concentrated aqueous hydrochloric acid in dioxane; the reaction time is reduced to only 10 min. Phenolphthalein or thymolphthalein can be used as indicators. A micromethod was suggested (Chakrabarty *et al.* 1970), consisting of titration with perchloric acid dissolved in acetic acid. In another modification (Durbetaki 1956; Masa, Vioque 1975), the sample is titrated directly with a solution of hydrobromic acid dissolved in acetic acid. The method was recommended by the AOCS (Firestone 1996). The methods of the epoxide determination are very easy and simple, but large amounts of organic solvents are necessary Now, epoxides are usually determined using various instrumental methods.

Determination of Double Bonds in Oxidized Lipids Using the Iodine Value

In the past, the content of double bonds in a fat or oil sample was evaluated almost exclusively with the use of the iodine value (IV) determination. The IV is the amount of halogen (expressed as iodine) absorbed, under conditions of the method, by 1 g of sample, and is expressed as a percentage of the sample weight. The IV is still sometimes used for monitoring lipid oxidation.

Several methods for determining the IV have been standardized, and the Wijs and Hanus methods are most widely used. In the Wijs method, iodine trichloride (ICl_3) is used as a halogenation agent, whereas in the Hanus method, iodine monobromide (IBr) is the halogenation agent. The halogenation agent is added in excess, and the unreacted halogen is converted into free iodine by reaction with a potassium iodide solution. Free iodine is titrated with a sodium thiosulfate solution, and the content of bound halogen is then calculated by comparing the free iodine content in the blank and after the reaction.

The sample of fat or oil is dissolved in chloroform or another suitable solvent. Because chloroform is toxic, cyclohexane may replace it, or it is omitted completely in some newer modifications in which only acetic acid is used as a solvent. The AOCS recommends cyclohexane as an alternative (Firestone 1996). The halogenation agent dissolved in acetic acid is then added, and the resulting reaction mixture is stored in the dark. An aqueous potassium iodide solution is then added. Free halogen is converted into free iodine. Excess water is added, and iodine is titrated with sodium thiosulfate solution, using starch as an indicator.

The method is often used for the analysis of fresh fat and oil samples, and both methods give good results. However, many lipid scientists use the same method for the analysis of oxidized oil. If the degree of oxidation of the sample is negligible, the common standard methods are applicable. The standard method is not satisfactory, however, for the analysis of oxidized samples. Hydroperoxides and some of their degradation products contain conjugated double bonds, and the IV does not give quantitative results in such a case. Another method was proposed for such cases, namely, the Woburn method (Von Mikusch and Frazier 1941), using a stronger solution of IBr, a greater excess of monobromiodide, and a longer reaction time. Other oxidation products also give erronous results and require the same more intensive halogenation. The presence of a hydroperoxide or another oxygen-containing group on a carbon atom adjacent to the double bond also causes moderately lower results. The difference between the standard procedure and the Woburn method may be ≥10% in samples at advanced stages of oxidation (Pokorný 1958).

The results are expressed in the same way in the case of oxidized lipids as for the standard procedure, except that the coefficient of variation (confidence interval) is higher. If such a procedure is used, it should specifically mentioned, together with other details of the procedure, which may be useful. The method requires large amounts of organic solvents, which carry a greater health risk; this should be taken into consideration when planning the analyses.

Concluding Remarks

Volumetric methods are very old, but the iodometric determination of the PV and the alkalimetric determination of the free fatty acid content (the acid value) are still useful and are recommended. Their disadvantage is the need to use large amounts of organic solvents. The modern alternatives using safer solvents are less accurate, and the titration is more difficult than with traditional standard procedures. The operator should decide which procedure to use. The other methods mentioned in this chapter would be better replaced by spectrophotometric or other instrumental methods.

References

Amer, M.M., Said, F., and Sayed Ahmad, A.K. (1960) A New Iodometric Method for the Determination of the Peroxide Value of Oils and Fats, *Egypt. Pharm. Bull. 42*, 271–276.

Amer, M.M., Sayed Ahmed, A.K., and Said, F. (1961) The Iodometric Determination of the Peroxide Value of Rancid Oils and Fats, *U.A.R.J. Pharm. Sci. 2*, 1–11.

Bogs, U. (1975) Elektrometrische Bestimmung der Peroxidzahl, *Pharmazie 30*, 5–6.

Chakrabarty, M.M., Bhattacharyya, D., and Kundu, M.K. (1970) A Micro-Titrimetric Method for the Determination of the Oxirane Functional Group, *Analyst 95*, 85–87.

Durbetaki, A.J. (1956) Direct Potentiometric Titration of Oxirane Oxygen by Hydrogen Chloride-Acetic Acid, *J. Am. Oil Chem. Soc. 33*, 221–223.

Fiedler, U. (1974) A Coulometric Method for the Determination of Low Peroxide Values of Fats and Oils, *J. Am. Oil Chem. Soc. 51*, 101–103.

Firestone, D., ed., (1996) *Official Methods and Recommended Practices of the American Oil Chemists' Society*, 4th ed., 3rd printing, AOCS Press, Champaign, IL.

Gutfinger, T., Peled, M., and Letan, A. (1976) Iodometric Determination of the Peroxide Value in Edible Oils, *J. Assoc. Off. Agric. Chem. 59*, 148–152.

Hara, S., Hasegawa, S., Suzuki, H., and Totani, Y. (1985) Potentiometric Determination of Low Peroxide Value of Lipids. III. Determination of Lipid Peroxides in Human Serum, *Yukagaku 34*, 263–287.

Hara, S., Washizu, O., and Totani, Y. (1982) Potentiometric Determination of Low Peroxide Values of Lipids, I. Detection Limit of Peroxides, *Yukagaku 31*, 1004–1008.

King, G. (1949) Estimation of Epoxides, *Nature 164*, 706–707.

Knight, H.B., and Swern, D. (1949) Reaction of Fatty Materials with Oxygen. IV. Determination of Functional Groups, *J. Am. Oil Chem. Soc. 26*, 366–370.

Kubota, Y., Mamuro, H., Kato, A., and Hashimoto, T. (1974) The Replacement Ratio of Air with Nitrogen under the Condition of Peroxide Value Measurement, *Yukagaku 23*, 114–116.

Lea, C.H. (1952) Methods for Determining Peroxides in Lipids, *J. Sci. Food Agric. 3*, 586–594.

Masa, M.P., and Vioque, E. (1975) Microdeterminación de Epoxiácidos en Aceites, *Grasas Aceites 26*, 78–83.

Paquot, C., and Hautfenne, A. (1987) *Standard Methods for the Analysis of Oils, Fats and Derivatives*, 7th ed., Blackwell, Oxford, pp. 199–200.

Pokorný, J. (1958) Analytical Investigation of Changes of Vegetable Oil During Oxidation, Sb. VŠCHT Praze, *Potrav. Technol. 2*, 181–219.

Pokorný, J., and Čmolík, J. (1961) Kinetics of the Reaction of Peroxides with Potassium Iodide, Sb. VŠCHT Praze, *Potrav. Technol. 5*, 163–176.

Popov, A., and Yanishlieva, N. (1968) Méthode Rapide de Détermination de la Stabilité des Savons, *Rev. Fr. Corps Gras 15*, 215–218.

Said, F., Amer, M.M., and Ahmad, A.K. (1964) The Determination of Both the Total and Labile Peroxide Value, *Fette Seifen Anstrichm. 66*, 1000–1006.

Sully, B.D. (1954) A Modified Iodometric Determination of Organic Peroxides, *Analyst 79*, 86–90.

Swern, D., Findley, T.W., Billen, G.N., and Scanlan, J.T. (1947) Determination of Oxirane Oxygen, *Anal. Chem. 19*, 414–415.

Szumilak, K., and Gudaszewski, T. (1985) Próba Udokladnienia Nadtlenków w Tłuszczu, *Tluszcze Jadalne 23*, 1–6.

Takagi, T., Mitsuno, Y., and Masumura, M. (1978) Determination of Peroxide Value by the Colorimetric Iodine Method with Protection of Iodide as Cadmium Complex, *Lipids 13*, 147–151.

Von Mikusch, J.D., and Frazier, C. (1941) Woburn Iodine Absorption Method. Measure of Total Unsaturation in the Presence of Conjugated Double Bonds, *Ind. Eng. Chem. Anal. Ed. 13*, 782–789.

Wheeler, D.H. (1932) Peroxide Formation as a Measure of Autoxidative Deterioration, *Oil Soap 9*, 89–97.

Yanishlieva, N., and Popov, A. (1972) Statistische Auswertung einer modifizierten Methode zur Bestimmung der Peroxidzahl von Lipiden, *Nahrung 16*, 121–122.

Yanishlieva, N.V., Popov, A., and Marinova, E.M. (1978) Eine modifizierte jodometrische Methode zur Bestimmung der Peroxidzahl in kleinen Lipidproben, *Dokl. Akad. Nauk Bulg. 31*, 869–872.

Chapter 3

Ultraviolet-Visible Spectrophotometry in the Analysis of Lipid Oxidation

Jan Pokorný[a], Štefan Schmidt[b], and Jana Parkányiová[a]

[a]Faculty of Food and Biochemical Technology, Institute of Chemical Technology, Prague, Czech Republic; and [b]Faculty of Chemical and Food Technology, Slovakian Technical University, Bratislava, Slovakia

Introduction

Spectrophotometric methods are among the oldest techniques for the analysis of oxidized lipids. These methods were developed several decades ago and have been used widely without substantial change since that time. Therefore, some very old references are cited. Spectrophotometric methods have the advantages of being simple, reproducible, and fast; the apparatus is relatively cheap, and various factors affecting the results have been thoroughly studied and are well known. The main disadvantage is that they are not sufficiently specific so that their combination with a preliminary chromatographic separation is sometimes necessary. Nevertheless, they can be used for a rapid control of the degree of lipid oxidation, as well as for a study of changes during oxidation under well-defined conditions. Standard procedures are available for most spectrophotometric methods, including data on the variance of results and the effect of interfering factors.

Compounds showing an absorption maximum in the ultraviolet (UV) region usually contain one or several conjugated C=C, C=O, or C=N double bonds. The position of the absorption maximum is shifted to higher wavelengths when the number of conjugated double bonds increases. In the advanced stages of lipid oxidation, the absorption maximum can be shifted even to the visible region. A group of methods, based on this phenomenon, is used in studies of lipid oxidation products especially as a way to monitor the progress of the reaction.

Formation of Active Compounds During Lipid Oxidation

Lipid oxidation starts with the formation of lipid free radicals, which can be studied by electron spin resonance, but also with the use of spectrophotometric methods. The primary lipid oxidation products, i.e., hydroperoxides, are studied with various spectrophotometric methods, even if the iodometric determination is more common. In some cases, the destruction of a colored compound, e.g. β-carotene, by hydroperoxides or their degradation products may be analyzed.

The most important group of spectrophotometric methods for the analysis of lipid oxidation, however, is that used to determine carbonyl oxidation products, especially aldehydes and ketones. These compounds react with suitable reagents producing colored products; sometimes, however, the colored product may be formed by the reaction of the reagent with a precursor of a carbonyl product. Colored compounds or natural pigments, such as carotenes or chlorophylls, may interfere with the analytical reaction, especially in the beginning of oxidation, when the concentration of the oxidation products is low. Purification of the material before the analysis is then recommended.

Spectrophotometry in the UV Region

Edible fats and oils do not absorb UV radiation at wavelengths higher than ~210 nm because they do not contain any conjugated dienoic systems or only traces. During the autoxidation, free lipid radicals are formed, and the original pentadienoic double bond systems are shifted into two isomeric derivatives containing two conjugated double bonds (Fig. 3.1). The conjugated dienoic compounds, dissolved in isooctane, absorb UV radiation with the maximum at 233 nm (AOCS 1997, Paquot and Hautfenne 1985, Yanishlieva and Popov 1973). The conjugated trienoic systems show maximum absorbance at 262 nm. With a further increase in conjugated double bonds, the maximum is shifted to still longer wavelengths (~30 nm/double bond). The advantage of direct UV absorption is that the reaction proceeds at ambient temperature. At higher temperatures, required by other methods, there is always a danger that hydroperoxides will decompose during the analysis, leading to the formation of carbonyl and other derivatives. The UV absorption is proportional to a peroxide value (PV) of up to ~200 mEq/kg in polyunsaturated oils oxidized at ambient temperature with free access to oxygen. The method can be used for the determination of conjugated fatty acids even in nonoxidized oils, and has been published as a standard method for the determination of conjugated dienoic and trienoic fatty acids by IUPAC (Paquot and Hautfenne 1985). Because the measurement could be affected by the presence of a hydroperoxide functional

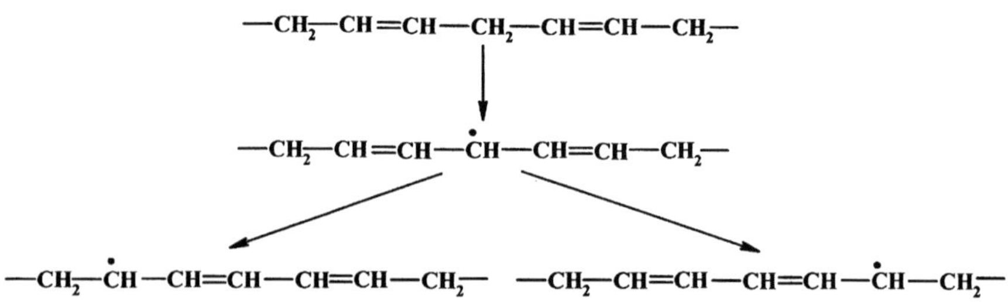

Fig. 3.1. Formation of conjugated double bond systems during the oxidation of polyunsaturated lipids.

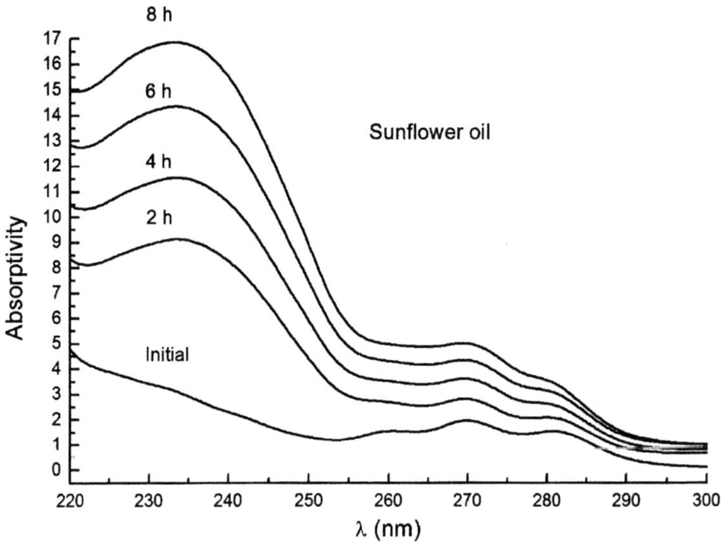

Fig. 3.2. Changes in ultraviolet spectra during oxidation of sunflower oil (*Source:* Š. Schmidt, unpublished results).

group, it was suggested to first convert the hydroperoxide group into a hydroxyl group by reaction with sodium borohydride (Fishwick and Swoboda 1977).

UV absorption is a suitable method for monitoring the course of lipid oxidation, particularly in those cases in which the peroxides formed in oxidizing lipids are very unstable so that conjugated diene formation begins to be a more reliable method for monitoring the course of oxidation than the PV. Absorptivity in the UV region between 220 and 320 nm can be measured following a standard IUPAC procedure (Paquot and Hautfenne 1987). Changes in absorptivity in the UV region were published and discussed by Sedláček between 1964 and 1972, and were correlated with the degree of sensory rancidity. Examples of oxidizing soybean and sunflower seed oils are shown (Sedláček 1968). A few other examples on the course of increasing absorption in vegetable oils due to lipid oxidation products having conjugated double bond systems are shown for sunflower oil (Fig. 3.2) and rapeseed oil (Fig. 3.3).

It is evident that other absorption maxima may appear at higher wavelengths in addition to the maximum of the –CH=CH-CH=CH- system located at 233 nm. The specific absorption coefficient at 233 nm corrected for absorption due to acid or ester groups (Paquot and Hautfenne 1985) is given by the formula $a = a_{233} - a_0$ where a_0 is 0.07 for esters, and 0.03 for soaps and fatty acids. The specific absorption coefficient for conjugated trienoic double bonds at 268 nm corrected for background absorption is given by the formula $a_3 = 2.8\ [a_{268} - 0.5(a_{262} + a_{274})] - a_0$ where a_0 is 0.07 for esters, and 0.03 for soaps and fatty acids. Other maxima appear due to the absorption of –CH=CH- bonds conjugated with carbonyl double bonds (Fig. 3.4) with an absorp-

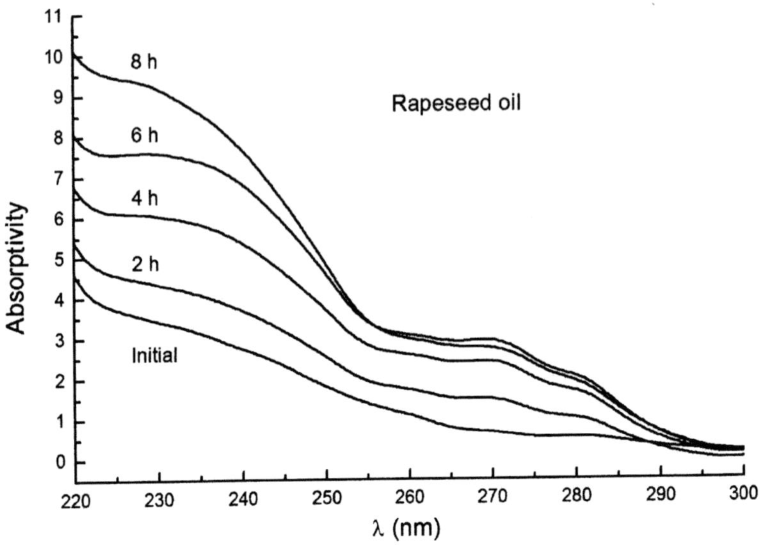

Fig. 3.3. Changes in ultraviolet spectra during oxidation of rapeseed oil (*Source:* Š. Schmidt, unpublished results).

tion maximum at 245 nm. In more extended double bond systems, the maximum is shifted further to the visible (Vis) region. It is well known that polyunsaturated oils become darker on oxidation.

The absorption peaks of the different conjugated double bond systems cannot be well differentiated and quantified by common UV spectrophotometry; nevertheless, UV absorption remains a useful method for monitoring the course of oxidation, not only for anhydrous fats and oils, but also for aqueous dispersions (Vossen *et al.* 1993) or chicken meat (Grau *et al.* 2000). The difference measurements are particularly sensitive.

Spectrophotometry in the Visible Region

As explained above, the absorption maximum of oxidized lipids depends on the number of conjugated C=C and C=O double bonds. In polyunsaturated edible oils, the

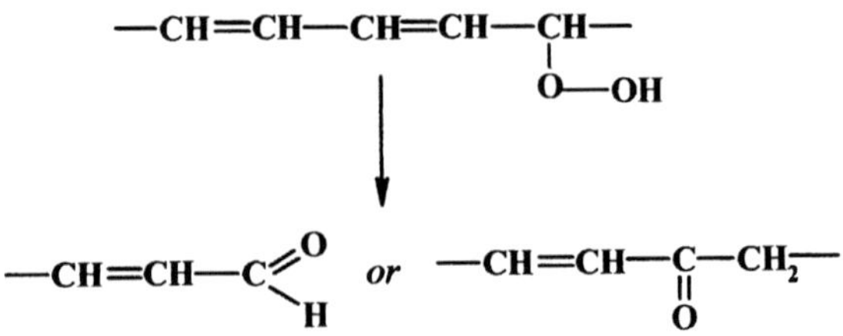

Fig. 3.4. Formation of conjugated systems containing C=C and C=O double bonds.

absorption due to conjugated double bond systems may be shifted to the visible region in advanced stages of oxidation, especially in oils containing linolenic, arachidonic, or still more unsaturated acids. The position of the maximum is still very close to the UV region. The color is deeper in lipids containing nitrogen, e.g., phospholipids or those contaminating proteins because the C=N double bonds stimulate the shift to longer wavelengths than the C=O bonds. The spectrophotometry in the visible region is important for the analysis of frying oils, but the brown color of frying oils is due mainly to tiny particles of burned fried material. Liquid oil, obtained after the removal of solid particles, is still deep brown, due to the presence of reaction products containing both oxygen and nitrogen (Koga *et al.* 1997). The formation of brown products by reactions of oxidized lipids with amino acids and proteins (often incorrectly called Maillard reactions) will be discussed in more detail in a later chapter. The determination of color changes is one of the important markers of frying oil quality.

Determination of Free Lipid Radicals Using Spectrophotometric Methods

Free lipid peroxy or alkoxy radicals can be determined on the basis of reactions resulting in the formation of colored, often fluorescent compounds. Luminol-enhanced chemiluminescence was used for the determination of alkylperoxy radicals in biological dispersions (Sawa *et al.* 1999). Phycoerythrin is highly reactive with free radicals. The phycoerythrin damage with peroxy free radicals resulting in color changes occurs 60 times more quickly than with other proteins, but more slowly than with ascorbic acid; however, the procedure is used mainly for materials of animal origin (DeLange and Glazer 1989). Another approach is to measure the decrease in coloration of a colored compound by reaction with free radicals. Alkyl, alkoxy, and peroxy radicals produced by lipid oxidation react with β-carotene, and the degree of bleaching is measured (Mortensen and Skibsted 1998). Fluorescence measurements, however, are more sensitive than color measurements and are therefore used much more frequently, especially in biological material. The most progressive methods for studying lipid free radicals are those based on electron spin resonance.

Determination of Hydroperoxides Using Spectrophotometric Methods

Lipid hydroperoxides are most often determined iodometrically by reaction with potassium iodide, and by titration of liberated iodine. It is also possible to determine free iodine spectrophotometrically (Ishimatsu *et al.* 1969). The color intensity is more reproducible, if iodine is first transformed into its cadmium complex (Takagi *et al.* 1978), but the toxicity of cadmium salts is a disadvantage. After another modification, aluminium chloride is added as a catalyst, and starch solution is added as a reagent; the violet coloration is then measured at 560 nm (Asakawa and Matsushita 1980). In the case of a low concentration of lipid hydroperoxides, the amount of liberated iodine is very low, and the results are easily influenced by traces of oxygen. Therefore, ferrous ions are added to the reaction medium to protect iodide against oxygen. The sharp peak of I_3^- ions, formed by reaction of the iodide ion with a molecule of free iodine, is then measured at 360 nm (Løvaas 1992). The spectrophotometry of iodine is reliable and very sensitive, but the method is not widely used.

Another spectrophotometric method is based on the reaction of lipid hydroperoxides with titanium tetrachloride in a chloroform-acetic acid medium. A yellow peroxotitanium salt is formed, and the absorbance is measured at 430 nm. The complex is separated from the fat phase either by extraction with rather concentrated hydrochloric acid (Janíček and Pokorný 1959) or by precipitation and transformation of the precipitate into peroxotitanium nitrate (Eskin and Frenkel 1976). The method is more specific than the iodometric method because only lipid hydroperoxides react with titanium salts. Cyclic and polymeric peroxides do not react. The method is very suitable for the investigation of oil history, particularly because it gives reliable information on the conditions of its oxidation. A serious inconvenience of the method is tedious work with aggressive chemicals.

Lipid hydroperoxides oxidize 10-L-methyl-carbamoyl-leucomethylene blue into methylene blue, which is measured (Yagi *et al.* 1986). Various other colored reagents sensitive to oxidation may be used. The reaction with xylenol orange is useful for the assessment of dairy products; therefore, it was standardized by the International Dairy Federation (IDF). However, a modified method based on the same mechanism, but using another solvent system, is preferred because it is more reliable (Nielsen *et al.* 2003). Edible oils can be analyzed with the use of methylene dichloride and ethanol as solvents even at very low PV, when large amounts of sample must be used (Navas *et al.* 2004).

An indirect spectrophotometric determination is based on the oxidation of ferrous salts to ferric salts by reaction with hydroperoxides, and by the reaction of ferric ions with *N,N'*-dimethyl-*p*-phenylene diamine (Vioque and Vioque 1962) or *N,N'*-di(2-naphthyl)phenylene-1,4-diamine (Ferracini and De Lima 1981). Other suitable reagents forming colored compounds with ferric ions are possible. The most widely used spectrophotometric method is the oxidation of ferrous salts by hydroperoxides, as described above, and the reaction of ferric salts with potassium isothiocyanate. Many modifications exist, suitable for particular food materials, e.g., milk lipids (Dieffenbacher and Lüthi 1986). The procedure is still frequently used for microdeterminations. Small levels of hydroperoxides are also studied by methods based on their luminescence as will be discussed in a separate chapter.

Determination of Carbonyl Oxidation Products Using Spectrophotometric Methods

Carbonyl compounds, mainly aldehydes and ketones, are the most reactive lipid oxidation products, formed by the decomposition of lipid hydroperoxides. They react easily with primary aromatic amines, such as *p*-anisidine or 2,4-dinitrophenylhydrazine, leading to the formation of intensely colored Schiff bases. The condensation reaction is more complicated when other reagents are used, e.g., 2-thiobarbituric acid. The reactions are not very specific because most analytical reagents react with a variety of structurally related carbonyl compounds. The reaction products with unsaturated carbonyl derivatives are more intensely colored than derivatives of saturated carbonyl compounds. Therefore, results obtained using spectrophotometric methods are not proportional to the amount of carbonyl compounds. Nevertheless, they give useful information on the degree of oxidation because the deviations from linearity are only moderate.

Kreis Test

The Kreis test is one of the oldest analytical methods for oxidized fats and oils; it was first published more than 100 years ago (Kreis 1899), and was modified by the same author in the succeeding years. A later modification (Pool and Prater 1945) was more widely used. The oxidation product measured in the Kreis test is epihydrin aldehyde (Fig. 3.5), which is an isomer of malonaldehyde. The active compound is released from oxidized lipid precursors in a strongly acidic medium. The sample reacts with a solution of phloroglucin and trichloroacetic acid in acetic acid after a short heating at 45°C, and the red solution is measured at 540 nm. The reaction mechanism is shown in Figure 3.5. Phloroglucin may be replaced by resorcin. The Kreis test was compared with the PV in oxidizing lipids (Watts and Major 1946); the Kreis test differed in that it is more affected by the linoleic acid content than is the PV. The procedure has been accepted as a standard method and is still included in several collections of standard methods, even if it is now used only rarely.

Benzidine Value

The benzidine value is a very simple, easy method for the evaluation of lipid oxidation. The method was developed ~50 years ago by G. Wode and co-workers, and the most easily available information was published a few years later (Holm *et al.* 1957). The method is based on the reaction of benzidine with a carbonyl group of

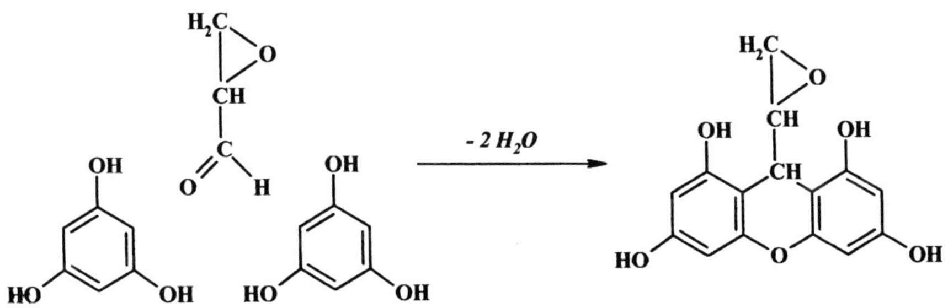

Fig. 3.5. Reaction of epihydrin aldehyde with phloroglucin.

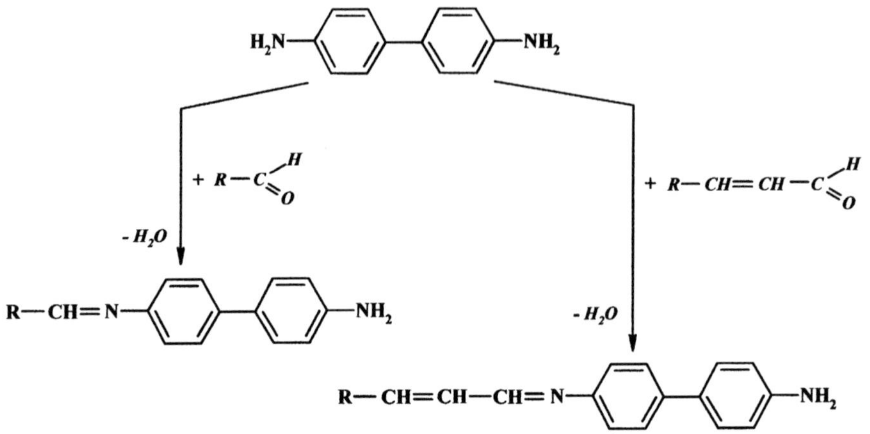

Fig. 3.6. Reactions of carbonyl compounds with benzidine.

oxidized lipids (Fig. 3.6). Because the reagent is added in great excess, only one of the two amine groups reacts. The absorbance of the condensation product is measured at the maximum at 350 nm. Condensation products with 2-alkenals and 2,4-alkadienals have higher absorbances because of longer-chain conjugated double bonds. The method was modified using a more suitable solvent mixture, one in which samples were more soluble (Pokorný and Janíček 1966). The absorbance is measured at 430 nm. Because of good repeatability and reproducibility, the benzidine method was tested by the Fat and Oil Commission of IUPAC, and accepted as a standard method (Paquot and Hautfenne 1985). However, benzidine was found to be a strong carcinogen so that the reagent was replaced by *p*-anisidine, and the benzidine value is no longer used. Of course, some older data are still found in the literature. The absolute values may not be entirely comparable with *p*-anisidine values.

p-*Anisidine Value*

When the carcinogenicity of benzidine was discovered, lipid scientists were immediately looking for a replacement. A compound with a similar structure, *p*-anisidine, was tested because it was not on the list of highly carcinogenic substances. The reaction mechanism (Fig. 3.7) is analogous to the case of the benzidine value. Unsaturated aldehydes give higher results than saturated aldehydes; therefore, the anisidine value (AV) does not give the real concentration of aldehydes, as was the case with the benzidine value, but it is a method of relative value. Anisidine reacts slowly even with hydroperoxides (Fig. 3.7). The reaction proceeds under the same conditions as the reaction with benzidine with similar results. Therefore, the procedure used for benzidine value determination was retained, and only benzidine was replaced by *p*-anisidine as a reagent. The repeatabilities and reproducibilities of the

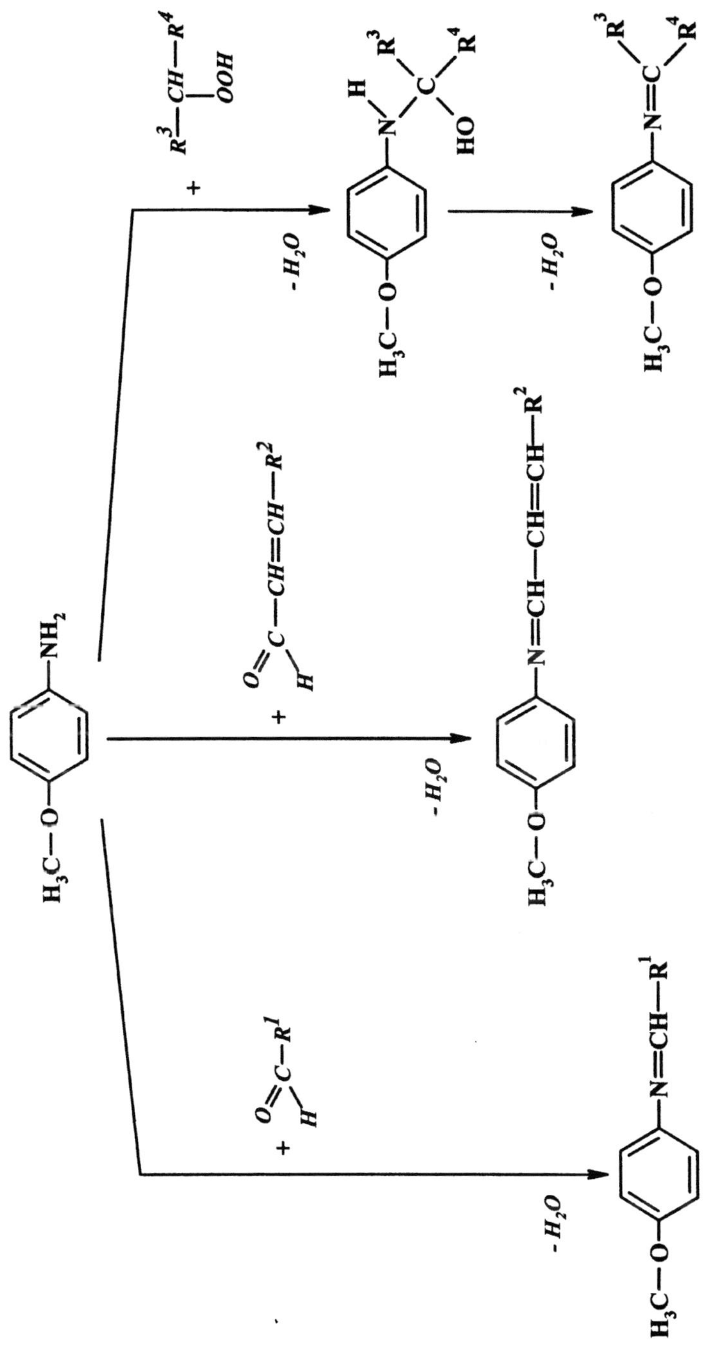

Fig. 3.7. Reactions of carbonyl compounds with *p*-anisidine.

two methods are very similar. The same procedure does not mean, however, that the data obtained by the two methods are identical. The structures of the two condensation products are comparable, but differences remain in the lengths of the conjugated double bond system. Therefore, the absolute values obtained by the two methods differ somewhat. In the case of low PV, the results obtained by the two methods are practically the same, but at high PV, benzidine values are ~20% higher than the corresponding AV values (Pardun 1974). The anisidine method was modified with the use of a chloroform solvent system to facilitate the solubilization of solid samples (Jirošuová 1975). When this modification is used, the absorbance may be measured at 430 nm. The AV can be determined also by flow injection analysis using 2-propanol as a solvent and monitoring the absorbance continuously at 350 nm (Labrinea *et al.* 2001).

The rancidity of oils and fats is due mainly to volatile carbonyl compounds, but *p*-anisidine also reacts with nonvolatile aldehydic and ketonic oxidation products. In spite of this difference, satisfactory correlation was observed between the AV and the content of volatiles in bleached oils (Doleschall *et al.* 2002). Bleached oils, however, contain only very low concentrations of hydroperoxides; because only small differences were found (Pardun 1974), the conclusion cannot be applied to more oxidized fats and oils.

The AV is particularly suitable for heated oils in which the hydroperoxides were mostly destroyed during the heating. It can be used for evaluation of blended or deodorized oils (heated to >200°C) and for materials extruded at the high temperature of 150°C (Wang *et al.* 2003). Deep frying is a technological procedure for which the *p*-anisidine value gives good information on the course of frying oil degradation. Reports of many experiments are available in the literature; therefore, we describe only a few typical applications. The heating temperature had a greater effect on the AV than other factors (Houhoula *et al.* 2002). The AV correlated with the overall odor intensity and with the content of individual aldehydes isolated from frying oils (Tompkins and Perkins 1999). The increase in AV during frying correlated with the increasing unsaturation of frying oil as seen in the difference between palm oil and a more unsaturated mixture of rapeseed and soybean oils (Rade *et al.* 1997). The AV increases during frying with the increasing unsaturation of the oil as seen in the difference between palm oil and a mixture of rapeseed and soybean oils (Rade *et al.* 1997). The higher content of unsaturated aldehydes in polyunsaturated edible oils clearly plays a role here. During decomposition of oxidized sunflower oil by heating in an oxygen-free atmosphere, statistical analysis using principal component analysis showed that the AV behaved differently from other methods used for monitoring the lipid oxidation, and that it could serve as an independent kinetic indicator of the course of oxidation (Heberger *et al.* 1999).

Due to new safety tests, even *p*-anisidine has been included on the list of toxic substances so that the determination of the *p*-anisidine value should be done carefully. The chemical should not come into contact with the skin. Nevertheless, we recommend the AV for monitoring the quality of refined edible oils.

TOTOX Value

The TOTOX (= TOTal OXidation products) value was introduced for the evaluation of refined oils. These oils contain small amounts of hydroperoxides before the refining, but during the deodorization step, oils are heated to <200°C under vacuum so that all of the hydroperoxides are destroyed. The nonvolatile carbonyl degradation products of hydroperoxides are rather unstable, and are easily oxidized again during further storage of refined oils. Hydroperoxides formed during further storage of deodorized oils contribute to the secondary oxidation that occurs. Thus, the stability of refined oils can be predicted by considering both hydroperoxides and carbonyl lipid oxidation products. Therefore, the TOTOX value includes a combination of both parameters: TOTOX = 2PV + AV. The reason for the multiplication of the PV by a factor of 2 is that the PV has a more pronounced effect on the stability of refined oil than the AV. Of course, the expression is only empirical and approximative. It is suitable for evaluating refined oils from a single factory or a single processor. It is valid for the evaluation of a specific product because the relation may not be the same for other fats and oils. Because the TOTOX value is very convenient and does not require any expensive equipment, it is still widely used in small factories in spite of its inaccuracy. The TOTOX value is suitable for quality assessment of refined edible oils. The method is not suitable for the evaluation of oils stored for long periods, when the peroxide content may reach values >5 mEq/kg.

Thiobarbituric Acid Value

The reaction with 2-thiobarbituric acid for the determination of lipid oxidation products has been used for >60 years. Thiobarbituric acid (TBA) was found to give a red pigment with malonaldehyde (the term malondialdehyde is also frequently used, but it is not correct because malonic acid is a bivalent compound so that to use the term malonaldehyde already assumes that both carboxyl groups of malonic acid have been replaced by aldehydic groups). Malonaldehyde is produced during the oxidation of polyunsaturated fatty acids or unsaturated aldehydes (Fig. 3.8). The sample was reacted with a strong acid to liberate malondialdehyde from its precursors, such as acetals. Malonaldehyde reacts with 2-thiobarbituric acid forming a conjugated double bond system (Fig. 3.9). At the absorption maximum of 535 nm the molecular absorptivity at the pH = 0.9 was equal to 1.56×10^5 (Sinnhuber and Yu 1958). In the reaction (Sinnhuber *et al.* 1958), one molecule of malonaldehyde reacts with two molecules of 2-TBA, forming intensely colored products (Fig. 3.9A).

In the original method, the volatile substances, including malonaldehyde, were distilled off and the distillate reacted with TBA in an acid medium. In the beginning, the TBA method was used in the analysis of animal tissues. It was observed that the reaction was specific for highly unsaturated fatty acids, e.g., TBA reacted with oxidized linolenic acid, but not with oxidized linoleic acid (Wilbur *et al.*

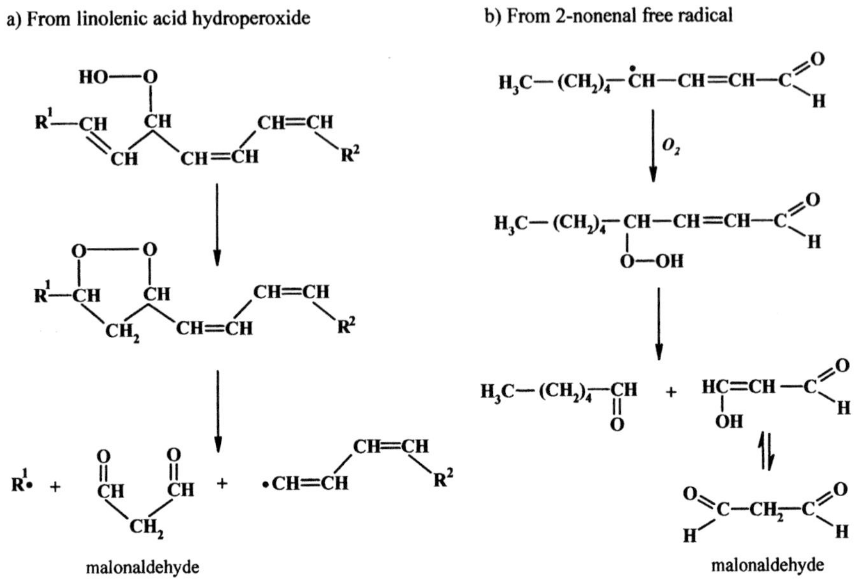

Fig. 3.8. Formation of malonaldehyde during lipid autoxidation.

1949). Surprisingly, the TBA method was then used for the assay of milk fat in which the content of trienoic fatty acids is very low. The earlier work was reviewed by Sidwell *et al.* (1955). Satisfactory correlation between the TBA test and the degree of rancidity was observed not only in butter, but also in meat and meat products (Dugan 1955) and in fishery products (Yu and Sinnhuber 1957). Malonaldehyde could be isolated in alkaline medium as well (Witas 1978). The distillate, obtained after the alkaline treatment, was then reacted with TBA in a strong acidic medium at pH 0.5. The distillation step was eliminated in a later modification by reacting lipids with TBA in an organic solvent solution in the presence of trichloroacetic acid (Dzikowski 1958). The TBA procedure was adapted to the flow-injection analysis of malonaldehyde in blood plasma (Ikatsu *et al.* 1992). The analysis was conducted at 95°C, the optimal time was 10 min, and the coefficient of variation was only 1.5%. The effect of different substances on the TBA value was discussed from the aspect of practical applications (Ward 1985).

Malonaldehyde was supposed to be present mainly as an acetal, and only 2% as free aldehyde. The heating in an acidic or alkaline medium was believed necessary for the hydrolysis of acetals into free malonaldehyde. In a series of samples, malonaldehyde was determined using a fluorometric method (Kikugawa *et al.* 1988), and it was observed that the content of malonaldehyde was much lower than that found using the TBA method. A similar observation resulted from the comparison of the TBA value and the determination of malonaldehyde using gas chro-

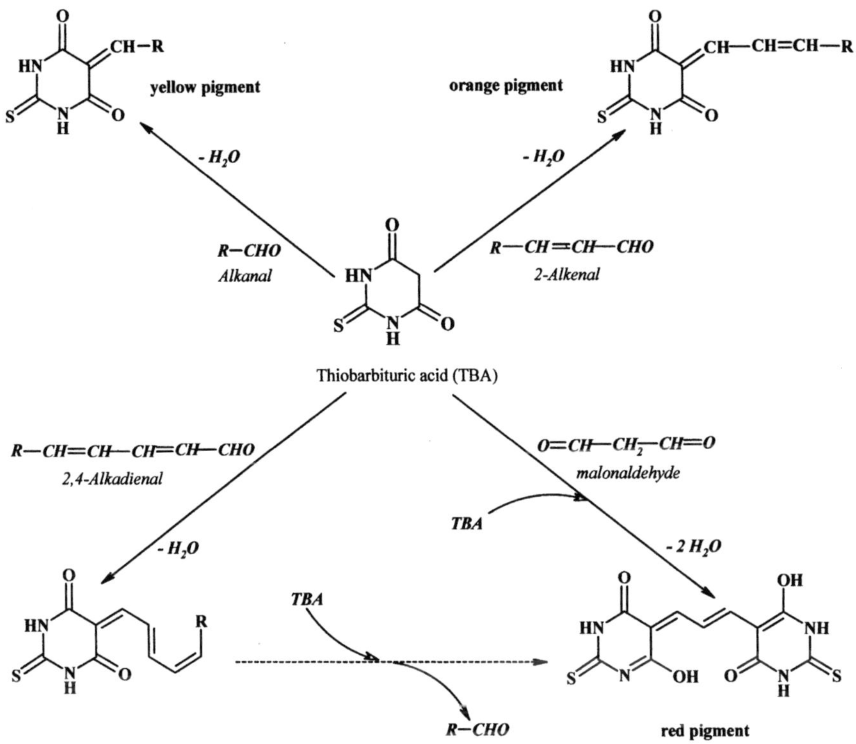

Fig. 3.9. Reactions of carbonyl compounds with 2-thiobarbituric acid.

matography (GC). It follows from these results that nonvolatile precursors other than acetals exist in oxidized lipids.

The reactivities of different carbonyl compounds and different TBA derivatives were tested, and the results published in a long series of papers reviewed in detail by Guillén Sans and Guzmán Chozas (1988). The reaction products of different carbonyl compounds with TBA were more complicated than the earlier researchers had anticipated; therefore, the reactivity of other lipid oxidation products was studied. The spectrum of the reaction product with oxidized lipids showed two main (and a few small) maxima, i.e., at 450 nm and at 530 nm (Fig. 3.10), and their ratios depended on the composition of the oxidized lipids (Marcuse 1970). The yellow absorbance maximum at 450 nm correlated better with the degree of rancidity than the red absorbance maximum at 530 nm (Marcuse and Pokorný 1994). The yellow maximum was unstable, increased rapidly to the maximum, and decreased if heating was continued. In contrast, the red maximum was relatively stable (Asakawa *et al.* 1975). Yellow pigments were first rapidly formed in the reaction of 2-alkenals with TBA (Kosugi *et al.* 1987), as early as after 15 min at 100°C, but they were later converted into orange pigments (maximum absorption

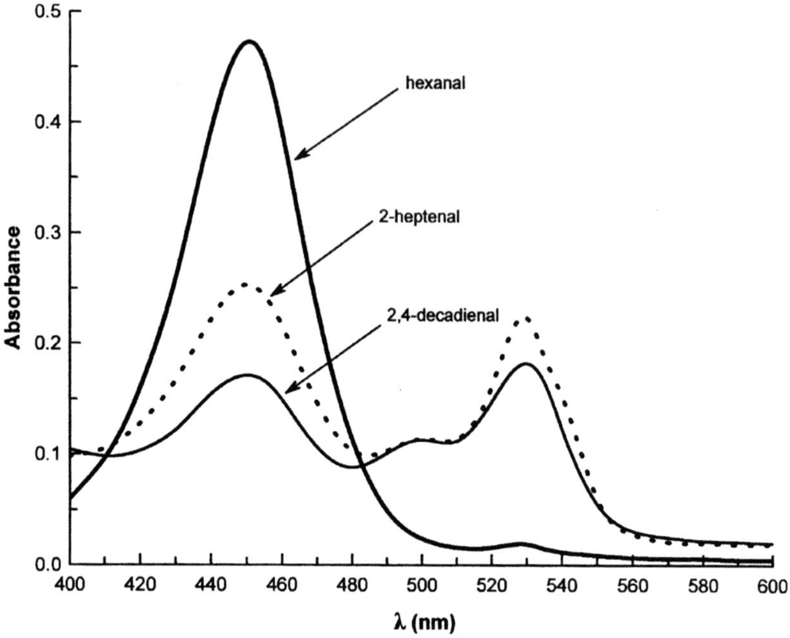

Fig. 3.10. Spectrum of reaction products between 2-thiobarbituric acid and aldehydes (*Source:* Z. Réblová, unpublished results).

at 490 nm), and after several hours into a red pigment (maximum at 532 nm). Precursors of the yellow and red pigments were colorless (Kosugi *et al.* 1987), and the last formed red pigment had the same retention time as the condensation product of TBA with malonaldehyde (Kakiuchi *et al.* 1989).

The reaction of the TBA with 2,4-alkadienals, which are typical oxidation products of polyunsaturated oils, resulted in an intense red coloration (Fig. 3.10), whereas the absorption maximum of alkanals was developed at 450 nm. Malonaldehyde is produced by decomposition of hydroperoxides produced from unsaturated aldehydes. A mechanism for the reaction with aldehydes in which retroaldolization of unsaturated aldehydes may occur during the treatment was suggested (Kosugi and Kikugawa 1986).

Lipid hydroperoxides also react with TBA, with the production of the same red pigment as that produced by the reaction of TBA with malonaldehyde. It was observed that lipid hydroperoxides dramatically increased the TBA value of a mixture of malonaldehyde, alkenal, and alkadienal (Kosugi and Kikugawa 1989). In less oxidized samples, containing only trace amounts of aldehydes, the reaction of hydroperoxide with TBA was nearly quantitative so that the TBA value could be used for the determination of peroxides. During the heating reaction, hydroperoxides and cyclic peroxides present in oxidized lipids decomposed into various prod-

TABLE 3.1
Effect of Lipid Peroxides on the TBA Value[a]

Lipid peroxide	Log A at 532 nm after 60 min	Log A at 450 nm in the maximum
3-Alkyl-1,2-dioxane	3.5	4.1
3-Alkyl-5-hydroxy-1,2-dioxane	3.5	3.1
3,5-Dialkyl-1,2-dioxolane	4.2	4.2
Hydroperoxyoctadecadienoic acid	3.3	4.0
Hydroperoxyoctadecatrienoic acid	4.8	4.0

[a]*Sources:* cyclic peroxides, Porter *et al.* 1976; hydroperoxides,Valentoý and Pokorný, unpublished results.

ucts reacting with TBA (Porter *et al.* 1976), and malonaldehyde likely could be formed from dienoic hydroperoxides, as well. The degree of interference of peroxides with the determination of TBA value is evident from the data shown in Table 3.1.

The decomposition of hydroperoxides and oxidized alkadienals is catalyzed by ions of transient valency metals, such as iron and copper. Therefore, the addition of these ions increased the TBA value by transforming hydroperoxides into compounds reacting with TBA (Jacobson 1993). Higher valency states are more active than lower valency states. Binding of trace metals by the addition of metal chelators to the reaction mixture decreased the rate of hydroperoxide decomposition, and consequently, the TBA value. EDTA was very efficient in this respect. The addition of the antioxidant di-*tert*-butylated hydroxytoluene, which inhibited the oxidation processes during the analysis, also decreased the TBA value (Kosugi *et al.* 1991).

All of the more recent modifications of the TBA method omit the distillation to simplify the procedure, and the sample is heated with the TBA solution in an acidic medium without the rather time-consuming and expensive distillation step. The IUPAC standard method is based on this same principle (Pokorný and Dieffenbacher 1989). Organic solvents may be replaced with water, which is cheaper and safer. The simple modifications of the TBA method are used, e.g., in the meat and poultry industries (Pikul *et al.* 1985). Phospholipids contribute to the amount of TBA-reactive substances in different degrees, depending on their composition (Pikul and Kummerow 1991).

On the basis of the above discussed experimental evidence, it is evident that it is not correct to evaluate the TBA value on the basis of malonaldehyde equivalents. Therefore, the TBA value is defined only as the absorbance obtained under defined conditions, and the compounds reacting under such conditions are called TBA Reactive Substances (TBARS). Naturally, malonaldehyde may still be used as a reference substance (available as the relatively stable tetraacetal; 1,1,3,3-tetraethoxypropane). Because many substances react with TBA, the TBA value has no absolute significance. It is not equivalent to a defined amount of an active compound. It has, of course, a relative value, useful for the comparison of a series of samples, such as the same products from a factory after storage under defined, always identical conditions.

As is clear from our discussion of the topic, TBA analysis should be replaced by other, more precise methods, particularly GC and HPLC, wherever possible. However, important advantages of the TBA method are that it is simple, rapid, relatively cheap, and suitable for running large series of analyses. Therefore, the TBA method is still often used in clinical experiments in which many samples are to be analyzed in an experiment. It is also suitable for the study of changes in the course of lipid oxidation in a single experiment or in a series of experiments conducted under the same conditions (such as analogous products from the same factory, treated and stored for approximately the same period of time). Samples of different composition or samples treated under different conditions are difficult to compare on the basis of the TBA test. In our opinion, the method should not be used for monitoring the degree of lipid oxidation or for the control of food quality.

Reaction with 2,4-Dinitrophenylhydrazine

The reaction of carbonyl groups with 2,4-dinitrophenylhydrazine (often abbreviated as DNPH, but here we use the abbreviation for the respective hydrazones) has been used in the analysis of organic compounds for >100 years. Carbonyl compounds (both aldehydes and ketones) react with 2,4-dinitrophenylhydrazine forming 2,4-dinitrophenylhydrazones (DNPH). The reaction mechanism is shown in Figure 3.11. When aldehydes and ketones were identified in oxidized fats and oils, one understands why the method was applied to the estimation of lipid oxidation.

The most widely used application is the procedure reported more than 50 years ago (Henick *et al.* 1954), but still used today without major changes. The procedure is based on the formation of DNPH from both aldehydes and ketones in the presence of a trichloroacetic acid catalyst. The resulting colored hydrazones can be measured spectrophotometrically in an alkaline solution in which the absorbance maximum is shifted to longer wavelengths. The authors reported the maximum absorption of the DNPH of saturated aldehydes at 432 nm, and of unsaturated aldehydes at 458 nm. They gave simple equations for the simultaneous determination of saturated (C_S) and unsaturated (C_U) carbonyl contents from the absorbances at the convenient wavelengths of 430 nm and 460 nm as follows: $C_U = (3.861\ A_{460} - 3.012\ A_{430})/0.854$ and

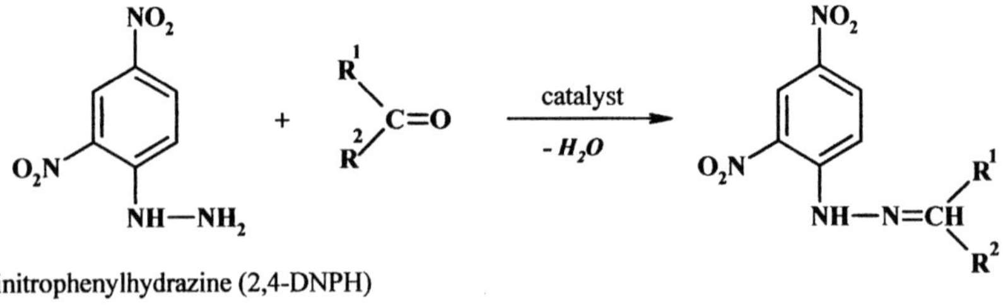

Fig. 3.11. Reaction of carbonylic lipid oxidation products with 2,4-dinitrophenylhydrazine.

$C_S = 3.861\ A_{460} - 2.170\ C_U$. In the presence of hydroperoxides, 2,4-dinitrophenylhydrazones can react with the hydroperoxide group forming isomeric DNPH, which may lead to higher results. In the presence of higher concentrations of hydroperoxides, it is advisable to reduce the hydroperoxides with potassium iodide before the reaction with 2,4-dinitrophenylhydrazine (Linow *et al.* 1964). Hydroperoxides can be reduced by triphenylphosphine also before the analysis (Suzuki and Maruta 1963). Carbonyl compounds are extracted from the reduced substrate with benzene and brought to reaction with 2,4-dinitrophenylhydrazine (toluene is now preferred). The yellow color due to the formation of DNPH was measured spectrophotometrically. A differentiation between different classes of carbonyl derivatives is possible by measurement at 430, 460, and 480 nm. Keto oxidation products, which are less reactive, may not react quantitatively under the reaction conditions used.

The absorption maxima of DNPH were found at 358 and 363 nm in an ethanolic solution, and at 440 nm in an alkaline solution. In the case of DNPH derived from aldehydes, fading of the maximum was observed, whereas the maximum of DNPH derived from ketones remained constant (Kumazawa and Oyama 1965). This difference in stabilities could help in distinguishing between the two classes of DNPH.

Another simple and fast method was suggested, based on the direct reaction of oxidized lipids with 2,4-dinitrophenylhydrazine, followed by the removal of the excess reagent using ion exchange chromatography, and the measurement of the residual absorption at 366 nm (Franzke and Baumgardt 1973). A simplified method was proposed (Endo *et al.* 2001) using 2-propanol as a solvent, and measuring the absorbance at 420 nm in an alkaline medium in which both saturated and unsaturated aldehydes and ketones have the same absorption coefficient. The use of 1-butanol as a solvent is also possible (Endo *et al.* 2003) because it is a good solvent for both lipids and DNPH.

The DNPH formed by the reaction of 2,4-dinitrophenylhydrazine with lipid oxidation products may be separated by HPLC, and the absorbance of the DNPH peaks can be measured. Both polar and nonpolar reaction products may be separated from one another, and the individual DNPH determined (Seppanen and Csallany 2001). We think this is the best way to utilize the advantages of the method because carbonyl compounds are better identified in the presence of other oxidation products. A comprehensive review was published (Schulte 2002) on the determination of carbonyl compounds in used frying oils by HPLC of DNPH.

Other Methods

The DNPH method was modified for the analysis of oils in the advanced stages of oxidation, when hydroperoxides are present in higher amounts than in the cases discussed above. It is suitable to use triphenyl phosphide to first reduce the hydroperoxides. They are reduced to hydroxyl derivatives that do not interfere with the subsequent determination of carbonyl functional groups. The carbonyl oxidation products are then determined after their conversion into DNPH (Chiba *et al.* 1989).

According to another procedure (Schwartz and Rady 1990), oxidized lipids are dissolved in cyclohexane, the solution is transmethylated, and carbonyls in the methyl ester fraction are reacted with 2,4-dinitrophenylhydrazine in the presence of monochloroacetic acid. The derivatives obtained are then fractionated by column chromatography on alumina, and the oxo acid fraction is measured between 320 and 420 nm. A rapid determination of carbonyl compounds in oxidized fats and oils was also devised on the basis of trichlorophenylhydrazone formation (White and Hammond 1983). The sample was dissolved in a mixture of cyclohexane and diethyl ether (99:1, v/v), the solution was passed through a Florisil column to remove the interfering hydrocarbons, and triacylglycerols and carbonyl compounds were then eluted with diethyl ether. The eluate was reacted with trichlorophenylhydrazine, and the hydrazones were isolated by chromatography on a Florisil column. The individual hydrazones were separated and quantified by GC on a nonpolar capillary column.

A simple and quick method for a quantitative determination of aldehydes in oxidized oils is based on the reaction of *N,N*-dimethyl-*p*-phenylene diamine (Fig. 3.12) in the presence of acetic acid. The Schiff bases produced in this reaction are determined by measuring the absorption at 400, 460, and 500 nm (Miyashita *et al.* 1991). The reaction time (10 min at 30°C) is much shorter than that of the conventional method using 2,4-dinitrophenylhydrazine as a reagent (30 min at 60°C).

Several other methods exist based on the reactions of carbonyl groups with different reagents. We tested a few of the methods given here, and our conclusion is that these methods are usually as suitable for determining the carbonyl group content as the methods discussed previously in more detail. Their common disadvantage is that they are used only rarely so that no comprehensive data on their application are found in the literature for comparison. Therefore, we suggest that only the 2-thiobarbituric acid value, the *p*-anisidine value, and the 2,4-dinitrophenylhydrazine procedure be used for the determination of carbonyl compounds in oxidized lipids.

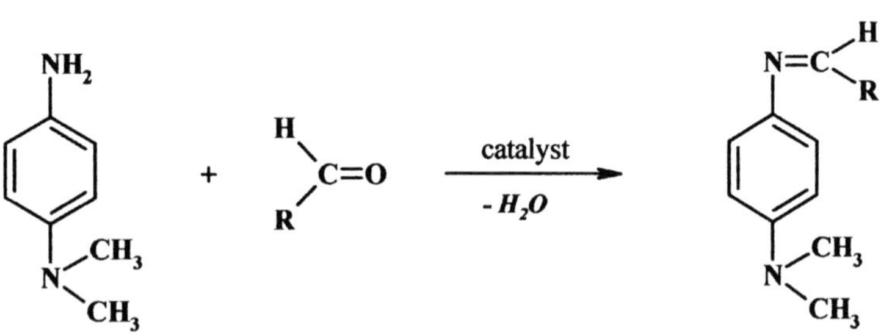

N,N-dimethyl-*p*-phenylenediamine (DPPD)

Fig. 3.12. Reaction of carbonylic lipid oxidation products with *N,N*-dimethyl-*p*-phenylendiamine.

Comparison of Different Spectrophotometric Methods for Assay of Lipid Oxidation

Many papers were published on the subject, but every comparison of different spectrophotometric methods provides only a relative value valid only for the specific materials and reaction or storage conditions. Their effect on each particular method is different so that the ratios of results will depend on the experimental or technological conditions. Therefore, the published data should be considered only as a way of estimating the values that could be expected, and it would be necessary to test the validity of methods with specific samples and under specific technological and analytical conditions.

Concluding Remarks

All spectrophotometric methods for the assay of lipid oxidation have advantages and disadvantages. They are very simple and inexpensive, requiring no specific equipment and no highly qualified operators. On the other hand, the spectrophotometric methods are not specific because many substances occurring in oxidizing lipids react under the test conditions. Their contribution to the final absorbance usually depends on the chemical structure of the oxidation products. The values obtained usually depend on the lipid composition, the stage of oxidation, the presence of other food components, and on the technological operations taking place. If it is possible, it is advisable to replace or couple spectrophotometric methods with more specific instrumental methods to achieve absolute results, corresponding to the real concentrations of oxidation products. Nevertheless, there are still large series of samples of similar character, which could be analyzed using a special spectrophotometric method; the results are valid, however, only in the particular series of samples, and cannot be compared with other series of other samples or samples treated under different conditions.

References

AOCS (1997) *Standard Methods and Recommended Practices*, 5th ed., AOCS Press, Champaign, IL.

Asakawa, T., and Matsushita, S. (1980) A Colorimetric Microdetermination of Peroxide Values Utilizing Aluminium Chloride as the Catalyst, *Lipids 15*, 965–967.

Asakawa, T., Nomura, Y., and Matsushita, S. (1975) On the Reacting Compounds in the TBA Method for the Determination of Lipid Oxidation, *Yukagaku 24*, 88–93.

Chiba, T., Takazawa, M., and Fujimoto, K. (1989) A Simple Method for Estimating Carbonyl Content in Peroxide-Containing Oils, *J. Am. Oil Chem. Soc. 66*, 1588–1591.

DeLange, R.J., and Glazer, A.N. (1989) Phycoerythrin Fluorescence/Based Assay for Peroxy Radicals: A Screen for Biologically Relevant Protective Agents, *Anal. Biochem. 177*, 300–306.

Dieffenbacher, A., and Lüthi, B. (1986) Die direkte kolorimetrische Bestimmung der Peroxidzahl in Milchprodukten, *Mitt. Gebiete Lebensm. Hyg. 77*, 544–553.

Doleschall, F., Kemény, Z., Recseg, K., and Kövári, K. (2002) A New Analytical Method to Monitor Lipid Peroxidation During Bleaching, *Eur. J. Lipid Sci. Technol. 104*, 14–18.

Dugan, L., Jr. (1955) Stability and Rancidity, *J. Am. Oil Chem. Soc. 32*, 605–609.

Dzikowski, J. (1958) Badanie zmian utleniajacych w tluszczach z kwasem 2-tiobarbiturowym w rozpuszczalnikach organicznych, *Rocz. Państw. Zakl. Hig. 9*, 461–468.

Endo, Y., Li, C. M., Tagiri-Endo, M., and Fujimoto, K. (2001) A Modified Method for the Estimation of Total Carbonyl Compounds in Heated and Frying Oils Using 2-Propanol as a Solvent, *J. Am. Oil Chem. Soc. 78*, 1021–1024.

Endo, Y., Tominaga, M., Tagiri-Endo, M., Kumozaki, M., Kouzui, H., Shiramasa, H., and Miyakoshi, K. (2003) A Modified Method to Estimate Total Carbonyl Compounds in Frying Oil Using 1-Butanol as a Solvent, *J. Oleo Sci. 52*, 353–358.

Eskin, N.A.M., and Frenkel, C. (1976) A Simple and Rapid Method for Assessing Rancidity of Oils Based on the Formation of Hydroperoxides, *J. Am. Oil Chem. Soc. 53*, 746–747.

Ferracini, V.L., and De Lima, C.G. (1981) Evaluation of N,N′-di(2-Naphthyl)phenylene-1, 4-Diamine as a Reagent for the Spectrophotometric Determination of Organic Peroxides, *Analyst 106*, 574–583.

Fishwick, M.J., and Swoboda, P.A.T. (1977) Measurement of Oxidation of Polyunsaturated Fatty Acids by Spectrophotometric Assay of Conjugated Derivatives, *J. Sci. Food Agric. 28*, 387–393.

Franzke, C., and Baumgardt, F. (1973) Schnellmethods zur Bestimmung von Carbonylverbindungen in Fetten (Heptanalzahl), *Nahrung 17*, 209–214.

Guillén Sans, R., and Guzmán Chozas, M. (1988) Reactividad del Ácidσ 2-Tiobarbitúrico con Compuestos Carbonílicos. Su Importancia en la Rancidez Oxidativa y en el Profil de Flavor de los Alimentos, *Grasas Aceites 39*, 185–189.

Grau, A., Guardiola, F., Boatella, J., Baucells, M.D., and Codony, R. (2000) Evaluation of Lipid Ultraviolet Absorption as a Parameter to Measure Lipid Oxidation in Dark Chicken Meat, *J. Agric. Food Chem. 48*, 4128–4135.

Heberger, K., Keszler, A., and Gude, M. (1999) Principal Component Analysis of Measured Quantities During Degradation of Hydroperoxides in Oxidized Vegetable Oils, *Lipids 34*, 83–92.

Henick, A.S., Benca, M.F., and Mitchell, J.H., Jr. (1954) Estimating Carbonyl Compounds in Rancid Fats and Foods, *J. Am. Oil Chem. Soc. 31*, 88–91.

Holm, L., Ekbom, K., and Wode, G. (1957) Determination of the Extent of Oxidation of Fats, *J. Am. Oil Chem. Soc. 34*, 606–609.

Houhoula, D.P., Oreopoulou, V., and Tzia, C. (2002) A Kinetic Study of Oil Deterioration During Frying and a Comparison with Heating, *J. Am. Oil Chem. Soc. 79*, 133–137.

Ikatsu, H., Nakajima, T., Murayama, N., and Korenaga, T. (1992) Flow-Injection Analysis for Malondialdehyde in Plasma with the Thiobarbituric Acid Reaction, *Clin. Chem. 30*, 2061–2062.

Ishimatsu, S., Inose, H., Okamura, M., and Goto, H. (1969) Estimation of Oil and Fat Peroxide Values, *Nippon Daigaku Yakugaku Kenkyu Hokoku 10*, 21–24.

Jacobson, G.A. (1993) Evaluation of Oxidized Lipids in Foods, *INFORM 4*, 811–819.

Janíček, G., and Pokorný, J. (1959) Colorimetric Determination of Fat Peroxides by Means of Titanium Chloride, *Sb. Vys. Sk. Chem.-Technol. Praze Oddil. Fak. Technol. Potravin 3*, 233–260.

Jiroušová, J. (1975) Modifizierte Bestimmung der Anisidinzahl bei oxydierten Fetten und Ölen, *Nahrung 19*, 319–325.

Kakiuchi, Y., Yamauchi, A., Hama, S., and Yamazaki, H. (1989) Analysis of Reaction Products in the Water Extraction-TBA Method for a Fats Deterioration Test, *Eisei Kagaku 35*, 291–296.

Kikugawa, K., Kato, T., and Iwata, A. (1988) Determination of Malonaldehyde in Oxidized Lipids by the Hantzsch Fluorometric Method, *Anal. Biochem. 174*, 512–521.

Koga, H., Tone, N., Muramoto, N., Sakurai, H., and Katayama, O. (1997) The Relation Between the Coloration of Frying Oil and the Deterioration Using Fried Model Materials, *Nippon Shokuhin Kagaku Kaishi 44*, 666–670.

Kosugi, H., Kato, T., and Kikugawa, K. (1987) Formation of Yellow, Orange, and Red Pigments in the Reaction of Alk-2-Enals with 2-Thiobarbituric Acid, *Anal. Biochem. 165*, 456–464.

Kosugi, H., and Kikugawa, K. (1986) Reaction of Thiobarbituric Acid with Saturated Aldehydes, *Lipids 21*, 537–542.

Kosugi, H., and Kikugawa, K. (1989) Enhancement of Red Pigment Formation in Thiobarbituric Reaction with a Combination of Alkenals, Alkadienals and Organic Hydroperoxides, *Yukagaku 38*, 224–230.

Kosugi, H., Kojima, T., and Kikugawa, K. (1991) Characteristics of the TBA Reactivity of Oxidized Fats and Oils, *J. Am. Oil Chem. Soc. 68*, 51–55.

Kreis, H. (1899) Über neue Farbenreaktionen fetter Öle, *Chemiker-Ztg. 23*, 802–803.

Kumazawa, H., and Oyama, T. (1965) Estimation of Total Carbonyl Content in Oxidized Oil by 2,4-Dinitrophenylhydrazine, *Yukagaku 14*, 167–171.

Labrinea, E.P., Thomaidis, N.S., and Georgiou, C.A. (2001) Direct Olive Oil Anisidine Value Determination by Flow Injection, *Anal. Chim. Acta 448*, 201–206.

Linow, F., Roloff, M., and Täufel, K. (1964) Analysis of Carbonyl Compounds in the Presence of Hydroperoxides After Reduction with Potassium Iodide, *Fette Seifen Anstrichm. 66*, 866–869 and 1052–1058.

Løvaas, E. (1992) A Sensitive Spectrophotometric Method for Lipid Hydroperoxide Determination, *J. Am. Oil Chem. Soc. 69*, 777–783.

Marcuse, R. (1970) Studien über den TBS Test, 1. Seine Temperaturabhängigkeit und Anwendung auf Vakuum-destillierte Fettoxydationsprodukte, *Fette Seifen Anstrichm. 72*, 635–640.

Marcuse, R., and Pokorný, J. (1994) Higher Correlation with Sensory Evaluation of Oxidative Rancidity by Modified TBA Test, *Fat Sci. Technol. 16*, 185–187.

Miyashita, K., Kanda, K., and Takagi, T. (1991) A Simple and Quick Determination of Aldehydes in Autoxidized Vegetable and Fish Oils, *J. Am. Oil Chem. Soc. 68*, 748–751.

Mortensen, A., and Skibsted, L.H. (1998) Reactivity of β-Carotene Towards Peroxy Radicals Studied by Laser Flash and Steady-State Photolysis, *FEBS Lett. 426*, 392–396.

Navas, J. A., Tres, A., Codony, R., Boatella, J., Bou, R., and Guardiola, F. (2004) Modified Ferrous Oxidation-Xylenol Orange Method to Determine Lipid Hydroperoxides in Fried Snacks, *Eur. J. Lipid Sci. Technol. 106*, 688–696.

Nielsen, N.S., Timm-Heinrich, M., and Jacobsen, C. (2003) Comparison of Wet Chemical Methods for Determination of Lipid Hydroperoxides, *J. Food Lipids 10*, 35–50.

Paquot, C., and Hautfenne, A. (1985) *IUPAC Standard Methods for the Analysis of Oils, Fats and Derivatives*, 7th ed., Blackwell, Oxford.

Paquot, C., and Hautfenne, A. (1987) *Standard Methods for the Analysis of Oils, Fats, and Derivatives,* 7th ed., Blackwell, Oxford.

Pardun, H. (1974) Beurteilung des Präoxydationsgrades bzw. der Oxydationsstabilität pflanzlicher Öle aufgrund ihrer Benzidinoder Anisidinzahl, *Fette Seifen Anstrichm. 76*, 521–528.

Pikul, J., and Kummerow, F.A. (1991) Thiobarbituric Acid Reactive Substance Formation as Affected by Distribution of Polyenoic Fatty Acids in Individual Phospholipids, *J. Agric. Food Chem. 39*, 451–457.

Pikul, J., Leszczynski, D.E., and Kummerow, F.A. (1985) Total Lipids, Fat Composition and Malonaldehyde Concentration in Chicken Liver, Heart, Adipose Tissue and Plasma, *Poult. Sci. 64*, 469–475.

Pokorný, J., and Dieffenbacher, A. (1989) Determination of 2-Thiobarbituric Acid Value: Direct Method, *Pure Appl. Chem. 61*, 1165–1170.

Pokorný, J., and Janíček, G. (1966) Modified Determination of Benzidine Value in Rancid Fats, *Sb. VSCHT Praze E 9*, 81–84.

Pool, M.F., and Prater, A.N. (1945) A Modified Kreis Test Suitable for Photocolorimetry, *Oil Soap 22*, 215–216.

Porter, N.A., Nixon, J., and Isaac, R. (1976) Cyclic Peroxides and the Thiobarbituric Assay, *Biochim. Biophys. Acta 441*, 506–512.

Rade, D., Mokrovčak, Z., and Štrucelj, D. (1997) Deep Fat Frying of French Fried Potatoes in Palm Oil and Vegetable Oil, *Food Technol. Biotechnol. 35*, 119–124.

Sawa, T., Nakao, M., Akaike, T., Ono, K., and Maeda, H. (1999) Alkylperoxyl Radical-Scavenging Activity of Various Flavonoids and Other Phenolic Compounds, *J. Agric. Food Chem. 47*, 397–402.

Schulte, E. (2002) Determination of Higher Carbonyl Compounds in Used Frying Fats by HPLC of DNPH Derivatives, *Anal. Bioanal. Chem. 372*, 644–648.

Schwartz, D.P., and Rady, A.H. (1990) Determination and Occurrence of Oxofatty Acids in Fats and Oils, *J. Am. Oil Chem. Soc. 67*, 635–641.

Sedláček, B.A.J. (1968) Studium der UV-Spektren oxydierter Fette, *Fette Seifen Anstrichm. 70*, 80–86.

Seppanen, C.M., and Csallany, A.S. (2001) Simultaneous Determination of Lipophilic Aldehydes by HPLC in Vegetable Oils, *J. Am. Oil Chem. Soc. 78*, 1253–1260.

Sidwell, C.G., Salwin, H., and Mitchell, J.H., Jr. (1955) Measurement of Oxidation in Dried Milk Products with Thiobarbituric Acid Test, *J. Am. Oil Chem. Soc. 32*, 13–16.

Sinnhuber, R.O., and Yu, T.C. (1958) 2-Thiobarbituric Acid Method for the Measurement of Rancidity in Fishery Products, 2. The Quantitative Determination of Malonaldehyde, *Food Technol. 12*, 9–12.

Sinnhuber, R.O., Yu, T.C., and Yu, T.C. (1958) Characterization of the Red Pigment Formed in the 2-Thiobarbituric Acid Determination of Oxidative Rancidity, *Food Res. 23*, 626–633.

Suzuki, Y., and Maruta, S. (1963) Determination of the Aldehydes and Ketones as Their 2,4-DNPH by Spectrophotometric Procedure, *Yukagaku 12*, 42–47.

Takagi, T., Mitsuno, Y., and Masumura, M. (1978) Determination of Peroxide Value by the Colorimetric Iodine Method with Protection of Iodide as Cadmium Complex, *Lipids 13*, 147–151.

Tompkins, C., and Perkins, E.G. (1999) The Evaluation of Frying Oils with the *p*-Anisidine Value, *J. Am. Oil Chem. Soc. 76*, 945–947.

Vioque, E., and Vioque, A. (1962) Microdeterminación Espectrofotométrica de Peróxidos en Grasas, *Grasas Aceites 13*, 203–206.

Vossen, R.C.R.M., Van Dam-Mieras, M.C.E., Hornstra, G., and Zwaal, R.F.A. (1993) Continuous Monitoring of Lipid Peroxidation by Measuring Conjugated Diene Formation in an Aqueous Liposome Suspension, *Lipids 28*, 857–861.

Wang, X., Wang, T., and Johnson, L.A. (2003) Chemical and Sensory Properties of Gas-Purged, Minimum-Refined, Extruded-Expelled Soybean Oil, *J. Am. Oil Chem. Soc. 80*, 923–926.

Ward, D.D. (1985) The TBA Assay and Lipid Oxidation: An Overview of the Relevant Literature, *Milchwissenschaft 40*, 583–588.

Watts, B. M., and Major, R. (1946) Comparison of a Simplified Quantitative Kreis Test with Peroxide Values of Oxidizing Fats, *Oil Soap 23*, 222–225.

White, P.J., and Hammond, E.G. (1983) Quantification of Carbonyl Compounds in Oxidized Fats as Trichlorophenylhydrazones, *J. Am. Oil Chem. Soc. 60*, 1769–1773.

Wilbur, K.M., Bernheim, F., and Shapiro, O.W. (1949) The Thiobarbituric Acid Reagent as a Test for the Oxidation of Unsaturated Fatty Acids by Various Agents, *Arch. Biochem. 24*, 305–313.

Witas, T. (1978) Determination of Oxidation Degree of Fish Oils by Thiobarbituric Acid Method with Alkaline Hydrolysis, *Nahrung 22*, 133–147.

Yagi, K., Kiuchi, K., Saito, Y., Miike, A., Kayahara, N., Tatano, T., and Ohishi, N. (1986) Use of a New Methylene Blue Derivative for Determination of Lipid Peroxides in Foods, *Biochem. Int. 12*, 367–371.

Yanishlieva, N., and Popov, A. (1973) La Spectrophotométrie Ultraviolette en Tant que Méthode d´Estimation de l´État d´Oxydation des Lipides Insaturés, *Rev. Fr. Corps Gras 20*, 10–26.

Yu, T. C., and Sinnhuber, R.O. (1957) 2-Thiobarbituric Acid Method for the Measurement of Rancidity in Fishery Products, *Food Technol. 11*, 104–108.

Chapter 4

Analysis of Nonvolatile Lipid Oxidation Compounds by High-Performance Size-Exclusion Chromatography

Gloria Márquez-Ruiz and M. Carmen Dobarganes

Instituto de la Grasa (CSIC), 41012 Sevilla, Spain

Introduction

High-performance size-exclusion chromatography (HPSEC) is a technique used to separate compounds according to their molecular size, normally related to their molecular weight (MW) provided that the compounds have a similar shape. This chapter deals with the use of HPSEC for the separation and quantitation of nonvolatile lipid oxidation compounds. A brief introduction on the basis of the technique and specific characteristics for lipid analysis will be followed by a description of the main methodologies based on HPSEC that have been developed to quantitate nonvolatile oxidation compounds. In the third part of the chapter, a review of the applications for the analysis of lipid oxidation compounds is included.

Originally, size exclusion chromatography was used mainly for characterization of high-MW molecules, either synthetic polymers or biopolymers, but applications soon expanded to other areas, largely supported by important technical advances. Fundamental developments and general applications of HPSEC during the last years were reported in various reviews (Balke *et al.* 1999 and 2000, Barth *et al.* 1994, 1996, and 1998, Barth and Boyes 1990 and 1992, Stulik *et al.* 2003).

Typically, stationary phases consist of macromolecules cross-linked to form a 3-dimensional network characterized by a specific pore size. The most important parameters influencing resolution are the pore volume, pore-size distribution, and particle size; improvements in the control of these parameters have contributed to the development of columns of high efficiency and separation capacity (Kulin *et al.* 1990). The migration of molecules between the stationary phase and the mobile phase occurs essentially by diffusion, and the elution order is inversely related to molecular size or weight. Thus, the larger molecules are excluded and emerge first, whereas the smaller molecules can diffuse into the pores of the gel, partially or completely, and elute later. It is possible to estimate the MW of unknown molecules by plotting retention volume vs. the logarithm of MW for a series of known standards. These plots provide accurate determinations of MW for molecules that adopt a conformation in solution similar to that of the standard (Stellwagen 1990, Stogiou *et al.* 2002). Developments in calibration methodologies, including direct calibration by standards and various instrumental methods [nuclear magnetic resonance (NMR), mass spectrometry (MS), light scattering] as well as universal calibration with viscometry detectors, were reported recently (Kostanski *et al.* 2004).

During recent years, HPSEC development in the areas of fractionation, detection, calibration, and resolution correction advanced considerably, especially in the field of polymer analyses. In particular, special emphasis was paid to new multidetector combinations that improve elucidation of molecular properties of polymers, including detection with Fourier Transform Infrared (FTIR) spectrometry with solvent-evaporation interfaces (Karami *et al.* 2001, Torabi *et al.* 2001). Some examples of the combinations used are the coupling of multiangle light scattering and differential refractometry detectors (Laguna *et al.* 2001), the combination of refractive and light scattering detectors with fluorescence spectroscopy (Mrkvickova *et al.* 2000), or the coupling of ultraviolet (UV) detection and on-line NMR spectroscopy and MS combined with on-line collection of the chromatographic eluent for subsequent FTIR (Ludlow *et al.* 1999).

General Characteristics of HPSEC for Lipid Analysis

The development of lipophilic gels applicable to the separation of organic compounds of MW range from 100 to 1500 encouraged researchers in the 1960s to apply the HPSEC technique to the analysis of lipids for the first time. For applications in fats and oils, HPSEC is appropriate for separating groups of compounds differing by at least 10–15% MW, which is necessary to achieve well-resolved peaks. Hence it is not applicable to compounds that are only slightly different in MW as a result of variable fatty acid composition or degree of unsaturation. This limitation constitutes, at the same time, the basis of the most specific and powerful applications in lipid analysis (Christie 1987 and 1995, Dobarganes and Márquez-Ruiz 1993, 1995, and 1998, Márquez-Ruiz and Dobarganes 1997).

To date, the stationary phases based on copolymers of styrene divinyl benzene have been the most promising and widely used for applications in organic media. Copolymers of styrene divinyl benzene are available over a wide range of pore sizes, but 50, 100, and 500 Å are essential porosities for low-MW separations (100–20,000 MW). Size exclusion columns are generally larger than those used in the other chromatographic modes, so that the amount of stationary phase and thus the effective pore volume available is increased. Packed columns are normally ~30 cm × 0.8 cm i.d., and they are often used in series. Thus, effective selection within a broad range is accomplished by the first column, and fractionation within a more defined range is achieved on the second or third column. Optimization of stationary phase particles is essential to enhance mass transfer given that resolution in HPSEC is dependent on diffusion. In this respect, development of spherically shaped, monosized particles contributed considerably to improving particle-size and pore-size distribution and hence efficiency and separation capacity (Kulin *et al.* 1990, Ugelstad *et al.* 1980). Particle sizes of 5 and 10 μm are the most commonly used.

A single solvent is used for the mobile phase, and HPSEC columns are normally filled with the same solvent used to dissolve the sample. Supports of copolymer of styrene divinylbenzene were designed to operate across a wide spectrum of solvents. Columns can even be transferred easily and rapidly between solvents of differing polarity without damage to the packed bed. Tetrahydrofuran (THF) is the most commonly used solvent, although toluene and dichloromethane are used for certain applications. Flow rates between 0.5 and 1.5 mL/min are the most usual, thus allowing performance of analysis in <30 min.

Detection with nonselective mass-sensitive detectors such as the refractive index detector, (RI), and evaporative light scattering detector (ELSD) are most common. The RI detector is limited to isocratic separations but provides good linearity for mass-sensitive measurements. The ELSD, on the other hand, has excellent gradient capabilities and a slightly superior sensitivity but provides nonlinear (exponential or slightly sigmoid) response curves because of underlying light scattering mechanisms (Abidi *et al.* 1999, Hansen and Artz 1995, Kaufmann *et al.* 2001). For quantitative purposes, a refractometer is simpler in that a single solvent is used, and linear responses are normally obtained in the ranges of interest.

Methodologies Based on HPSEC for Analysis of Nonvolatile Oxidation Compounds

The literature reflects the main problem encountered in the evaluation of lipid oxidation, which is the need to use more than one measurement to obtain a complete picture of the oxidation state because the methods normally applied focus on partial and distinct aspects of the oxidation process (Frankel 1993, 1998a, and 1998b, Frankel and Meyer 2000, Rossell 1994). In addition, most of the methods commonly used to evaluate oxidation are indices or do not provide information on the amounts of compounds formed. The first compounds formed in the autoxidation of unsaturated lipids, which proceeds *via* a free radical mechanism, are hydroperoxides; further oxidation, decomposition, and polymerization reactions result in a complex mixture of volatile and nonvolatile oxidation compounds of different polarity and MW. Although volatile compounds are of extraordinary importance for sensory properties and consumer acceptance, nonvolatile products are by far the most abundant and nutritionally relevant compounds because they are retained in food and hence are part of the diet.

The methodologies described below are unique in that they provide quantitative data on all of the groups of nonvolatile compounds formed during oxidation, independently of the stage of the process. This is attained by virtue of the HPSEC separation of lipid molecule groups differing in MW, after a preliminary separation by adsorption chromatography. Figure 4.1 gives a schematic view of all of the methodologies proposed; they are described in the following sections.

Combination of Adsorption Chromatography and HPSEC

Silica Column and HPSEC. In 1988, the first methodology based on the combination of adsorption chromatography using silica columns and HPSEC for quantita-

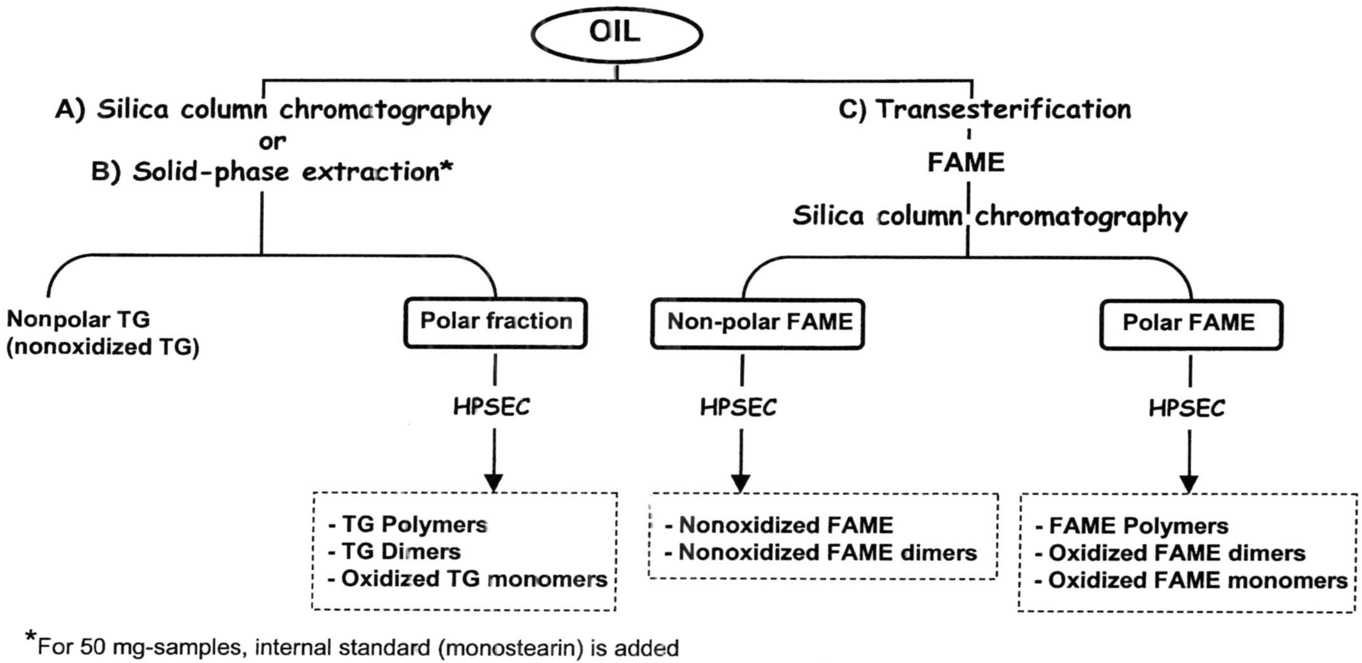

Fig. 4.1. Schematic representation of methodologies based on HPSEC for quantitation of nonvolatile oxidation compounds. Abbreviations: TG, triacylglycerols; FAME, fatty acid methyl esters.

tion of oxidized and hydrolytic compounds was published (Dobarganes *et al.* 1988). Starting from 1 g of fat or oil, nonpolar and polar fractions are eluted with 150 mL of a mixture hexane:diethyl ether 90:10 and 150 mL diethyl ether, respectively (see Fig. 4.1A). Nonpolar and polar fractions are evaporated under reduced pressure, and separation is verified by thin-layer chromatography (TLC) (Fig. 4.2). After gravimetric determination of polar compounds, the polar fractions are further analyzed by HPSEC, using two columns (100 and 500 Å), packed with a porous, highly cross-linked styrene divinylbenzene copolymer (particle size: 5 μm), connected in series. THF serves as the mobile phase, at a flow rate of 1 mL/min and a refractive index detector is used. HPSEC analysis requires a total run time of only 15 min. Concentrations are calculated from peak areas and gravimetric determination of the polar fraction. The methodology was recently standardized by the IUPAC, with slight modifications (Dobarganes *et al.* 2000) for the analysis of virgin and refined oils, and for used frying fats.

By virtue of the previous separation of the most abundant fraction of the sample, i.e., nonpolar or nonoxidized triacylglycerols (TG), application of HPSEC to the concentrated fraction of polar compounds allows the separation and quantitation of different groups of oxidized and hydrolytic compounds that differ substantially in MW: TG polymers (TGP), TG dimers (TGD), oxidized TG monomers (oxTGM), diacylglycerols (DG), monoacylglycerols (MG), and free fatty acids (FFA), in that order of elution. This methodology allows applications in a broad range of samples, from unused, nonoxidized oils to thermally oxidized oils, thus covering samples with different degrees of oxidation. This is illustrated in Figures 4.3 and 4.4, which show typical HPSEC profiles of total oils (directly injected in HPSEC) and polar fractions of unused oils and used frying oils, respectively.

The advantages of the HPSEC analysis on the minor polar fraction devoid of nonoxidized TG include quantitation of two important groups of compounds, i.e., oxTGM and DG, which constitute oxidation and hydrolysis markers, respectively. For analysis of lipid oxidation, quantitation of oxTGM is of great relevance because this group of compounds comprises all monomeric TG containing at least one oxidized fatty acyl group. Oxygenated functions may be peroxide groups, the primary oxidation products formed, or those characteristic of decomposition and/or further reactions of hydroperoxides, e.g., epoxy compounds, alcohols, or ketones. Therefore, quantitation of oxTGM constitutes an excellent global measurement of the nonvolatile oxidation formed from the initiation reactions until the advanced stages of oxidation.

Even though polymerization compounds (TGD and TGP) can be otherwise quantitated by direct analysis of the total oil by HPSEC, a substantial increase in sensitivity is achieved in the polar fraction due to the effect of concentration. Quantitation of TGD and TGP completes the information obtained on the oxidation state because their increase indicates the onset of the accelerated oxidation stage. In addition, the improvement in quantitation of polymerization compounds benefits analysis of used frying fats and oils because these are major compounds under conditions of oxidation at high temperatures. Applications of this methodol-

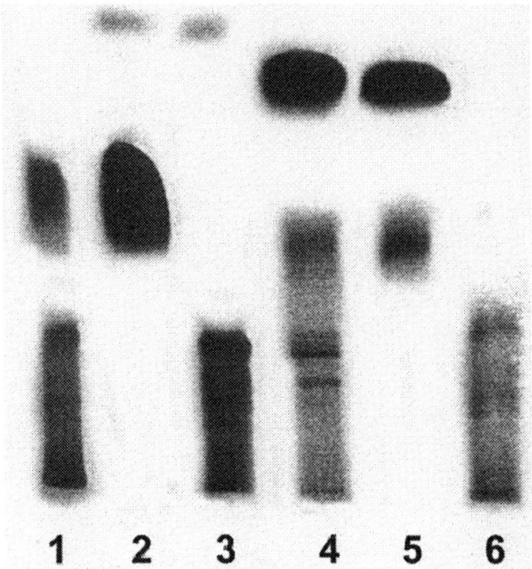

Fig. 4.2. Efficiency of silica column chromatography fractionation checked by thin-layer chromatography: 1, total sample; 2, nonpolar TG; 3, polar fraction; 4, total FAME; 5, nonpolar FAME; 6, polar FAME. Abbreviations: TG, triacylglycerols; FAME, fatty acid methyl esters.

ogy in different aspects of lipid oxidation, as well as the significance of the groups of compounds quantitated for each application, are described later in this chapter.

Other authors have introduced certain modifications in this analytical procedure for particular uses, such as the improvement of resolution in the range of lowest MW by replacement of two styrene/divinylbenzene copolymer columns with pore sizes of 100 and 500 Å by three columns of 100, 50 and 50 Å (Hopia *et al.* 1992) and the improvement of resolution in the high-molecular-mass range by using 500, 500 and 100 Å columns (Gomes 1992) or 3-μm mixed-bed styrene/divinylbenzene copolymer columns (Abidi *et al.* 1999). On the other hand, in an effort to improve quantitative determinations, peaks corresponding to the

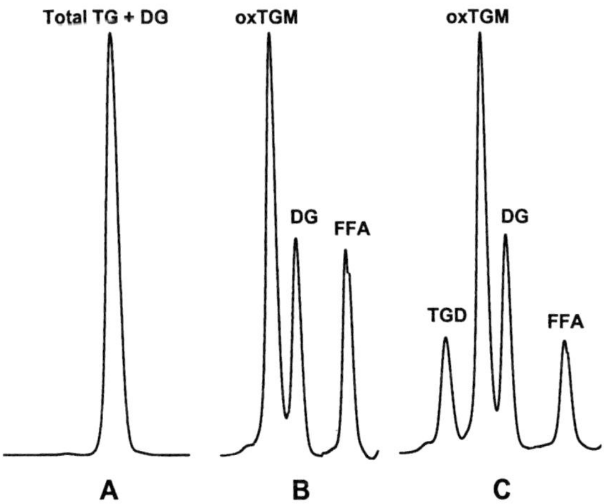

Fig. 4.3. HPSEC chromatograms of total virgin olive oil (A) and polar fractions of virgin olive oil (B) and of refined olive oil (C). Abbreviations: TG, triacylglycerols; TGD, triacylglycerol dimers; oxTGM, oxidized triacylglycerol monomers; DG, diacylglycerols; FFA, free fatty acids (this peak also includes polar unsaponifiable compounds).

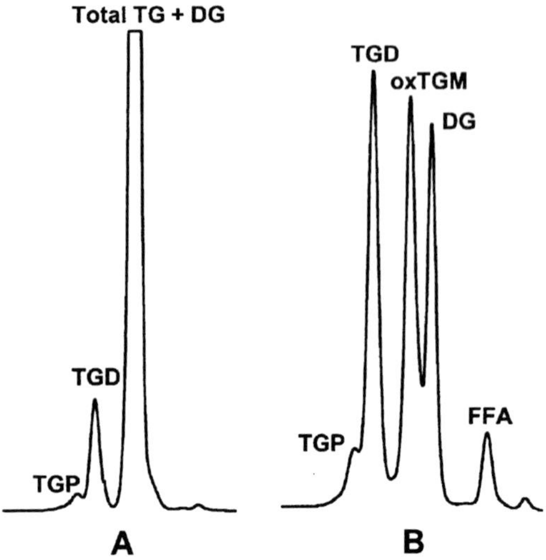

Fig. 4.4. HPSEC chromatograms of total used frying oil (A) and its polar fraction (B). Abbreviations: TG, triacylglycerols; TGP, triacylglycerol polymers; TGD, triacylglycerol dimers; oxTGM, oxidized triacylglycerol monomers; DG, diacylglycerols; FFA, free fatty acids (this peak also includes polar unsaponifiable compounds).

groups of compounds present in the polar fraction of refined oils were collected by preparative gel permeation chromatography for potential use as standards in HPSEC analyses (Gomes and Caponio 1999). In the selection of other detectors, the possibilities of obtaining quantitative data with ELSD were explored (Abidi *et al.* 1999, Hopia *et al.* 1992). Dual viscometric/refractometric detection was used recently for simultaneous determination of MW and concentrations in the analysis of used frying oils (Abidi and Warner 2001).

Solid Phase Extraction and HPSEC. Silica column separation can be replaced by solid phase extraction (SPE) using silica gel cartridges (Sébédio *et al.* 1986) or NH_2 cartridges (Hopia *et al.* 1992), thus requiring lower amounts of sample and solvents, and shortening analysis time. This alternative possibility was proposed in a method developed for 50-mg samples, based on the use of SPE (silica gel cartridges) for the separation of nonpolar and polar fractions, and the addition of an internal standard for quantitation purposes (Márquez-Ruiz *et al.* 1996a). Briefly, 2 mL of the sample solution, containing 50 mg of oil and 1 mg of monostearin, used as an internal standard, is placed on the silica cartridge for SPE (Fig. 4.1B). Monostearin is used as the internal standard because MG are normally in negligible, not detectable amounts in fats and oils. The nonpolar fraction is eluted with 15 mL of hexane:diethyl ether 90:10. A second fraction containing polar compounds and the internal standard is eluted with 15 mL of diethyl ether. Nonpolar and polar fractions are evaporated under reduced pressure and redissolved in 1 mL of THF for further analyses, i.e., TLC to determine the efficiency of the separation and HPSEC. Fractions of polar compounds are analyzed by HPSEC under the same conditions previously described. Precision, accuracy, and recovery data were determined. Samples containing levels of polar compounds ranging from 3.7 to 24.3%

were analyzed by both this procedure (SPE-HPSEC method) and that based on gravimetric determination of the polar fraction (silica column-HPSEC method); no significant differences were found between the mean values obtained. However, the lower the concentration of the polar compound, the lower the SD found for the internal standard method. In view of these results, it seems that the SPE-HPSEC method is useful for samples within a wide range of polar compounds and especially adequate for samples of low oxidation level. As an example, Figure 4.5 includes representative HPSEC chromatograms obtained using the SPE-HPSEC method, which show the evolution of nonvolatile oxidation compounds during oxidation. Further comments on this figure are included below.

Transesterification and Combination of Adsorption Chromatography and HPSEC. Information complementary to that obtained by the methodologies outlined above can be attained by analyzing specifically the oxidized fatty acyls included in TG molecules, following the procedure shown in Figure 4.1C, which starts from fatty acid methyl esters (FAME) (Márquez-Ruiz *et al.* 1990).

Briefly, FAME are obtained by transesterification of 1 g of sample with sodium methoxide and hydrochloric acid/methanol. FAME are recovered quantitatively and separated by silica column chromatography, using 150 mL hexane/diethyl ether (88:12, vol/vol) to elute the nonpolar fraction and 150 mL diethyl ether to obtain the polar fraction. Analyses of the nonpolar and polar fractions are performed by HPSEC, under the chromatographic conditions described above. The combined chromatographic analysis permits quantitation of five groups of FAME (see Fig. 4.6). The most abundant nonoxidized FAME and the nonpolar or nonoxi-

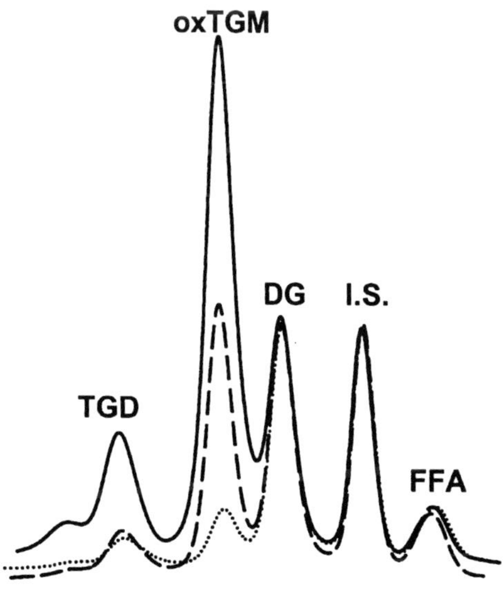

Fig. 4.5. HPSEC chromatograms of polar fractions of sunflower oil during oxidation at 25°C: at the starting point (dotted line), during the early oxidation stage (dashed line), and at the end of the induction period (solid line). Abbreviations: TGD, triacylglycerol dimers; oxTGM, oxidized triacylglycerol monomers; DG, diacylglycerols; I.S., internal standard; FFA, free fatty acids (this peak also includes polar unsaponifiable compounds).

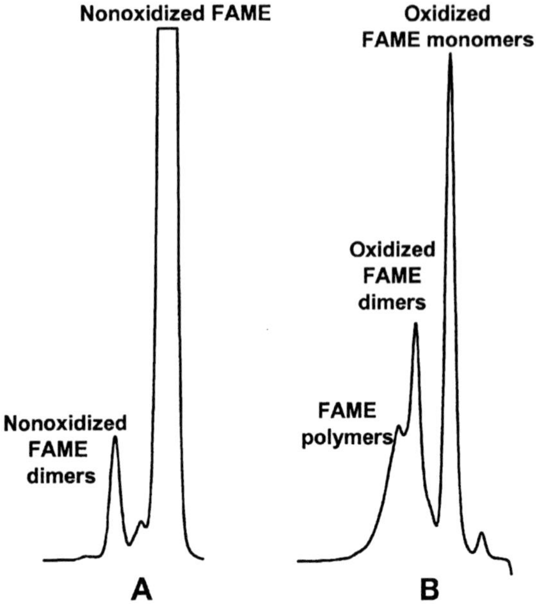

Fig. 4.6. HPSEC chromatograms of non-polar FAME (A) and polar FAME (B) fractions of used frying high-oleic sunflower oil. Abbreviation: FAME, fatty acid methyl esters.

dized FAME dimers, linked by C-C bonds and lacking extra oxygenated functions in their structure, are quantitated in the first fraction (A). FAME polymers, oxidized FAME dimers, and oxidized FAME monomers are determined in turn in the polar fraction (B). Therefore, global quantitation of the compounds eluted in the second fraction provides a measurement of the total oxidized fatty acyl groups included in TG molecules. Isolation of exclusively nonoxidized FAME in the nonpolar fraction can be alternatively achieved using hexane/diethyl ether (95:5, vol/vol) to elute this fraction. Then, the peak of TG dimers in the polar fraction would include both the nonpolar and oxidized types of dimers. Given that quantitation is based on gravimetric determinations and considering the high contribution of unchanged fatty acyl groups in oxidized TG molecules, it is important to note that applications of this methodology present certain limitations in sensitivity for samples of low oxidation level.

Applications in Lipid Oxidation

This section discusses the main applications of HPSEC, essentially of the methodologies described above, for quantitation of nonvolatile oxidation compounds in different aspects of lipid oxidation.

Oxidative Status of Crude and Refined Oils. Refining of crude fats and oils is a common process in edible oils; its purpose is to remove unwanted minor components (phospholipids, FFA, color, and volatile components). Applications of the methodologies described above have added valuable information on the quality evaluation and oxidative status of crude and refined oils. Figure 4.3 presents the HPSEC chromatograms of a virgin olive oil (A), and the polar fractions of the virgin (B) and refined olive oil (C). Because the fraction of polar compounds in crude and refined fats and oils is low (TG normally account for >95% of the sample), no minor compounds can be detected in the entire sample. In contrast, four peaks can be well resolved in the polar fraction, i.e., TGD, oxTGM, DG, and FFA. Quantitation of these groups of compounds in crude oils and samples taken at different stages of the refining process indicated the following: (i) TGD are formed during the deodorization step, whereas they are absent or detected in very low amount in crude oils; (ii) oxTGM and DG remain at levels similar to those present in the crude oils because they are nonvolatile under refining conditions and are neither formed nor eliminated in any of the other processing stages; and (iii) FFA decrease with respect to the crude oils, as expected, due to the neutralization step (Dobarganes *et al.* 1990). Thus, oxTGM measurement is an excellent marker of the oxidative status of the refined oil as well as of the original crude oil. Similarly, DG is an indicator of the degree of hydrolysis in the crude and refined oils. Overall information on the quality of refined oils, which would affect its subsequent performance during storage and utilization for food preparation to a different extent, can be obtained. This analytical approach was adopted in recent years to study the influence of the refining conditions on fats and oils quality (De Greyt *et al.* 1997, Dobarganes *et al.* 1990, Gomes and Caponio 1997 and 1998, Gomes *et al.* 2003a, Hopia 1993a, Ruiz-Méndez *et al.* 1997).

Specifically, for the important issue of characterization of virgin olive oil, the methodology has provided useful parameters, as illustrated in Figure 4.3, i.e., the absence of TGD in the case of virgin olive oils, and the occurrence of TGD plus a high ratio of DG/FFA in the case of refined oils (Dobarganes *et al.* 1989, Pérez-Camino *et al.* 1993). In the last decade, Gomes and co-workers used this analytical approach and published a large number of articles on the quality of virgin and refined olive oils (Caponio *et al.* 2003a, Gomes 1992 and 1995; Gomes *et al.* 1997, 2002a, 2002b, and 2003b).

Oxidation Kinetics. Trilinolein (LLL) was used as a model unsaturated TG to evaluate modifications during oxidation and gain insight into the oxidation kinetics (Márquez-Ruiz 1996b and 2003a). Through quantitation of the three groups of nonvolatile oxidation compounds formed, i.e., oxidized LLL monomers, LLL dimers, and LLL polymers, the influence of both temperature (25, 60, and 100°C) and the addition of α-tocopherol on oxidation kinetics was examined. In samples with antioxidant, the considerable delay in the formation of nonvolatile oxidation compounds allowed the clear distinction of two stages (see Fig. 4.7 for illustration of the oxidation progress pattern): first, a period characterized by slow progress of oxidation, or induction period (IP) and, second, an accelerated oxidation stage. The end of the IP could therefore be defined as the time point at which a notable shift

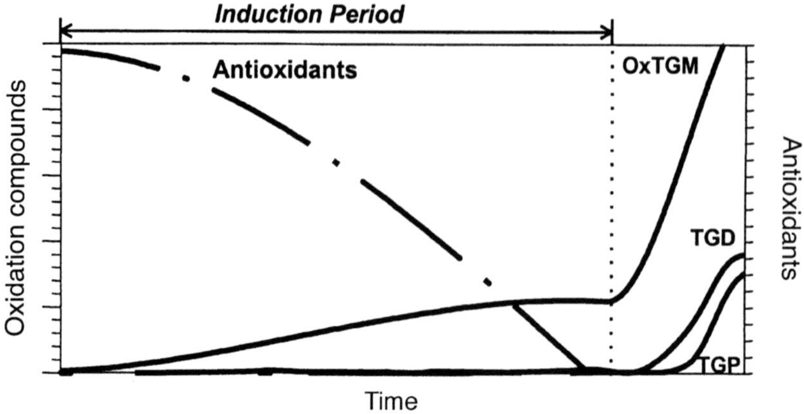

Fig. 4.7. General oxidation profile found in model triacylglycerols and oils oxidized at low temperatures. Abbreviations: oxTGM, oxidized triacylglycerol monomers; TGD, triacylglycerol dimers; TGP, triacylglycerol polymers.

in oxidation rate was observed. The only group of compounds increasing during the early oxidation stage was the group of oxidized LLL monomers, comprised mainly of hydroperoxides at that early stage. At the end of the IP, oxidation was accelerated, as shown by the sharp increase in oxidized LLL monomers, significant formation of polymerization products, and exhaustion of antioxidants. In general, increases of ~1% in dimer concentrations indicated the start of the accelerated phase at all temperatures tested.

As expected, the main effect of the increase in temperature was the decrease in the IP, but an additional and very important observation was the influence of temperature on the amounts of primary oxidation compounds (oxidized LLL monomers) accumulated at the end of the induction period, which decreased as the temperature increased, indicating that the slope of the initial linear stage of oxidation depended on temperature. This was probably related to the effect of temperature on antioxidant degradation. Therefore, polymerization started at very different levels of primary oxidation products, depending on temperature. Such differences were clearly reflected in the ratio of oxidized monomers-to-polymerization compounds obtained at 25, 60, and 100°C. For example, for similar levels of total oxidation compounds (27.9–29.0%), that ratio was ~20:1, 9:1, and 3:1, respectively.

The kinetic parameters were calculated from the experimental data, considering that oxidized LLL monomers are, in practice, the only products formed during the early stages of oxidation, and do not participate in other side reactions during this period. Values found for the reaction order did not differ from 0 when the antioxidant was present, thus indicating that the increase in oxidized LLL monomers was linear during the induction period. In addition, the influence of temperature on the oxidation rate during the induction period was examined on the basis of the Arrhenius law, and a linear relation was obtained between ln IP and 1/T (T = absolute temperature) in samples with α-tocopherol added at 25, 60, and 100°C, thus reflecting that assays at 60 or 100°C could be useful in foreseeing the IP at room temperature.

Oxidation during Storage of Oils and Foods. Quantitation of nonvolatile oxidation compounds by the silica column-HPSEC or the SPE-HPSEC methodologies proved to be relevant measurement of oxidation for practical applications, i.e., evaluation of oxidative stability and examination of the evolution of oxidation during storage of oils and foods.

To illustrate the changes observed during oxidation, three HPSEC chromatograms of polar fractions of sunflower oil oxidized at 25°C were overlapped in Figure 4.5 as follows: (i) at the starting point (dotted line); (ii) during the early stage of oxidation or IP (dashed line), and (iii) at the end of the IP (solid line). Changes in the groups of compounds resolved are well illustrated, given that polar fractions were dissolved in equal volumes of solvent. Thus, the peak corresponding to monostearin, used as an internal standard (methodology described above), is of similar magnitude in all chromatograms. Values for the starting oil, expressed as wt% on oil, were 0.9% oxTGM and 0.5% TGD, levels commonly found in refined oils. OxTGM comprised the only group of compounds increasing during the early stage of oxidation, as illustrated in the polar fraction by the dashed line, separated from a sample that contained 2.9% oxTGM and was not still rancid. Finally, the polar fraction of a sample taken right after the end of the IP (solid line), when tocopherols were exhausted, had a considerable amount of oxTGM (17.7% on oil) and a significant increase in TGD (1.4% on oil), thus indicating that the sample was entering the period of advanced oxidation.

Quantitation of oxTGM was proposed as a measurement of oxidation in fats, oils, and foods; it was found that this group of compounds experienced the largest increase during early oxidation, before detection of rancidity (Pérez-Camino *et al.* 1990 and 1991). Similar results were found in studies in which TG mixtures and various edible oils were autoxidized (Hopia 1993b, Hopia *et al.* 1993).

In the recently published results of a 3-y long study, the evolution of oxidation in sunflower oils differing in the degree of unsaturation was followed in detail to determine changes in nonvolatile oxidation compounds during long-term storage at room temperature, under well-controlled conditions (Martín-Polvillo *et al.* 2004). Sunflower oils differing exclusively in degree of unsaturation, namely, conventional high-linoleic sunflower oil (HLSO), genetically modified high-oleic sunflower oil (HOSO), and a 1:1 mixture of the two were used. The oxidation pattern found was the same as that obtained for LLL and is represented in Figure 4.7. As expected, the IP length depended on the degree of unsaturation, although the oil mixture prepared using equal amounts of HLSO and HOSO had values markedly closer to those found for the most unsaturated oil. Overall, two important findings stood out. First, the amounts of oxTGM accumulated at the end of the IP increased as the unsaturation degree of the oil increased, and was more than double for HLSO compared with HOSO. Second, the relation between the IP values obtained at room temperature was similar to that between the oil stability indices as determined by Rancimat at 100°C, thus reflecting the utility of this determination to predict the shelf-life of oil.

The results obtained using the analytical methodology were compared with the data provided through determination of peroxide value (PV) and UV absorption at 270 nm (K_{270nm}), two indices commonly used to evaluate primary and secondary oxidation products, respectively. PV determination is widely applied to evaluate the extent of oxidation in fats, oils, and food lipids. As a measurement of hydroperoxide formation, it has been recognized as a useful index for the early stages of oxidation. PV reaches a maximum during the progress of oxidation followed by a decrease when the rate of decomposition of hydroperoxides exceeds the rate of their formation at more advanced stages, which varies according to the degree of unsaturation and the storage conditions (Frankel 1998a). Results obtained for PV were compared with those of oxTGM, which, as already stated, are comprised primarily of hydroperoxides during the early stage of oxidation. The relation between the two determinations for all oil samples within the early oxidation stage (up to TGD concentrations of ~1%) showed an excellent correlation that was independent of the degree of unsaturation of the oil. Once oxidation accelerated, this relation became complex because secondary oxidation products were formed. Thus, TG-containing oxygenated functions other than hydroperoxide (e.g., epoxy, keto, or hydroxy) begin to contribute to the amount of oxTGM. Hydroperoxide functions are present not only in primary oxidation compounds but are also involved in dimeric linkages of polymerization compounds (Dobarganes and Márquez-Ruiz 1996). K_{270nm} constitutes a measurement of conjugated trienes as well as ethylenic diketones and conjugated ketodienes produced from polyunsaturated lipids. Even though this index does not provide quantitative data, it has been used traditionally to evaluate secondary oxidation products. Only slight changes in K_{270nm} were detected during the induction period; however, once oxidation accelerated and tocopherol was exhausted, a significant increase was observed, which was parallel to the formation of polymerization compounds.

The results obtained in that study showed that the evolution of oxidation in the sunflower oils tested was very similar to that observed earlier in LLL model systems. Thus, similar kinetic considerations were applied, with the conclusion that an increment in the reaction constant k occurred as the degree of unsaturation increased.

It is important to note that in the case of highly unsaturated oils, polymerization is very rapid at low temperatures because of the high instability of unsaturated hydroperoxides; hence, the simple determination of polymers by direct application of HPSEC was used satisfactorily for quality evaluation of commercial fish oil capsules (Sagredos 1992, Shukla and Perkins 1991) and routine assessment of fish oil quality (Burkow and Henderson 1991). In this context, advantages of polymer determination vs. the thiobarbituric acid reactive substance value and polyene index to monitor oxidation of fish oils during storage were also reported (Márquez-Ruiz *et al.* 2000).

With respect to the evolution of nonvolatile oxidation compounds during the storage of foods, in the course of an extensive project, chips were prepared industrially with conventional HLSO, HOSO, and palm olein (PO), and stored at room temperature for up to 6 mon. Initial values for oxTGM in chips indicated that frying performance had been excellent; after storage for 25 wk, only HLSO chips showed a considerable rise specifically in oxTGM, and changes from initial values were significant even at 15 wk. The rest of the nonvolatile oxidation compounds quantitated remained at the initial levels. Quite in contrast, HOSO and PO samples presented roughly the same oxidation levels as initially after 25 wk, thus showing a notable shelf-life at room temperature (Martín-Polvillo *et al.* 1996). Interestingly, these results were in excellent agreement with parallel sensory assessments by a panel test, showing that HLSO chips were distinctly rancid from wk 17, whereas HOSO and PO behaved similarly, maintaining fruity characteristics for odor and taste for >6 mon (Raoux *et al.* 1996).

Further studies on oxidative stability at moderate temperature of fried potatoes prepared in HLSO and HOSO and differing in initial levels of nonvolatile oxidation compounds revealed some insight into the different behavior of α-tocopherol depending on the temperature, and underlined the importance of the remaining level of natural antioxidants in food oils to stop rapid initiation of oxidation during storage (Márquez-Ruiz *et al.* 1999a).

Evaluation of the oxidative status of oils extracted from commercialized products by quantitation of oxTGM, TGD and TGP recently came into general use, and the products tested include snack foods, fried and bakery products (Piispa *et al.* 1996), and, more recently, margarines (Caponio *et al.* 2002a and 2003b, Caponio and Gomes 2004), the oils used for covering canned fish and vegetable foodstuffs (Caponio *et al.* 2002b and 2003c, Gomes *et al.* 1998), bouillon cubes, and condiments (Caponio *et al.* 2001a and 2002c).

Oxidation in Dispersed Lipids. Lipid oxidation in systems in which the fat or oil is dispersed as droplets in emulsions or encapsulated in dried products is poorly understood. In oil-in-water emulsions, lipid droplets are dispersed in a continuous water phase, stabilized by proteins, phospholipids, or surfactants. Some examples of the numerous foods constituted by oil-in-water food emulsions include milk, mayonnaises, salad dressings, infant foods, creams, and soups (Frankel 1998c). In turn, through the process of oil microencapsulation, natural or formulated oil-in-water emulsions are dried to obtain a powdery ingredient in which oil droplets are surrounded by a matrix of proteins and/or carbohydrates intended to protect sensitive oils, mask or preserve flavors and aromas (Balassa and Fanger 1971, Gibbs *et al.* 1999, Shahidi and Han 1993). The most relevant formulated microencapsulated oils are infant formulas, flavoring additives, pigments, and microencapsulated fish oils; the last-mentioned are used as functional ingredients in a growing number of dairy and bakery products. In addition, there is a large group of microencapsulated oils obtained from drying natural foods such as milk powders, dried eggs, or dehydrated soups and sauces. Finally, part of the lipids in some natural products, such as dried nuts, may also be found as an encapsulated fraction.

The evaluation of oxidation in dispersed lipids is even more complicated than it is in bulk lipids because of the influence of a multitude of additional variables. In emulsions, the nature of the droplet membrane or interface, the interactions between ingredients, and the partitioning of reactants and products of oxidation among the oil, aqueous and interfacial regions may play an important role (Genot *et al.* 2003, Velasco *et al.* 2002). In the case of dried microencapsulated oils, pH, water activity, particle size, globule size distribution, and additional factors derived from the presence of the other matrix components and the heterogeneous lipid distribution are also involved (Velasco *et al.* 2003).

Oxidation in Emulsions. The action of antioxidants in emulsions is one of the important aspects that warrant special interest currently because it is not predictable from their activity in bulk oils (Frankel *et al.* 1994, Huang *et al.* 1994). Studies on the evolution of oxidation and the efficiency of phenolic antioxidants in sunflower oil-in-water emulsions through quantitation of nonvolatile compounds by SPE-HPSEC were published recently (Velasco *et al.* 2004). Two groups of phenolic antioxidants that are structurally similar were tested: (i) α-tocopherol and its water-soluble analog, Trolox; and (ii), gallic acid and its ester derivatives, propyl gallate and dodecyl gallate.

By comparing the time intervals within which oxidation accelerated, it is clear that α-tocopherol was more effective than its hydrophilic analog, Trolox (Fig. 4.8). Thus, oxidation accelerated at 10–14 h for Trolox and 26–32 h for α-tocopherol in emulsions stabilized by Tween-20. The protection conferred by α-tocopherol was ~3 times higher than that provided by its polar counterpart in both emulsions stabilized by Tween-20 and those containing sodium caseinate and lactose (these latter data not shown). Similarly, it was found that dodecyl gallate, the least polar and most lipophilic compound among the gallic acid ester derivatives tested, was the most effective because oxidation accelerated at 3–4 h for propyl gallate and 8–10 h for dodecyl gallate (Fig. 4.8). The lower activity of hydrophilic vs. lipophilic antioxidants in oil-in-water emulsions has been related to their partition behavior between the different phases and attributed in part to their tendency to concentrate in the aqueous phase and/or higher affinity for certain emulsifiers (Frankel 1998c). On the other hand, results obtained in this work showed that substantial amounts of α-tocopherol coexisted with significant polymerization; this was indicative of the heterogeneity of oxidation, i.e., differences of oxidation rate in oil droplets.

Quite a different use of the HPSEC technique to evaluate the action of antioxidants in emulsions consisted in the detection of hydroperoxy TG and hydroperoxy cholesterol esters after reduction, using the postcolumn fluorometric diphenyl-1-

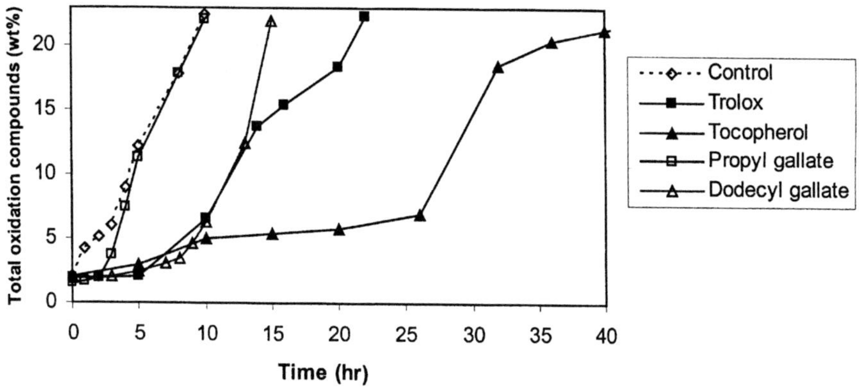

Fig. 4.8. Effect of antioxidants on formation of total nonvolatile oxidation compounds (sum of oxidized triacylglycerol monomers, dimers, and polymers) in sunflower oil-in-water emulsions oxidized under accelerated conditions (Cu-catalyzed, 40°C).

pyrenylphosphine oxidation principle (Hartvigsen *et al.* 2000). In that study, the addition of Panodan TR DATEM emulsifier to fish oil-enriched mayonnaises reduced formation of hydroperoxy TG during storage, whereas opposite effects were found for gallic acid and EDTA.

Oxidation in Dried Microencapsulated Oils. One important factor influencing oxidation in foods in which there is a noncontinuous lipid phase, as in dried microencapsulated oils, is the coexistence of a portion of the lipid phase, which is hexane-extractable and usually called free, surface, or nonencapsulated oil, and a portion of noncontinuous lipid phase, also known as encapsulated oil, wherein lipids are in droplets and whose extraction requires previous disruption of the matrix structure (Fritsch 1994). This leads to difficulty in interpreting the real oxidation status from data obtained from the analysis of the total lipids extracted. For example, external oxidation (in the surface oil) might induce rancidity even if the encapsulated oil has a low oxidation level, whereas otherwise, rancidity might not be detected until the oxidized encapsulated oil is released. Moreover, evolution of oxidation in the noncontinuous or dispersed lipid phase may become very complex because lipid droplets are isolated from each other in the matrix and, consequently, different oxidation rates can occur in different droplets.

Recently, an analytical approach consisting of the quantitation of nonvolatile oxidation compounds by SPE-HPSEC in the surface and encapsulated lipid phases, extracted separately and quantitatively, allowed detection of important differences in the progress of oxidation in both phases, characterized by distinct oxidation patterns (Márquez-Ruiz *et al.* 2000 and 2003b, Velasco *et al.* 2000). A representative example of the results obtained for the surface and encapsulated fractions is shown in Figure 4.9, in this case for dried microencapsulated sunflower oils stored at

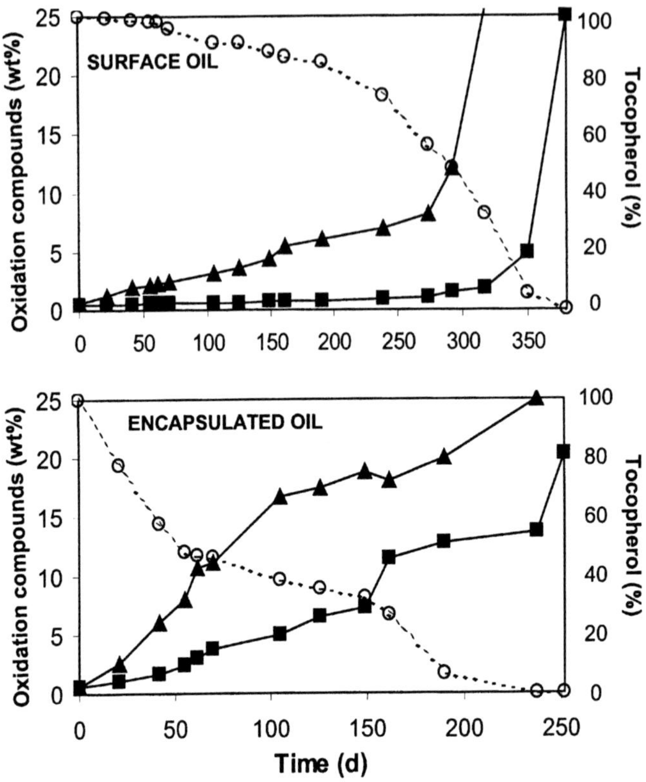

Fig. 4.9. Evolution of oxidized triacylglycerol monomers (▲), triacylglycerol dimers plus polymers (■) and tocopherol content (○) in dried microencapsulated oils stored at 25°C.

25°C. Oxidation was more rapid in the encapsulated oil fraction even though, theoretically, the more accessible, external or surface oil was not protected by the matrix and was more exposed to oxidation. In other experiments, using infant formulas, either the inverse or similar oxidation rates were observed, indicating that the great number of variables influencing oxidation in these systems exerts a crucial role in the relative oxidation rate of the surface and encapsulated fractions. This notion was already pointed out by Fritsch in a paper (Fritsch 1994) that stressed the fact that lipid distribution is of paramount importance in food oxidation and is still too often ignored.

Interestingly, the oxidation profile of this surface fraction was similar to that obtained for bulk oils (Fig. 4.7), typical of lipids in continuous phase (monophasic lipid systems), and characterized by increase of hydroperoxides during the induction period (measured as oxTGM) and a clear end of the induction period marked by the initiation of polymerization and exhaustion of tocopherols (here at ~350 d). However, the oxidation profile of encapsulated oil was rather unusual, i.e., high polymer values were found in samples with high levels of tocopherol remaining. For example, after 150 d, encapsulated oil fractions contained as much as 8% polymers and still had ~30% residual tocopherol. At this point, samples did not have objectionable flavor when tested as intact samples despite the high level of oxidation, but rancidity could be detected when the encapsulated oil was released. It was also notable that both polymer increase and tocopherol loss showed a "shifting" or "uneven" profile. Overall results reflected the coexistence of oil globules with a wide range of oxidation status, likely including some at low stages of oxidation and still protected by the presence of tocopherol, and others devoid of antioxidants and well within the advanced oxidation stage. Therefore, analysis of the noncontinuous phase of microencapsulated oils provided a profile typical of a mixture of oils showing different oxidation rates. The heterogeneity of oxidation observed in these studies is of great importance because this phenomenon may occur in a large number of foods and biological systems constituted by dispersed oils.

Oxidation at High Temperatures. The main culinary process that involves oxidation at high temperature is frying. During frying, thermal, oxidative, and hydrolytic reactions take place; thus, a complex mixture of new compounds is formed. Quantitation of the polar compounds formed by means of silica columns was proposed by IUPAC for quality control of used frying oils (IUPAC 1987, Waltking and Wessels 1981); it is currently included in some European regulations that limit polar compounds for human consumption to ~25% (Firestone 1996). In addition, quantitation of polymerized TG seemed also to be valuable in the area of heated and used frying oils because polymerized TG are major compounds among the degradation compounds formed. In fact, good correlations were found between amounts of polymers and polar compounds (Gere 1982, Perrin *et al.* 1985, Schulte 1982). Analysis of polymerized TG by HPSEC stands out for its simplicity because it is necessary only to dilute the oil or fat in the appropriate solvent; the chromatographic determination is short and performed with a single solvent. As a consequence, after two interlaboratory tests carried out in 1986-87, the IUPAC Commission on Oils, Fats and Derivatives adopted a method for the determination of polymerized TG in used frying fats and oils for samples containing not <3% polymers (IUPAC 1992, Wolff *et al.* 1991). The method proposes a single column of 30 cm × 0.77 cm i.d. packed with a high-performance spherical gel made of copolystyrene divinyl benzene of 5 mm, THF as the mobile phase, and a refractive index detector with a sensitivity at full scale at least 1×10^{-4} of the refractive index. Sample concentration suggested was 50 mg/mL for an injection valve with a 10 μL loop. The analysis time is ~10 min at a flow rate of 1 mL/min. At present, it is a commonly used method for the analysis of used frying oils and fats (Gertz 2000, Gertz and Kochlar 2002, Hansen *et al.* 1994, Kiatsrichart *et al.* 2003, Lampi *et al.* 1999, Masson *et al.* 1997, Neff *et al.* 2003, Soheili *et al.* 2002a and 2002b).

In this context, a new index called OSET (oxidative stability at elevated temperature) was described recently for estimating the stabilizing activity of additives at simulated frying temperature. The test consists of accelerating polymerization with acid-catalyzed silica gel; oils or fats are heated for 2 h at 170°C in the presence of the additive tested, and HPSEC is used to evaluate dimeric and polymeric TG contents (Gertz and Kochlar 2002).

The methodology based on silica column-HPSEC has proved to be an excellent alternative for the evaluation of used frying oils. In addition to providing both the determination of total polar compounds and polymerized TG, broader knowledge is gained of the different groups of nonvolatile oxidation compounds and hydrolytic products formed (Dobarganes 1998, Dobarganes *et al.* 1988 and 1999, Dobarganes and Márquez-Ruiz 1996). In fact, very different patterns of polar compound distribution were obtained for samples with a similar content of total polar compounds (Dobarganes *et al.* 1988). As shown in Figure 4.4, the advantages offered by the combined technique are clearly reflected in the HPSEC profiles obtained by simply injecting the entire oil samples and those corresponding to the polar compound fractions, i.e., increased sensitivity in polymer quantitation, overcoming the limitation of the IUPAC method to a minimum of 3% content for analyses of total samples by HPSEC (IUPAC 1992), and differentiation of thermally oxidized compounds (oxTGM, TGD and TGP) from hydrolytic products (DG and FFA). The contribution of both types of products to the amount of polar compounds helps determine the real nutritional relevance of the 25% polar compound limitation established because thermally oxidized products are those associated with negative physiologic effects, whereas hydrolytic compounds are products naturally released from lipolysis in the gut before absorption.

Results obtained using this methodology in recent years have contributed to improved knowledge of important issues in the frying process: (i) Composition of oils absorbed by the fried food and lipid interchange between frying oil and food (Jorge *et al.* 1996a, Pérez-Camino *et al.* 1992, Pozo-Díaz *et al.* 1995, Sébédio *et al.* 1996). (ii) Contribution of hydrolysis among the compounds formed in the frying process (Arroyo *et al.* 1995, Dobarganes *et al.* 1993, Masson *et al.* 1997, Pérez-Camino *et al.* 1992). (iii) Thermal stability and frying performance of oils from genetically modified sunflower seeds (Dobarganes *et al.* 1993, Márquez-Ruiz *et al.* 1999b, Sébédio *et al.* 1996). (iv) Action of the main variables involved in the frying process, i.e., length of heating, temperature, surface-to-oil volume ratio and degree of unsaturation (Jorge *et al.* 1996b). (v) Differences between continuous and discontinuous frying (Jorge *et al.* 1996a). (vi) Relations between loss of antioxidants and formation of degradation compounds, in connection with the degree of unsaturation (Barrera-Arellano *et al.* 1999 and 2002, Verleyen *et al.* 2001 and 2002). (vii) Effect of the addition of dimethylpolysiloxane to frying oil (Jorge *et al.* 1996a and 1996b). (viii) Evaluation of performance of different oils during discontinuous frying (Abidi *et al.* 2003, Abidi and Warner 2001, Arroyo *et al.* 1995, Bastida *et al.* 2001a and 2001b, Houhoula *et al.* 2003, Masson *et al.*

1997, 1999 and 2002). (ix) Effect of oil replenishment during frying on oil quality (Cuesta *et al.* 1993, Cuesta and Sánchez-Muniz 1998, Romero *et al.* 1995, 1998, and 1999). In addition, changes resulting from other treatments at high temperature, such as microwave heating, were evaluated (Caponio *et al.* 2001b and 2002d).

On the other hand, through application of the methodology on the FAME of used frying oils (Fig. 4.1C), broader information on the degradation of used frying oils can be obtained because the amount of affected fatty acyl groups is measured exclusively (Jorge *et al.* 1997, Márquez-Ruiz *et al.* 1990 and 1995a). Figure 4.6 shows FAME nonpolar and polar fractions of used frying HOSO at the limit of rejection (24.3% polar compounds). Total altered FAME accounted for 8.8% on total oil, distributed in 2.9% oxidized FAME monomers, 1.8% oxidized FAME dimers, 3.0% nonpolar FAME dimers and 1.1% FAME polymers.

As illustrated in this example, evaluation of used frying fats collected by Food Inspection Services in Spain using both the silica column-HPSEC and transesterification-silica column-HPSEC procedures showed that samples with polar compound levels around the limit for rejection (21.1–27.6% polar compounds) gave values of total altered FAME from 8.1 to 11.3% and, among them, substantial amounts of oxidized FAME monomers (~30 mg/g oil) (Márquez-Ruiz *et al.* 1995a). In this group, a wide array of oxygenated compounds with unknown nutritional implications is included (Dobarganes and Márquez-Ruiz 2003, Márquez-Ruiz and Dobarganes 1996). Also, results offered some insight into the complexity of the TGP structure, by comparing TG and FAME dimer and polymer values. In general, the low FAME polymers-to-TG polymers ratios in contrast to the FAME dimers-to-TG dimers ratios revealed the considerable contribution of dimeric linkages to the structures of trimeric and higher oligomeric TG (Márquez-Ruiz *et al.* 1995a).

By virtue of the analytical approach starting from FAME, the digestibility of the five groups of FAME quantitated could be determined (Márquez-Ruiz *et al.* 1992a, 1993, and 1995b). The high digestibility coefficients found for oxidized fatty acid monomers indicated that such oxidized compounds are of utmost importance from the nutritional standpoint, supported also by their quantitative relevance in the diet. On the other hand, the digestibility of nonoxidized fatty acids was negatively dependent on the global alteration level of the dietary oil. This finding was later attributed to impaired hydrolysis of TGP and TGD, which include in part nonoxidized fatty acids, due to the difficulties involved in the pancreatic lipase action on complex glyceridic molecules (Márquez-Ruiz *et al.* 1992b and 1998). For these latter experiments, quantitation of the complex mixtures of partial glycerides obtained after lipolysis, including nonhydrolyzed TGD and TGP, was essential and was successfully achieved by direct application of HPSEC. In recent years, Sánchez-Muniz and co-workers have continued to investigate the *in vitro* and *in vivo* digestibility of used frying oils and fats (Arroyo *et al.* 1996, González-Muñoz *et al.* 1996, 1998 and 2003, Sánchez-Muniz *et al.* 1999 and 2000).

Concluding Remarks

The simplicity and high reproducibility of HPSEC in combination with adsorption chromatography have contributed significantly to the development of new analytical methodologies in lipid analysis. After the initial application in the area of frying fats and oils, the potential of the technique for oxidation studies was clearly demonstrated in the last decade. In particular, the following applications stand out:

1. In contrast to the evaluation of oxidation status through complementary analytical indices, HPSEC in combination with adsorption chromatography permits accurate quantitation of the total primary and secondary oxidation compounds in a single analysis.
2. Determination of the main groups of oxidation compounds is of indisputable utility to clarify the oxidation profile in different lipid systems and to gain information on kinetic parameters.
3. Furthermore, application of the methodologies described above even permits the detection of the main differences between oxidation in a continuous lipid phase (monophasic lipid systems) and a noncontinuous lipid phase.

Among future applications, its use as a complementary preparative technique to concentrate oxidation compounds is foreseen. In this respect, HPSEC might constitute an excellent analytical tool to facilitate the analysis of specific oxidation compounds by other chromatographic techniques.

Acknowledgment

This work was funded in part by Ministerio de Ciencia y Tecnología (Project 2001-0505).

References

Abidi, S.L., and Warner, K. (2001) Molecular-Weight Distributions of Degradation Products in Selected Frying Oils, *J. Am. Oil Chem. Soc. 78*, 763–769.

Abidi, S.L., and Rennick, K.A. (2003) Determination of Nonvolatile Components in Polar Fractions of Rice Bran Oils, *J. Am. Oil Chem. Soc. 80*, 1057–1062.

Abidi, S.L., Kim, I.H., and Rennick, K.A. (1999) Determination of Nonvolatile Components of Heated Soybean Oils Separated with High-Efficiency Mixed-Bed Polystyrene/Divinylbenzene Columns, *J. Am. Oil Chem. Soc. 76*, 939–944.

Arroyo, R., Cuesta, C., Sanchez-Montero, J.M., and Sanchez-Muniz, F. (1995) High Performance Size Exclusion Chromatography of Palm Olein Used for Frying, *Fat Sci. Technol. 95*, 292–296.

Arroyo, R., Sanchez-Muniz, F.J., Cuesta, C., Burguillo, F.J., and Sanchez-Montero, J.M. (1996) Hydrolysis of Used Frying Palm Olein and Sunflower Oil Catalyzed by Porcine Pancreatic Lipase, *Lipids 31*, 1133–1139.

Balassa, L.L., and Fanger, G.O. (1971) Microencapsulation in the Food Industry, *CRC Crit. Rev. Food Technol. July*, 245–265.

Balke, S.T., Mourey, T.H., and Schunk, T.C. (1999) Size Exclusion Chromatography: Practical Methods for Quantitative Results, *Polym. React. Eng. 7*, 429–452.

Balke, S.T., Mourey, T.H., and Karami, A. (2000) Evaluating Size Exclusion Chromatography Fractionation, *Int. J. Polym. Anal. Charact. 6*, 13–33.

Barrera-Arellano, D., Ruiz-Méndez, M.V., Márquez-Ruiz, G., and Dobarganes, M.C. (1999) Loss of Tocopherols and Formation of Degradation Compounds in Triacylglycerol Model Systems Heated at High Temperature, *J. Sci. Food Agric. 79*, 1923–1928.

Barrera-Arellano, D., Ruiz-Méndez, M.V., Velasco, J., Márquez-Ruiz, G., and Dobarganes, M.C. (2002) Loss of Tocopherols and Formation of Degradation Compounds at Frying Temperatures in Oils Differing in Degree of Unsaturation and Natural Antioxidant Content, *J. Sci. Food Agric. 82*, 1696–1702.

Barth, H.G., and Boyes, B.E. (1990) Size Exclusion Chromatography, *Anal. Chem. 62*, 381R–394R.

Barth, H.G., and Boyes, B.E. (1992) Size Exclusion Chromatography, *Anal. Chem. 64*, 428R–442R.

Barth, H.G., Boyes, B.E., and Jackson, C. (1994) Size Exclusion Chromatography, *Anal. Chem. 66*, 595R–620R.

Barth, H.G., Boyes, B.E., and Jackson, C. (1996) Size Exclusion Chromatography, *Anal. Chem. 68*, 445R–466R.

Barth, H.G., Boyes, B.E., and Jackson, C. (1998) Size Exclusion Chromatography and Related Separation Techniques, *Anal. Chem. 70*, 251R–278R.

Bastida, S., and Sańchez-Muniz, F.J. (2001a) Thermal Oxidation of Olive Oil, Sunflower Oil and a Mix of Both Oils During Forty Discontinuous Domestic Fryings of Different Foods, *Food Sci. Technol. Int. 7*, 15–21.

Bastida, S., and Sańchez-Muniz, F.J. (2001b) Polar Content vs. TAG Oligomer Content in the Frying-Life Assessment of Monounsaturated and Polyunsaturated Oils Used in Deep-Frying, *J. Am. Oil Chem. Soc. 79*, 447–451.

Burkow, I.C., and Henderson, R.J. (1991) Isolation and Quantification of Polymers from Autoxidized Fish Oils by High-Performance Size-Exclusion Chromatography with an Evaporative Mass Detector, *J. Chromatogr. 552*, 501–506.

Caponio, F., and Gomes, T. (2004) Examination of Lipid Fraction Quality of Margarine, *J. Food Sci. 69*, 97–100.

Caponio, F., Gomes, T., Bilancia, M.T., and Delcuratolo, D. (2001a) Quality of the Fat Fraction of Stock Cube and Condiment, *Riv. Ital. Sostanze Grasse 78*, 593–596.

Caponio, F., Pasqualone, A., Bilancia, M.T., Sacco, D., Delcuratolo, D., and Gomes, T. (2001b) Measure of Thermo-Oxidation of Vegetable Oils Microwave-Heated in Model Systems, *Ind.-Aliment. 40*, 628–632.

Caponio, F., Gomes, T., Delcuratolo, D., and Bilancia, M.T. (2002a) Quality of the Lipid Fraction of Vegetable Margarines, *Riv. Ital. Sostanze Grasse 79*, 59–62.

Caponio, F., Gomes, T., and Summo, C. (2002b) Use of HPSEC Analysis of Polar Compounds in the Ascertainment of the Degradation Level of Oils Utilised as Liquid Medium in Canned Tuna, *Riv. Ital. Sostanze Grasse 79*, 285–288.

Caponio, F., Gomes, T., and Delcuratolo, D. (2002c) Qualitative and Quantitative Characterisation of the Lipid Fraction of Bouillon Cubes, *Eur. Food Res. Technol. 215*, 200–203.

Caponio, F., Pasqualone, A., and Gomes, T. (2002d) Effects of Conventional and Microwave Heating on the Degradation of Olive Oil, *Eur. Food Res. Technol. 215*, 114–117.

Caponio, F., Gomes, T., Summo, C., and Pasqualone, A. (2003a) Influence of the Type of Olive-Crusher Used on the Quality of Extra Virgin Olive Oils, *Eur. J. Lipid Sci. Technol. 105*, 201–206.

Caponio, F., Gomes, T., and Bilancia, M.T. (2003b) Measurement of Degradation of the Lipid Fraction in Margarines, *Eur. Food Res. Technol. 216*, 83–87.

Caponio, F., Gomes, T., and Summo, C. (2003c) Assessment of the Oxidative and Hydrolytic Degradation of Oils Used as Liquid Medium of In-Oil Preserved Vegetables, *J. Food Sci. 68*, 147–151.

Christie, W.W. (1987) *High-Performance Liquid Chromatography and Lipids. A Practical Guide*, Pergamon Press, Oxford.

Christie, W.W. (1995) Size Exclusion Chromatography of Lipids, *Lipid Technol. 7*, 7–18.

Cuesta, C., and Sánchez-Muniz, F.J. (1998) Quality Control During Repeated Fryings, *Grasas Aceites 49*, 310–318.

Cuesta, C., Sánchez-Muniz, F.J., Garrido-Polonio, C., López-Varela, S., and Arroyo, R. (1993) Thermoxidative and Hydrolytic Changes in Sunflower Oil Used in Fryings with a Fast Turnover of Fresh Oil, *J. Am. Oil Chem. Soc. 70*, 1069–1073.

De Greyt, W.F., Kellens, M.J., and Huyghebaert, A.D. (1997) Polymeric and Oxidized Triglyceride Content of Crude and Refined Vegetable Oils, *Fett/Lipid 99*, 287–290.

Dobarganes, M.C. (1998) Formation and Analysis of High-Molecular Weight Compounds in Frying Fats and Oils, *OCL 5*, 41–47.

Dobarganes, M.C., and Márquez-Ruiz, G. (1993) Size Exclusion Chromatography in the Analysis of Lipids, In: *Advances in Lipid Methodology–Two* (Christie, W.W., ed.), pp. 113–137, The Oily Press, Dundee, Scotland.

Dobarganes, M.C., and Márquez-Ruiz, G. (1995) High-Performance Size-Exclusion Chromatography Applied to the Analysis of Edible Fats, In: *New Trends in Lipid and Lipoprotein Analysis* (Sébédio, J.L., and Perkins E.G., eds.), pp. 81–92, AOCS Press, Champaign, IL.

Dobarganes, M.C., and Márquez-Ruiz, G. (1996) Dimeric and Higher Oligomeric Triglycerides, In: *Deep Frying: Chemistry, Nutrition and Practical Applications* (Perkins, E.G., and Erickson, M.D., eds.), pp. 89–111, AOCS Press, Champaign, IL.

Dobarganes, M.C., and Márquez-Ruiz, G. (1998) Analytical Evaluation of Fats and Oils by Size-Exclusion Chromatography, *Analysis 26*, M61–M65.

Dobarganes, M.C., and Márquez-Ruiz, G. (2003) Oxidized Fats in Foods, *Curr. Opin. Clin. Nutr. Metab. Care 6*, 157–163.

Dobarganes, M.C., Pérez-Camino, M.C., and Márquez-Ruiz, G. (1988) High Performance Size Exclusion Chromatography of Polar Compounds in Heated and Non-Heated Fats, *Fat Sci. Technol. 90*, 308–311.

Dobarganes, M.C., Pérez-Camino, M.C., and Márquez Ruiz, G. (1989) Application of Minor Glyceridic Component Determination to the Evaluation of Olive Oil, In: *Actes du Congres International Chevreul pour l'Etude des Corps Gras. Premier Congrès Eurolipid*, Vol. 2, pp. 578–584, Angers, France.

Dobarganes, M.C., Pérez-Camino, M.C., Márquez-Ruíz, G. and Ruíz-Méndez, M.V. (1990) New Analytical Possibilities for Quality Evaluation of Refined Oils. In: *Edible Fats and Oils: Basic Principles and Modern Practices* (Erickson, D.R., ed.) pp. 427–429, AOCS Press, Champaign, IL.

Dobarganes, M.C., Márquez-Ruiz, G., and Pérez-Camino, M.C. (1993) Thermal Stability and Frying Performance of Genetically Modified Sunflower Oils, *J. Agric. Food Chem. 41*, 678–681.

Dobarganes, M.C., Márquez-Ruiz, G., Berdeaux, O., and Velasco, J. (1999) Determination of Oxidized Compounds and Oligomers by Chromatographic Techniques, In: *Frying of*

Foods (Boskou, D., and Elmadfa, I., eds.), pp. 143–161, Technomic Publishing Company, Lancaster, PA.

Dobarganes, M.C., Velasco, J., and Dieffenbacher (2000) The Determination of Polar Compounds, Polymerized Triacylglycerols, Oxidized Triacylglycerols and Diacylglycerols in Fats and Oils, *Pure Appl. Chem. 72*, 1563–1575.

Firestone, D. (1996) Regulation of Frying Fats and Oils, In: *Deep Frying: Chemistry, Nutrition and Practical Applications* (Perkins, E.G., and Erickson, M.D., eds.), pp. 323–334, AOCS Press, Champaign, IL.

Frankel, E.N. (1993) In Search of Better Methods to Evaluate Natural Antioxidants and Oxidative Stability in Food Lipids, *Trends Food Sci. Technol. 4*, 220–225.

Frankel, E.N. (1998a) Methods to Determine Extent of Oxidation, In: *Lipid Oxidation*, pp. 79–98, The Oily Press, Dundee, UK.

Frankel, E.N. (1998b) Foods, In: *Lipid Oxidation*, pp. 187–225, The Oily Press, Dundee, UK.

Frankel, E.N. (1998c) Oxidation in Multiphase Systems, in *Lipid Oxidation*, pp. 161–186, The Oily Press, Dundee, UK.

Frankel, E.N., and Meyer, A.S. (2000) The Problems of Using One-Dimensional Methods to Evaluate Multifunctional Food and Biological Antioxidants, *J. Sci. Food Agric. 80*, 1925–1941.

Frankel, E.N., Huang, S.-W., Kanner, J., and German, J.B. (1994) Interfacial Phenomena in the Evaluation of Antioxidants: Bulk Oils vs. Emulsions, *J. Agric. Food Chem. 42*, 1054–1059.

Fritsch, C.W. (1994) Lipid Oxidation—the Other Dimensions, *INFORM 5*, 423–431.

Genot, C., Meynier, A., and Riaublanc, A. (2003) Lipid Oxidation in Emulsions, In: *Lipid Oxidation Pathways* (Kamal-Eldin, A., ed.), pp. 190–244, AOCS, Champaign, IL.

Gere, A. (1982) Studies of the Changes in Edible Fats During Heating and Frying, *Die Nahrung 26*, 923–932.

Gertz, C. (2000) Chemical and Physical Parameters as Quality Indicators of Used Frying Fats, *Eur. J. Lipid Sci. Technol. 102*, 566–572.

Gertz, C., and Kochlar, S.P. (2002) New Practical Aspects About Deep Frying Process, *INFORM 13*, 386–389.

Gibbs, B.F., Kermarsh, S., Alli, I., and Mulligan, C.N. (1999) Encapsulation in the Food Industry: A Review, *Int. J. Food Sci. Nutr. 50*, 213–224.

Gomes, D. (1992) Oligopolymer, Diglyceride and Oxidized Triglyceride Contents as Measures of Olive Oil Quality, *J. Am. Oil Chem. Soc. 69*, 1219–1223.

Gomes, D. (1995) A Survey of the Amounts of Oxidized Triglycerides and Triglyceride Dimers in Virgin and Lampante Olive Oils, *Fat Sci. Technol. 97*, 368–372.

Gomes, T., and Caponio, F. (1997) A Study of Oxidation and Polymerization Compounds During Vegetable Oil Refining, *Riv. Ital. Sostanze Grasse 75*, 97–100.

Gomes, T., and Caponio, F. (1998) Evaluation of the State of Oxidation of Olive-Pomace Oils. Influence of the Refining Process, *J. Agric. Food Chem. 46*, 1137–1142.

Gomes, T., and Caponio, F. (1999) Effort to Improve the Quantitative Determination of Oxidation and Hydrolysis Compound Classes in Edible Vegetable Oils, *J. Chromatogr. 844*, 77–86.

Gomes, T., Caponio, F., Baiano, A., and de Pilli, T. (1997) Investigation on the Degree of Oxidation and Hydrolysis of Refined Olive Oils. An Approach for Better Product Characterisation, *Ital. J. Food Sci. 9*, 277–285.

Gomes, T., Caponio, F., Baiano, A., and de Pilli, T. (1998) Measurement of Oxidative and Hydrolytic Degradation of Olive Oils Used for Covering Preserved Foodstuffs, *Riv. Ital. Sostanze Grasse 75*, 77–82.

Gomes, T., Caponio, F., Delcuratolo, D., and Bilancia, M.T. (2002a) Oxidation, Polymerisation and Hydrolysis Substances in Olive Oil. Their Use in the Assessment of Quality, *Riv. Ital. Sostanze Grasse 79*, 211–214.

Gomes, T., Caponio, F., Baiano, A., de Pilli, T., Bilancia, M.T., and Delcuratolo, D. (2002b) The Percent Amount of Virgin Oils in the Commercial Class of Olive Oil. A Possibility to Better Define Quality Characteristics, *Riv. Ital. Sostanze Grasse 79*, 161–164.

Gomes, T., Caponio, F., and Delcuratolo, D. (2003a) Fate of Oxidized Triglycerides During Refining of Seed Oils, *J. Agric. Food Chem. 51*, 4647–4651.

Gomes, T., Caponio, F., and Delcuratolo, D. (2003b) Non-Conventional Parameters for Quality Evaluation of Refined Oil with Special Reference to Commercial Class Olive Oil, *Food Chem. 83*, 403–408.

Gonzalez-Muñoz, M.J., Tulasne, C., Arroyo, R., and Sánchez-Muniz, F.J. (1996) Digestibility and Absorption Coefficients of Palm Olein—Relationships with Thermal Oxidation Induced by Potato Frying, *Fett/Lipid 98*, 104–108.

Gonzalez-Muñoz, M.J., Bastida, S., and Sánchez-Muniz, F.J. (1998) Short-Term *In Vivo* Digestibility of Triglyceride Polymers, Dimers and Monomers of Thermoxidized Palm Olein Used in Deep-Frying, *J. Agric. Food Chem. 46*, 5188–5193.

Gonzalez-Muñoz, M.J., Bastida, S., and Sánchez-Muniz, F.J. (2003) Short-Term *In Vivo* Digestibility Assessment of a Highly Oxidized and Polymerized Sunflower Oil, *J. Sci. Food Agric. 83*, 413–418.

Hansen, S.L., and Artz, W.E. (1995) The Evaporative Light-Scattering Detector, *INFORM 6*, 170–176.

Hansen, S.L., Myers, M.R., and Artz, W.E. (1994) Nonvolatile Components Produced in Triolein During Deep-Fat Frying, *J. Am. Oil Chem. Soc. 71*, 1239–1243.

Hartvigsen, K., Hansen, L.F., Lund, P., Bukhave, K., and Holmer, G. (2000) Determination of Neutral Lipid Hydroperoxides by Size Exclusion HPLC with Fluorometric Detection. Application to Fish Oil Enriched Mayonnaises During Storage, *J. Agric. Food Chem. 48*, 5842–5849.

Hopia, A. (1993a) Analysis of High Molecular Weight Autoxidation Products Using High Performance Size Exclusion Chromatography: I. Changes During Processing, *Food Sci. Technol. 26*, 568–571.

Hopia, A. (1993b) Analysis of High Molecular Weight Autoxidation Products Using High Performance Size Exclusion Chromatography: I. Changes During Autoxidation, *Food Sci. Technol. 26*, 563–567.

Hopia, A.I., Piironen, V.I., Koivistoinen, P.E., and Hyvoenen, L.E.-T. (1992) Analysis of Lipid Classes by Solid-Phase Extraction and High-Performance Size-Exclusion Chromatography, *J. Am. Oil Chem. Soc. 69*, 772–776.

Hopia, A., Lampi, A., Piironen, V., Hyvönen, L., and Koivistoinen, P. (1993) Application of High-Performance Size-Exclusion Chromatography to Study the Autoxidation of Unsaturated Triacylglycerols, *J. Am. Oil Chem. Soc. 70*, 779–784.

Houhoula, D.P., Oreopoulou, V., and Tzia, C. (2003) The Effect of Process Time and Temperature on the Accumulation of Polar Compounds in Cottonseed Oil During Deep-Fat Frying, *J. Sci. Food Agric. 83*, 314–319.

Huang, S.-W., Frankel, E.N., and German, J.B. (1994) Antioxidant Activity of α- and γ-Tocopherols in Bulk Oils and in Oil-in-Water Emulsions, *J. Agric. Food Chem. 42*, 2108–2114.

IUPAC (1987) Standard Method 2.507: Determination of Polar Compounds in Frying Fats, In: *Standard Methods for the Analysis of Oils, Fats and Derivatives*, 7th ed., International Union of Pure and Applied Chemistry, Blackwell, Oxford.

IUPAC (1992) Standard Method 2.508: Determination of Polymerized Triglycerides in Oils and Fats by High Performance Liquid Chromatography, In: *Standard Methods for the Analysis of Oils, Fats and Derivatives*, 7th ed., International Union of Pure and Applied Chemistry, Blackwell, Oxford.

Jorge, N., Márquez-Ruiz, G., Martín-Polvillo, M., Ruiz-Méndez, M.V., and Dobarganes, M.C. (1996a) Influence of Dimethylpolysiloxane Addition to Edible Oils: Performance of Sunflower Oil in Discontinuous and Continuous Laboratory Frying, *Grasas Aceites 47*, 20–25.

Jorge, N., Márquez-Ruiz, G., Martín-Polvillo, M., Ruiz-Méndez, M.V., and Dobarganes, M.C. (1996b) Influence of Dimethylpolysiloxane Addition to Edible Oils: Dependence on the Main Variables of the Frying Process, *Grasas Aceites 47*,14–19.

Jorge, N., Guaraldo-Goncalves, L.A., and Dobarganes, M.C. (1997) Influence of Fatty Acid Composition on the Formation of Polar Glycerides and Polar Fatty Acids in Sunflower Oils Heated at Frying Temperatures, *Grasas Aceites 48*, 17–24.

Karami, A., Balke, S.T., and Schunk, T.C. (2001) Quantitative Interpretation of Fourier-Transform Infrared Spectroscopic Data from a Size-Exclusion Chromatography Solvent-Evaporation Interface, *J. Chromatogr. 911*, 27–37.

Kaufmann A., Ryser B., and Suter B. (2001) HPLC with Evaporative Light Scattering Detection for the Determination of Polar Compounds in Used Frying Oils, *Eur. Food Res. Technol. 213*, 372–376.

Kiatsrichart, S., Brewer, M.S., Cadwallader, K.R., and Artz, W.E. (2003) Pan-Frying Stability of Nusun Oil, a Mid-Oleic Sunflower Oil, *J. Am. Oil Chem. Soc. 80*, 479–483.

Kostanski, L.K., Keller, D.M., and Hamielec, A.E. (2004) Size-Exclusion Chromatography—A Review of Calibration Methodologies, *J. Biochem. Biophys. Methods 58*, 159–186.

Kulin, L.I., Flodin, P., Elligsen, T., and Ugelstad, J. (1990) Monosized Polymer Particles in Size-Exclusion Chromatography. 1. Toluene as Solvent, *J. Chromatogr. 514*, 1–9.

Laguna, M.T.R., Medrano, R., Plana, M.P., and Tarazona, M.P. (2001) Polymer Characterization by Size-Exclusion Chromatography with Multiple Detection, *J. Chromatogr. 919*, 13–19.

Lampi, A.-M., Dimberg, L.H., and Kamal-Eldin, A. (1999) A Study on the Influence of Fucosterol on Thermal Polymerization of Purified High Oleic Sunflower Triacylglycerols, *J. Sci. Food Agric. 79*, 573–579.

Ludlow, M., Louden, D., Handley, A., Taylor, S., Wright, B., and Wilson, I.D. (1999) Size-Exclusion Chromatography with On-Line Ultraviolet, Proton Nuclear Magnetic Resonance and Mass Spectrometric Detection and On-Line Collection for Off-Line Fourier Transform Infrared Spectroscopy, *J. Chromatogr. 857*, 89–96.

Márquez-Ruiz, G., and Dobarganes, M.C. (1996) Nutritional and Physiological Effects of Used Frying Fats, In: *Deep Frying: Chemistry, Nutrition and Practical Applications* (Perkins, E.G., and Erickson, M.D., eds.), pp. 160–182, AOCS Press, Champaign, IL.

Márquez-Ruiz, G., and Dobarganes, M.C. (1997) Analysis of Lipid Oxidation Products by Combination of Chromatographic Techniques, In: *New Techniques and Applications in Lipid Analysis* (McDonald, R.E., and Mossoba, M.M., eds.), pp. 216–233, Technomic Publishing Company, Lancaster, PA.

Márquez-Ruiz, G., Pérez-Camino, M.C., and Dobarganes, M.C. (1990) Combination of Adsorption and Size-Exclusion Chromatography for the Determination of Fatty Acid Monomers, Dimers and Polymers, *J. Chromatogr. 514*, 37–44.

Márquez-Ruiz, G., Pérez-Camino, M.C., and Dobarganes, M.C. (1992a) Digestibility of Fatty Acid Monomers, Dimers and Polymers in the Rat, *J. Am. Oil Chem. Soc. 69*, 930–934.

Márquez-Ruiz, G., Pérez-Camino, M.C., and Dobarganes, M.C. (1992b) *In Vitro* Action of Pancreatic Lipase on Complex Glycerides from Thermally Oxidized Oils, *Fat Sci. Technol. 94*, 307–312.

Márquez-Ruíz, G., Pérez-Camino, M.C., and Dobarganes, M.C. (1993) Evaluation of Hydrolysis and Absorption of Thermally Oxidized Olive Oil in Non-Absorbed Lipids in the Rat, *Ann. Nutr. Metabol. 37*, 121–128.

Márquez-Ruiz, G., Tasioula-Margari, M., and Dobarganes, M.C. (1995a) Quantitation and Distribution of Altered Fatty Acids in Frying Fats, *J. Am. Oil Chem. Soc. 72*, 1171–1176.

Márquez-Ruiz, G., Pérez-Camino, M.C., and Dobarganes, M.C. (1995b) Assessments on the Digestibility of Oxidized Compounds from [1-^{14}C]-Linoleic Acid Using a Combination of Chromatographic Techniques, *J. Chromatogr B 675*, 1–8.

Márquez-Ruiz, G., Jorge, N., Martín-Polvillo, M., and Dobarganes, M.C. (1996a) Rapid, Quantitative Determination of Polar Compounds in Fats and Oils by Solid-Phase Extraction and Exclusion Chromatography using Monostearin as Internal Standard, *Grasas Aceites 749*, 55–60.

Márquez-Ruiz, G., Martín-Polvillo, M., and Dobarganes, M.C. (1996b) Quantitation of Oxidized Triglyceride Monomers and Dimers as a Useful Measurement for Early and Advanced Stages of Oxidation, *Grasas Aceites 47*, 48–53.

Márquez-Ruiz, G., Guevel, G., and Dobarganes, M.C. (1998) Application of Chromatographic Techniques to Evaluate Enzymatic Hydrolysis of Oxidized and Polymeric Triglycerides by Pancreatic Lipase "*In Vitro*," *J. Am. Oil Chem. Soc. 75*, 119–126.

Márquez-Ruiz, G., Martín-Polvillo, M., Jorge, N., Ruiz-Méndez, M.V., and Dobarganes M.C. (1999a) Influence of Used Frying Oil Quality and Natural Tocopherol Content on Oxidative Stability of Fried Potatoes, *J. Am. Oil Chem. Soc. 76*, 421–425.

Márquez-Ruiz, G., Garcés, R., León-Camacho, M., and Mancha, M. (1999b) Thermoxidative Stability of Triacylglycerols from Mutant Sunflower Seeds, *J. Am. Oil Chem. Soc. 76*, 1169–1174.

Márquez-Ruiz, G., Velasco, J., and Dobarganes, M.C. (2000) Evaluation of Oxidation in Dried Microencapsulated Fish Oils by Combination of Adsorption and Size Exclusion Chromatography, *Eur. Food Res. Technol. 211*,13–18.

Márquez-Ruiz, G., Martín-Polvillo, M., and Dobarganes, M.C. (2003a) Effect of Temperature and Addition of Alpha-Tocopherol on the Oxidation of Trilinolein Model Systems, *Lipids 38*, 233–240.

Márquez-Ruiz, G., Velasco, J., and Dobarganes, M.C. (2003b) Oxidation in Dried Microencapsulated Oils, In: *Lipid Oxidation Pathways* (Kamal-Eldin, A., ed.), pp. 245–264, AOCS, Champaign, IL.

Martín-Polvillo, M., Márquez-Ruiz, G., Jorge, N., Ruiz-Méndez, M.V., and Dobarganes, M.C. (1996) Evolution of Oxidation During Storage of Crisps and French Fries Prepared with Sunflower Oil and High Oleic Sunflower Oil, *Grasas Aceites 47*, 54–58.

Martín-Polvillo, M., Márquez-Ruiz, G., and Dobarganes, M.C. (2004) Oxidative Stability of Sunflower Oils Differing in Unsaturation Degree During Long-Term Storage at Room Temperature, *J. Am. Oil Chem. Soc. 81*, 577–583.

Masson, L., Robert, P., Romero, N., Izaurieta, M., Valenzuela, S., Ortiz, J., and Dobarganes, M.C. (1997) Comportamiento de Aceites Poliinsaturados en la Preparación de Patatas Fritas para Consumo Inmediato: Formación de Nuevos Compuestos y Comparación de Métodos Analíticos, *Grasas Aceites 48*, 273–281.

Masson, L., Robert, P., Izaurieta, M., Romero, N., and Ortiz, J. (1999) Fat Deterioration in Deep Fat Frying of French Fries Potatoes at Restaurant and Food Shop Sector, *Grasas Aceites 50*, 460–468.

Masson, L., Robert, P., Dobarganes, M.C., Urra, C., Romero, N., Ortiz, J., Goicoechea, E., Pérez, P., Salame, M., and Torres, R. (2002) Stability of Potato Chips Fried in Vegetable Oils with Different Degree of Unsaturation. Effect of Ascorbyl Palmitate During Storage, *Grasas Aceites 53*, 190–198.

Mrkvickova, L., Pokorna, V., and Pecka, J. (2000) Study of Polymer Complexes by Size Exclusion Chromatography Coupled with Light Scattering Combination with Fluorescence Spectroscopy, *Macromol. Symp. 162*, 227–233.

Neff, W.E., Warner, K., and Eller, F. (2003) Effect of γ-Tocopherol on Formation of Nonvolatile Lipid Degradation Products During Frying of Potato Chips in Triolein, *J. Am. Oil Chem. Soc. 80*, 801–806.

Pérez-Camino, M.C., Márquez-Ruiz, G., Ruiz-Méndez, M.V., and Dobarganes, M.C. (1990) Determinación Cuantitativa de Triglicéridos Oxidados para la Evaluación Global del Grado de Oxidación en Aceites y Grasas Comestibles, *Grasas Aceites 41*, 366–370.

Pérez-Camino, M.C., Márquez-Ruiz, G., Ruiz-Méndez, M.V. and Dobarganes, M.C. (1991) Lipid Oxidation in Fats and Fatty Foods. Quantitative Determination of Oxidized Triglycerides, in *Proceedings of Eur. Food Chem. VI* (Baltes, W., Eklund T., Fenwick R., Pfannhauser, W., Ruiter, A., and Thier, H.P., eds.), Vol. 2, pp. 569–574, Lebensmittelchemische Gesellschaft, Frankfurt.

Pérez-Camino, M.C., Márquez-Ruiz, G., Ruiz-Méndez, M.V., and Dobarganes, M.C. (1992) Lipid Changes During Frying of Frozen Prefried Foods, *J. Food Sci. 56*, 1644–1648.

Pérez-Camino, M.C., Ruiz-Méndez, M.V., Márquez Ruiz, G., and Dobarganes, M.C. (1993) Aceites de Oliva Vírgenes y Refinados: Diferencias en Componentes Menores Glicerídicos, *Grasas y Aceites 44*, 91–96.

Perrin, J.L., Perfetti, P., Dimitriades, C., and Naudet, M. (1985) Etude Analytique Approfondie d'Huiles Chauffées I. Techniques Analytiques et Essais Préliminaires, *Rev. Franc. Corps Gras 32*, 151–158.

Piispa, E., Hyvonen, L.E.T., and Hopia, A. (1996) Characterization of Quality of Fat in Processed Foods by Fatty Acid Analysis and High-Performance Size-Exclusion Chromatography, *Fett/Lipid 98*, 257–260.

Pozo-Díaz, R.M., Masoud-Musa, T.A., Pérez-Camino, M.C., and Dobarganes, M.C. (1995) Intercambio Lipídico durante la Fritura de Patatas Prefritas Congeladas en Aceite de Girasol Alto Oleico, *Grasas Aceites 46*, 85–91.

Raoux, R., Morin, O., and Mordret, F. (1996) Sensory Assessment of Stored French Fries and Crisps Fried in Sunflower and High Oleic Sunflower Oils, *Grasas Aceites 47*, 63–74.

Romero, A., Cuesta, C., and Sánchez-Muniz, F.J. (1995) Quantitation and Distribution of Polar Compounds in an Extra Virgin Olive Oil Used in Fryings with Turnover of Fresh Oil, *Fett Wiss. Technol. 97*, 403–407.

Romero, A., Cuesta, C., and Sánchez-Muniz, F.J. (1998) Effect of Oil Replenishment During Deep-fat Frying of Frozen Foods in Sunflower Oil and High-Oleic Acid Sunflower Oil, *J. Am. Oil Chem. Soc. 75*, 161–167.

Romero, A., Cuesta, C., and Sánchez-Muniz, F.J. (1999) Does Frequent Replenishment with Fresh Monoenoic Oils Permit the Frying of Potatoes Indefinitely? *J. Agric. Food Chem. 47*, 1168–1173.

Rossell, J.B. (1994) Measurements of Rancidity, In: *Rancidity of Foods* (Allen, J.C., and Hamilton, R.J., eds.), pp. 22–53, Chapman & Hall, Glasgow.

Ruiz-Méndez, M.V., Márquez-Ruiz, G., and Dobarganes, M.C. (1997) Relationships Between Quality of Crude and Refined Edible Oils Based on Quantitation of Minor Glyceridic Compounds, *Food Chem. 60*, 549–554.

Sagredos, A.N. (1992) On the Quality and Contamination of Fish Oil Capsules, *Fett Wiss. Technol. 94*, 101–111.

Sánchez-Muniz, F.J., Bastida, S., and González-Muñoz, M.J. (1999) Column and High-Performance Size Exclusion Chromatography Applications to the *In Vivo* Digestibility Study of a Thermoxidized and Polymerized Olive Oil, *Lipids 34*, 1187–1192.

Sánchez-Muniz, F.J., Arroyo, R., Sánchez-Montero, J.M., and Cuesta, C. (2000) *In Vitro* Digestibility Study of Thermal Oxidized Palm Oleins, *Food Sci. Technol. Int. 6*, 449–456.

Schulte, V.E. (1982). Gelchromatographische Bestimmung Polymerisierter Triglyceride, *Fette Seifen Anstrichm. 84*, 178–180.

Sébédio, J.L., Septier, C., and Grandgirard, A. (1986) Fractionation of Commercial Frying Oil Samples Using Sep-Pak Cartridges, *J. Am. Oil Chem. Soc. 63*, 1541–1543.

Sébédio, J.L., Dobarganes, M.C., Márquez, G., Wester, I., Christie, W.W., Dobson, G., Zwobada, F., Chardigny, J.M., Mairot, Th., and Lahtinen, R. (1996) Industrial Production of Crisps and Prefried French Fries Using Sunflower Oils, *Grasas Aceites 47*, 5–13.

Shahidi, F., and Han, X.-Q. (1993) Encapsulation of Food Ingredients, *Crit. Rev. Food Sci. Nutr. 33*, 501–547.

Shukla, V.K.S., and Perkins, E.G. (1991) The Presence of Oxidative Polymeric Materials in Encapsulated Fish Oils, *Lipids 26*, 23–26.

Soheili, K.C., Artz, W.E., Tippayawat, P. (2002a) Pan-Heating of Low-Linolenic Acid and Partially Hydrogenated Soybean Oils, *J. Am. Oil Chem. Soc. 79*, 287–290.

Soheili, K.C., Preeyanooch, T., and Artz, W.E. (2002b) Comparison of a Low-Linolenic and a Partially Hydrogenated Soybean Oil Using Pan-Fried Hash Browns, *J. Am. Oil Chem. Soc. 79*, 1197–1200.

Stellwagen, E. (1990) Gel-filtration, *Methods Enzymol. 182*, 317–328.

Stogiou, M., Kapetanaki, C., and Iatrou, H. (2002) Examination of the Universality of the Calibration Curve of Size Exclusion Chromatography by Using Polymers Having Complex Macromolecular Architectures, *Int. J. Polym. Anal. Charact. 7*, 273–283.

Stulik, K., Pacakova, V., and Ticha, M. (2003) Some Potentialities and Drawbacks of Contemporary Size-Exclusion Chromatography, *J. Biochem. Biophys. Methods 56*, 1–13.

Torabi, K., Karami, A., Balke, S.T., and Schunk, T.C. (2001) Quantitative FTIR Detection in Size-Exclusion Chromatography, *J. Chromatogr. 910*, 19–30.

Ugelstad, J., Moerk, P.C., Herder Kaggerud, K., Ellingsen, T., and Berge, A. (1980) Swelling of Oligomer-Polymer Particles—New Methods of Preparation of Emulsions and Polymer Dispersions, *Adv. Colloid Interface Sci. 13*, 101–140.

Velasco, J., Dobarganes, M.C., and Márquez-Ruiz, G. (2000) Oxidation of Free and Encapsulated Oil Fractions in Dried Microencapsulated Oils, *Grasas Aceites 51*, 441–448.

Velasco, J., Dobarganes, M.C., and Márquez-Ruiz, G. (2002) Oxidación en Sistemas Lipídicos Heterofásicos: Emulsions Aceite en Agua, *Grasas Aceites 53*, 239–247.

Velasco, J., Dobarganes, M.C., and Márquez-Ruiz, G. (2003) Variables Affecting Lipid Oxidation in Dried Microencapsulated Oils, *Grasas Aceites 54*, 304–314.

Velasco, J., Dobarganes, M.C., and Márquez-Ruiz, G. (2004) Antioxidant Activity of Phenolic Compounds in Model Oil-in-Water Emulsions Containing Sunflower Oil, Sodium Caseinate and Lactose, *Eur. J. Lipid Sci. Technol. 106*, 325–333.

Verleyen, T., Kamal-Eldin, A., Dobarganes, C., Verhe, R., Dewettinck, K., and Huyghebaert, A. (2001) Modeling of α-Tocopherol Loss and Oxidation Products Formed During Thermoxidation in Triolein and Tripalmitin Mixtures, *Lipids 36*, 719–726.

Verleyen, T., Kamal-Eldin, A., Mozuraityte, R., Verhe, R., Dewettinck, K., Huyghebaert, A., and De Greyt, W. (2002) Oxidation at Elevated Temperatures: Competition Between α-Tocopherol and Unsaturated Triacylglycerols, *Eur. J. Lipid Sci. Technol. 10*, 228–233.

Waltking, A.E., and Wessels, M. (1981) Chromatographic Separation of Polar and Non-Polar Components of Frying Fats, *J. Assoc. Off. Anal. Chem. 64*, 1329–1330.

Wolff, I.P., Mordret, F.X., and Dieffenbacher, A. (1991) Determination of Polymerized Triglycerides in Oils and Fats by High-Performance Liquid Chromatography, *Pure Appl. Chem. 63*, 1163–1171.

Chapter 5

Analysis of Lipid Oxidation Products by NMR Spectroscopy

Taina I. Hämäläinen[a] and Afaf Kamal-Eldin[b]

[a]Department of Chemistry, University of Helsinki, FIN-00014 Helsinki, Finland, and [b]Department of Food Science, Swedish University of Agricultural Sciences (SLU), 750 07 Uppsala, Sweden

Introduction

Nuclear magnetic resonance (NMR) spectroscopy is the preeminent tool for identifying the structure of organic molecules and a versatile analytical technique. The development of this technique, since Bloch and Purcell discovered the NMR phenomenon in 1945, has been fast, particularly over the past 20 years. In recent years, the importance of NMR spectroscopy has grown in lipid chemistry, and we suspect that this magnificent technique has the potential to become a valuable tool for studying lipid oxidation.

In lipid chemistry, NMR applications range from sophisticated structure elucidation to routine quality control. This versatility stems from the several advantages of NMR spectroscopy. One of these is the fast and simple sample preparation. Moreover, NMR spectroscopy is nondestructive, and both qualitative and quantitative data can be obtained. NMR spectroscopy is suitable for the analysis of pure compounds as well as mixtures, and it offers the possibility of automatization. However, although software is also being developed for the automatic interpretation of NMR spectra (Griffiths 2000), chemists are still needed for this task. Furthermore, one major drawback of NMR spectroscopy is that it is costly; however, the availability of NMR spectrometers has grown recently. Yet another difficulty, particularly in the early days, is the relatively low sensitivity of NMR spectroscopy. Continuous hardware advances have, nonetheless, increased the sensitivity of NMR spectroscopy to the point at which submilligram amounts of medium-weight molecules can be analyzed routinely (Shapira *et al.* 2004).

This chapter is divided into two parts. The first part demonstrates how NMR spectroscopy can be used as a tool for structural analysis; it includes a brief introduction to the principles of NMR spectroscopy, which is intended to help the reader to follow the examples given. For those who are planning to use NMR spectroscopy in their own studies, the IUPAC recommendations (Harris *et al.* 1997) and the ASTM (American Society for Testing and Materials) standard (reprinted in Günther 1996b) are recommended as references for standard practice for data presentation relating to high-resolution NMR spectroscopy. The second part of this chapter focuses on the analysis of compounds that are formed during lipid oxidation. It concentrates on high-resolution NMR spectroscopy and on nonbiological samples. The main emphasis is on the NMR spectra of individual pure compounds because this knowledge is essential for applying as well as developing new NMR applications for mixtures of lipid oxidation products. One purpose of the chapter is to map out how much of this type of fundamental knowledge is available, and to determine the potentiality and requirements for the utilization of NMR spectroscopy in the analysis of lipid oxidation.

Principles of Nuclear Magnetic Resonance Spectroscopy

The study of the interaction of electromagnetic radiation with matter is called spectroscopy. NMR spectroscopy employs radio frequency pulses to induce transitions between magnetic energy levels of atomic nuclei. The record of these transitions is an NMR spectrum.

NMR spectroscopy arises from a property of nuclei known as *spin*. Nuclear spin is characterized by a nuclear spin quantum number I. Thus, only those nuclei possessing this property are detectable. The atomic nuclei that chemists are concerned with in lipid chemistry include proton (^1H), carbon-13 (^{13}C), and phosphorus-31 (^{31}P), which all have a spin quantum number I of $^1/_2$. These half-spin nuclei can occupy two different energy levels when they are immersed in a static magnetic field. In NMR measurements, transitions between the energy levels are induced by irradiating the nuclei with a radio frequency ν_I corresponding to the energy difference (ΔE) of these levels. Transitions occur only when this resonance condition ($\Delta E = h\nu_I$, where h is Planck's constant) is fulfilled. A transition to a higher energy level is called absorption and the reverse transition is called emission.

At present, most NMR spectra are run on 200–600 MHz spectrometers by the pulse Fourier NMR technique. In this technique, all of the nuclei of one species (e.g., all of the protons of the sample) are irradiated by a radio frequency pulse simultaneously; the system is followed when it returns to its equilibrium state and an emission spectrum is recorded. The NMR resonance frequencies depend on both the nature of the atomic nuclei and the strength of the external magnetic field. For example, when the field strength is 4.70 T, ^1H resonates at 200 MHz and ^{13}C at 50.3 MHz, whereas when the field strength is 14.09 T, ^1H resonates at 600 MHz and ^{13}C at 150.9 MHz (note that the frequency of a spectrometer refers to the ^1H resonance frequency). To be able to compare the frequencies measured on spectrometers operating at different magnetic field strengths, the position of the signal in an NMR spectrum is given by the chemical shift δ. The δ scale used for chemical shifts is dimensionless and the quantity is reported in ppm (the correct usage is $\delta = 5.00$ or δ 5.00, not $\delta = 5.00$ ppm). The standard orientation of spectra is with low radio frequency (high applied magnetic field) to the right. The low δ values on the right are often described as being *upfield* and the high δ values on the left are said to be *downfield.* The chemical shifts δ are determined relative to an internal reference. The primary internal reference for proton and carbon spectra in non-aqueous solutions is tetramethylsilane (TMS). The position of the TMS resonance is defined as δ 0.00. NMR sample is dissolved, preferably in a solvent that does not give rise to NMR signals. The most commonly used solvents are $CDCl_3$, $(CD_3)_2CO$ (acetone-d_6), $(CD_3)_2SO$, C_6D_6, and D_2O. The sample size for a routine ^1H NMR spectrum is ~1–10 mg and for a ^{13}C NMR spectrum ~50–100 mg in 0.7 mL of deuterated solvent. However, with a longer acquisition time or with smaller sample volume, a high-quality ^1H NMR spectrum can be obtained from microgram amounts of material and a ^{13}C NMR spectrum from <1 mg. Moreover, the low-nanogram range can be achieved for ^1H NMR spectroscopy utilizing a stopped-flow capillary HPLC-NMR technique (Krucker *et al.* 2004); the whole carbon skeleton of a medium-weight compound can be mapped out from micromoles of sample (Chauret *et al.* 1996) utilizing a high-sensitivity ^{13}C probe, known as nanoprobe.

NMR Spectroscopy as a Tool for Structural Analysis

Spectral Parameters. NMR spectroscopy is a useful tool for structure analysis of organic compounds, because the NMR frequencies depend not only on the external magnetic field and the nature of the atomic nuclei but also on the chemical environment of the particular nuclei. Therefore, protons that experience a different chemical environment (i.e., chemically nonequivalent protons) resonate at slightly different frequencies; thus, it is possible to assign certain characteristic chemical shift ranges to the protons of the various functional groups in organic compounds. For example, the signals of aliphatic methyl protons typically have a chemical shift between 0.6 and 1.9 ppm relative to TMS in $CDCl_3$, whereas a typical chemical shift value for olefinic protons usually lies between 4.5 and 8 ppm (Fig. 5.1A). In the same way, ^{13}C NMR spectroscopy exploits the different chemical environments that exist for carbon atoms within different types of organic compounds. In glycerol trioleate, for instance, differences in the signals for olefinic carbon atoms are produced by structural changes up to 11 atomic centers away (Gunstone 1993a). Moreover, ^{13}C NMR spectroscopy is more informative because it has better resolution than ^1H NMR spectroscopy. The chemical shifts of most functional groups in organic compounds spread normally over 15 ppm in ^1H NMR spectrum compared with over 200 ppm in the ^{13}C NMR spectrum. For example, the ^1H NMR signals of a mixture of two methyl linoleate hydroperoxides are spread just over 7 ppm; some are clearly coinciding or overlapping because the number of resolved groups of signals is ~15 (Fig. 5.1A). The ^{13}C NMR signals of the hydroperoxides are spread over 160 ppm; only two of the 38 carbon signals (19 from each isomer) overlap, the spectrum consisting of 37 singlets (Fig. 5.1B).

The splitting (or number of lines, i.e., the multiplicity) particularly of the ^1H resonance signal provides additional information on the structure of the molecule. It tells which atomic nuclei are close enough to interact with each other, and the

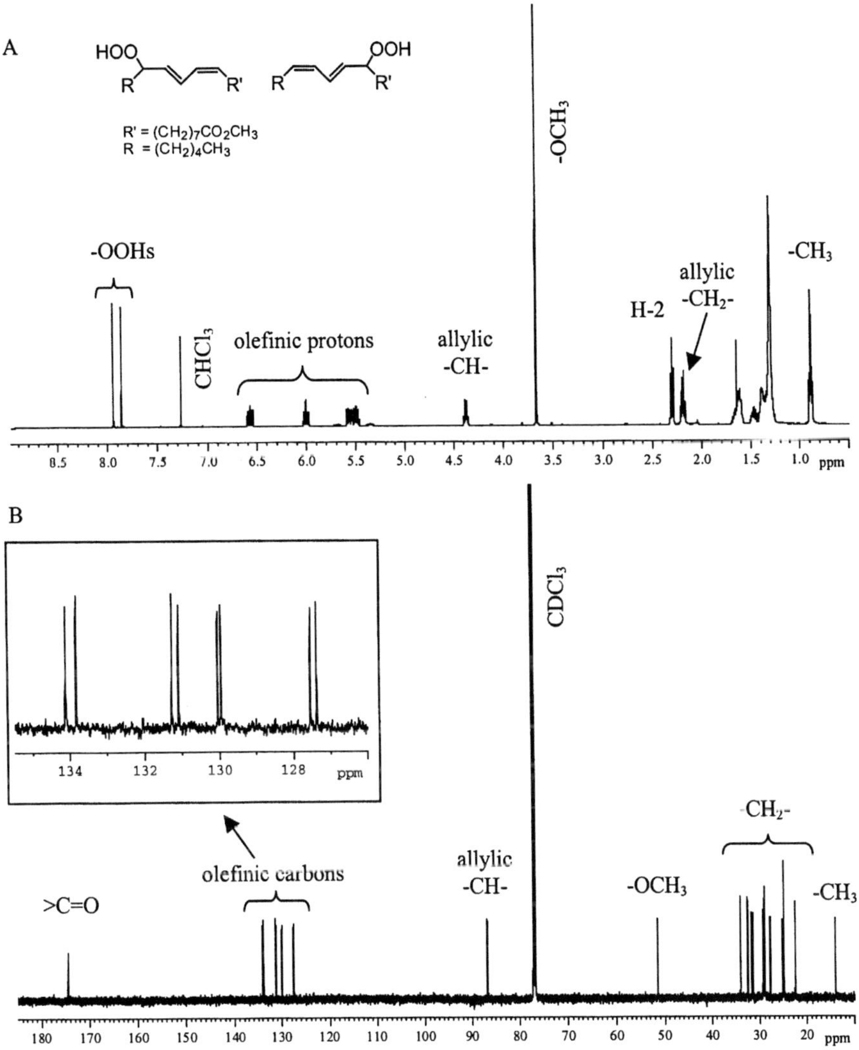

Fig. 5.1. (A) 500 MHz ^1H NMR spectrum and (B) 125 MHz ^{13}C NMR spectrum of two methyl linoleate hydroperoxides in $CDCl_3$ (10 mg/0.7 mL).

magnitude of the interaction gives information about the relative position of the nuclei. The nuclei interact with other nuclei through the bonding electrons; when the nuclei are nonequivalent, this can cause splitting of the corresponding line in the spectrum according to the following: (i) the number and distance of the coupled nuclei; (ii) their orientation toward each other; and (iii) the character of the bonds between the two coupling nuclei. In the case of σ-bonds, these interactions (spin-spin couplings) are typically observed only when the nuclei are one, two (geminal coupling), or three (vicinal coupling) bonds away. Long-range couplings across a larger number of bonds typically occur only if more polarizable π-bond systems are involved (e.g., H-C=C-C-H). This interaction through the bonding electrons is called indirect spin-spin, J, or scalar coupling, and the strength of the interaction is given by a spin-spin coupling constant J. (The notation $^{n}J_{AB}$ is used to represent a coupling over n bonds between nuclei A and B, where A and B need not be nuclei of the same species.) The coupling constants J are independent of the external field and are therefore expressed in hertz (Hz). The splitting patterns and intensity distribution of the signals in multiplets can often be explained by simple rules; thus they give more detailed insight into the structure. In first-order spectra, the signal of one or several equivalent protons is a multiplet with a multiplicity of $M = n + 1$, if the proton(s) couples with n other equivalent protons, and the relative intensities of the individual multiplet lines follow the same patterns as the nth binominal coefficients. The process of removing the spin-spin splitting between the spins is called decoupling. To compensate for the low natural abundance of carbon-13 (1.1%), ^{13}C NMR spectra are usually run with broadband heteronuclear decoupling, i.e., elimination of all ^{1}H-^{13}C couplings, which results in sharp singlets for all ^{13}C signals (Fig. 5.1B). The ^{13}C-^{13}C couplings are not a concern because the possibility of such couplings is low.

A third spectral parameter measured routinely from a ^{1}H NMR spectrum is the signal intensity, i.e., the area of the signal obtained by integration. Because the signal intensity is directly proportional to the relative number of nuclei inducing the signal, it can be used in a way that provides quantitative information of analytical value. Although ^{13}C NMR spectra are normally run in a manner that does not give quantitative integrals, this can be accomplished with an appropriate pulse program (e.g., with inversed gated decoupling techniques) or by running the spectrum in the presence of a relaxation agent (e.g., some paramagnetic salt) and using appropriate pulse delay.

Multidimensional NMR Spectroscopy and Hyphenated Techniques. The assignment of the various chemical shifts in a 1D NMR spectrum of a particular compound is based on spectral parameters (chemical shift δ, coupling constant J, and intensity) and it can be aided by decoupling experiments, by the use of chemical shift reagents, by studies of deuterated analogs, and by available knowledge built up over the years. Extensive studies on a series of compounds have made it possible to determine the effect of various functional groups on other functionalities (empirical correlations and additivity rules); today, computer software is available for calculating and predicting chemical shifts. When these methods and existing data are of little help to the assignment of the 1D NMR spectrum, more advanced multidimensional NMR techniques can be employed. By the use of homonuclear and heteronuclear correlations, multidimensional NMR spectroscopy offers ways not only to confirm assignments but also to make further assignments in a simple and direct manner. Examples of 2D NMR techniques that are based on homonuclear (e.g., ^1H to ^1H) chemical-shift correlations are ^1H-^1H correlated spectroscopy (COSY) for the spin-spin coupled pairs of protons that normally are three bonds apart, but which can be optimized for long-rage couplings (long-range COSY or COSYLR), ^1H-^1H total correlation spectroscopy (TOCSY), which can be used to identify entire spin-coupled networks in a molecule, and incredible natural abundance (INADEQUATE), which correlates directly attached carbons and is used for direct tracing out of the carbon skeletons of molecules. Heteronuclear (e.g., ^1H to ^{13}C) chemical-shift correlation spectroscopy includes techniques such as heteronuclear single quantum coherence (HSQC) and heteronuclear multiple bond correlation (HMBC), which offer a way to identify all directly bonded and long-range (2–3 bonds away) carbon-proton pairs, respectively. Today, these above-mentioned 2D and even 3D NMR techniques such as HSQC-TOCSY can be run routinely. A combination of several 2D NMR techniques can provide connectivity information for separate spin systems that allows the chemist to piece together an overall chemical structure for a molecule often even without prior knowledge of the specific shift assignments and, most importantly, with a high degree of certainty.

In recent years, much effort has been invested in the development of hyphenated NMR techniques, to the extent that HPLC-NMR has become an established technique in analytical science involving ^1H NMR observation either in continuous or stopped-flow mode. Although, to our knowledge, this technique has not yet been applied to the analysis of lipid oxidation products, mixtures of pharmaceuticals, drug metabolites, and natural products in plant extracts have been analyzed by HPLC-NMR-mass spectrometry (MS) (Wilson 2000). The characterization and identification of ecdysteroids by reversed-phase HPLC combined with on-line UV, Fourier transform infrared (FTIR), ^1H NMR, and time-of-flight (TOF) MS (Louden *et al.* 2001) serves as an example of the utilization of the technique in the field of lipid chemistry. Current development on HPLC-NMR focuses on real-time 2D NMR spectroscopy identification of analytes undergoing continuous chromatographic separation (Shapira *et al.* 2004) as well as coupling capillary separation techniques with NMR spectroscopy (Rapp *et al.* 2003). The advantages for 2D techniques for structure elucidation are apparent, and the capillary separation aims at cutting the consumption of expensive fully deuterated eluents, which has been one of the major drawbacks of the hyphenated HPLC-NMR techniques. Capillary LC-NMR online separation was applied successfully to a mixture of unsaturated fatty acid methyl esters and in that study, the stereochemical assignment of the *cis* and *trans* double bond configuration could easily be accomplished (Rapp *et al.* 2003). Evidently, the separation of the individual components of mixtures will set the limits and thus the usefulness of the hyphenated NMR techniques in the analysis of lipid oxidation.

*Structure Elucidation of 13-Hydroxy-9-*cis*,11-*trans*-octadecadienoic Acid Methyl Ester.* Hämäläinen *et al.* (2002) fully assigned the ^1H and ^{13}C NMR spectra of six hydroxy derivatives of 9-*cis*,11-*trans*-conjugated linoleic acid (9*c*,11*t*-CLA) methyl ester hydroperoxides. The assignment of the various chemical shifts

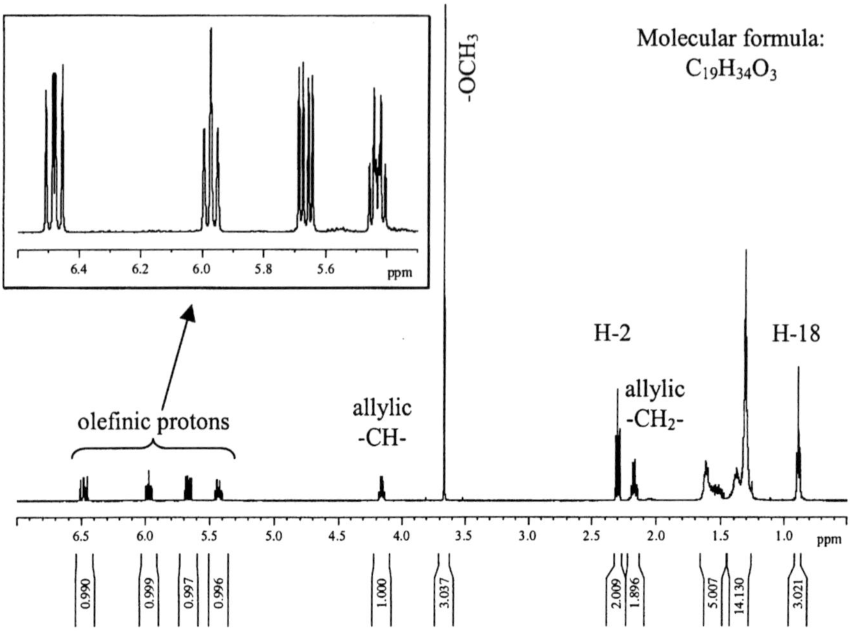

Fig. 5.2. 500 MHz ^{1}H NMR spectrum of Me 13-OH-9*c*,11*t* with integrals and enlarged olefinic region.

was accomplished unequivocally by using a combination of 2D NMR techniques. To demonstrate the utility, simplicity, and beauty of NMR spectroscopy, the detailed interpretation of the ^{1}H and ^{13}C NMR spectra of one of the hydroxy compounds, namely, 13-(*R,S*)-hydroxy-9-*cis*,11-*trans*-octadecadienoic acid methyl ester (Me 13-OH-9*c*,11*t*), is discussed below.

Easily recognizable groups based on chemical shifts, signal integrals, and multiplicity in the ^{1}H NMR spectrum of the 13-hydroxy isomer are the olefinic, allylic methine (-CH-), methoxy (-OCH_3), H-2, allylic methylene (-CH_2-), and H-18 proton signals (Fig. 5.2). The four multiplets in the olefinic region, for which the signal integration corresponds to one proton each, are characteristic of a conjugated diene structure with two lower field "inner" positioned olefinic protons at δ_H 6.48 and 5.97, and two higher field "outer" positioned olefinic protons at δ_H 5.67 and 5.43. The stereochemistry of the conjugated diene structure can be resolved from the splitting of these olefinic signals because the geometry of a double bond can be determined on the basis of the values of the vicinal coupling constants ^3J. The coupling constant of 15.2 Hz measured from the multiplet at δ_H 6.48 is characteristic for olefinic protons that have a *trans* geometry, whereas the coupling constant of 11.0 Hz measured from the multiplet at δ_H 5.97 is characteristic of a *cis* double bond. In addition, the ^3J values reveal that the "inner" positioned olefinic proton at δ_H 6.48 couples with the "outer" positioned olefinic proton at δ_H 5.67, whereas the

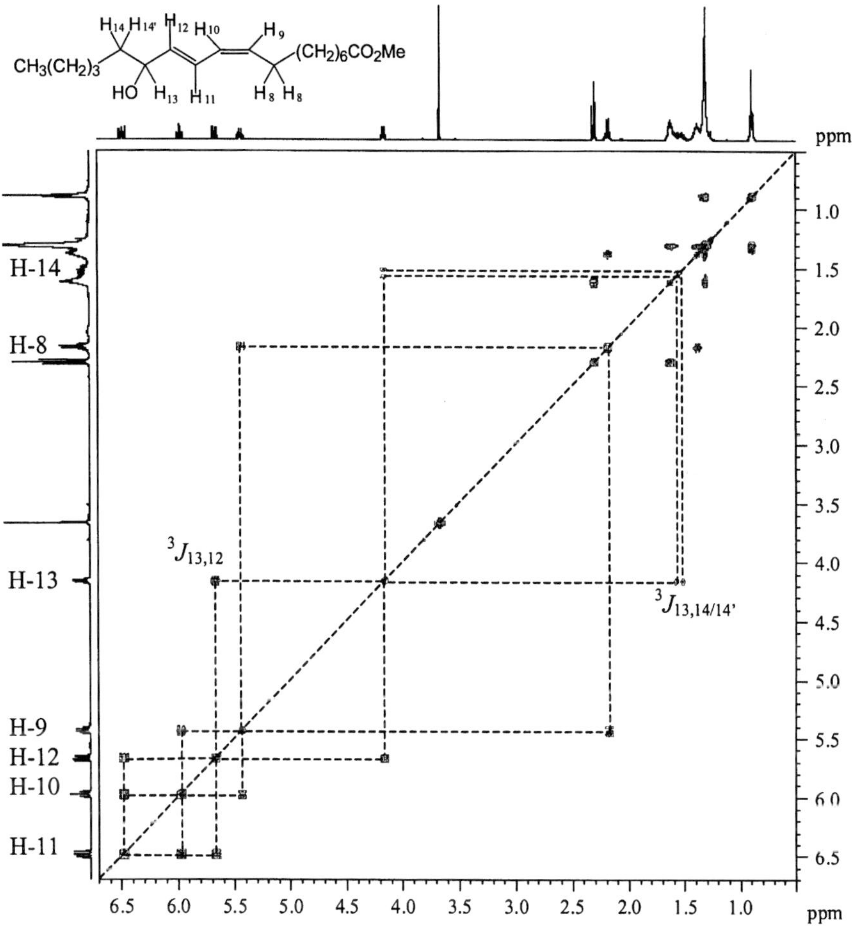

Fig. 5.3. 500 MHz ^1H-^1H correlated spectroscopy (COSY) spectrum of Me 13-OH-9*c*,11*t*. The diagonal and cross peaks joined by dashed lines indicate the olefinic and allylic proton-to-proton chemical shift correlations.

"inner" positioned olefinic proton at δ_H 5.97 couples with the "outer" positioned olefinic proton at δ_H 5.43.

The ^1H-^1H COSY experiment offers a way to identify the spin-spin coupled pairs of protons that are three bonds apart. The position of the hydroxy group was determined by GC-MS and this information can now be used to aid the assignment of the spectrum. The allylic methine proton, that is, H-13, resonates at δ_H 4.16, and the off-diagonal correlation cross peaks reveal the location of its coupling partners, H-14 and H-12 (Fig. 5.3). Because the methyl group protons resonate at δ_H 0.89, it is not a vinylic methyl group; thus, the structure contains a 9-*cis*,11-*trans* diene instead of a 14-*trans*,16-*cis* diene system. Therefore, the resonance at δ_H 5.67 arises from H-12, and the two higher-order multiplets at ~δ_H 1.57 and 1.51 from H-14 and H-14′. (The geminal protons H-14 and H-14′ are diastereotopic and resonate at different frequencies. Because geminal couplings are generally –10 to –20 Hz, and $|\Delta\nu| \approx 30$ Hz, the signals are higher-order multiplets. Resonance signals are of first order if $|\Delta\nu|$ is larger than 5–10 times J.) The former signal (H-14) is partly overlapping with the H-3 signal and this COSY spectrum demonstrates how 2D NMR spectroscopy offers a way of resolving overlapping signals and estimating their chemical shift values. As mentioned above, the ^3J values revealed the coupling partner of H-12; thus, the resonance at δ_H 6.48 is H-11 and thereafter it is simple to assign the remaining two olefinic resonance signals. Moreover, referring to the correlations on the COSY spectrum, these assignments can be confirmed as follows: δ_H 6.48 (H-11), 5.97 (H-10), 5.67 (H-12), and 5.43 (H-9). In addition, further assignments can be made on the basis of the cross peak signals in the COSY spectrum: allylic methylene protons (H-8) at δ_H 2.17, homoallylic methylene protons (H-7) at δ_H 1.45–1.35, and H-3 at δ_H 1.60. The multiplet at δ_H 1.35–1.25 contains the overlapping signals of the yet unassigned methylene groups (H-4, H-5, H-6, H-15, H-16, and H-17); the hydroxy group resonance signal lies somewhere underneath the H-3 and H-14 resonance signals based on signal integration (Fig. 5.2).

As explained above, the splitting patterns and intensity distribution of the signals in the multiplets can be predicted by simple rules in first-order spectra. Hence, the determination of the multiplicity provides support for the previous assignment of the chemical shifts of various protons. Proton H-11, for instance, has two chemically nonequivalent vicinal hydrogen atoms (H-12 and H-10), and this would lead to multiplicity of $M = (1+1)(1+1) = 4$ with signal intensities of 1:1:1:1. Indeed, at first glance, the multiplet at δ_H 6.48 (H-11) appears (Fig. 5.2, enlarged region) as a double doublet (dd). However, at closer inspection, the lines are not smooth (Fig. 5.4A), and because the structure contains a π-bond system, this can be expected to be caused by long-range coupling. When the spectral resolution is improved by windowing, the finer structure of the multiplet (Fig. 5.4B) is revealed. This additional splitting, as determined by a long-range COSY (COSYLR) experiment (Fig. 5.5), is due to coupling of proton H-11 with H-13 and H-9, which are four bonds apart (^4J); thus, the multiplicity of the H-11 signal is $M = (1+1)(1+1)(1+1)(1+1) = 16$ and it corresponds to a doublet of doublet of doublet of doublets (dddd). However, as can be seen from Figure 5.4B, the H-11 multiplet appears as if it consists of only 12 lines (signal intensities: 1:2:1:1:2:1:1:2:1:1:2:1), because the inner lines of each two adjacent doublets are coalescing and appear as a triplet.

The position of the double bond system in the alkyl chain (i.e., whether the structure is a *cis,trans* or a *trans,cis* isomer) can be confirmed by a 2D TOCSY experiment that establishes which one of the allylic resonances is closest to the methyl group. A TOCSY spectrum shows the protons belonging to the same coupled spin system; by using a suitable mixing time in the pulse program, it is possible to control how far the magnetization is transferred within the spin system. Figure 5.6 shows an expanded

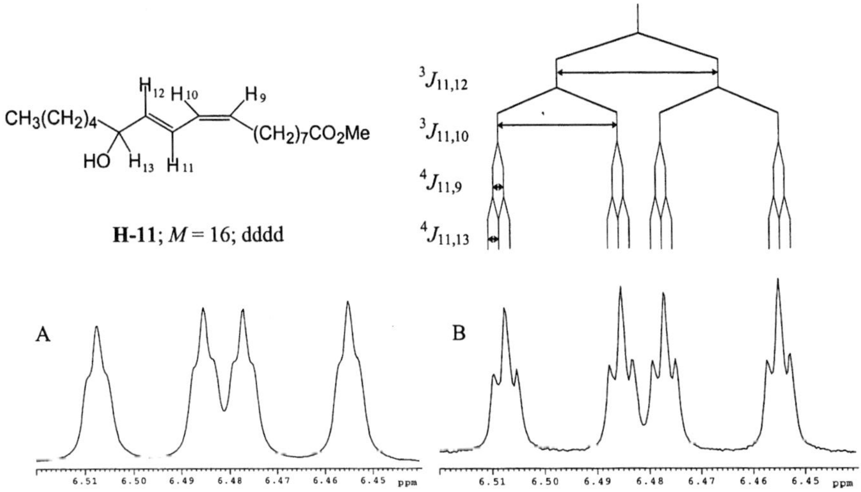

Fig. 5.4. A part of the 500 MHz ^{1}H NMR spectrum of Me 13-OH-9*c*,11*t* showing the H-11 resonance signal, and its multiplicity; (A) line broadening (lb) = 0.30 Hz, no zero filling; (B) lb = –0.10 Hz, zero filling.

plot of a trace of a 2D TOCSY spectrum of the 13-hydroxy isomer run with the relatively long mixing time of 140 ms. The methyl group protons correlate with the allylic methine proton but not with the allylic methylene protons, and thus the position of the double bond system must be 9-*cis*,11-*trans*. This is in accord with the methyl group resonance correlating more strongly with the protons in the *trans* double bond than with the more remote protons in the *cis* double bond.

Now the full assignment of the ^{1}H NMR (500 MHz, $CDCl_3$, 27°C) spectrum of Me 13-OH-9*c*,11*t* (99%, 69 mM) can be written as follows (reproduced from Hämäläinen *et al.* 2002): δ_H 6.48 (dddd, 1 H, H-11, $^3J_{11,12}$ = 15.2 Hz, $^3J_{11,10}$ = 11.1 Hz, $|^4J_{11,9}| \approx |^4J_{11,13}| \approx$ 1.1 Hz), 5.97 (ddtd, 1 H, H-10, $^3J_{10,9} \approx$ 11.0 Hz, $|^4J_{10,8}| \approx$ 1.4 Hz, $|^4J_{10,12}| \approx$ 0.6 Hz), 5.67 (dd, 1 H, H-12, $^3J_{12,13}$ = 6.9 Hz), 5.43 (dt, 1 H, H-9, $^3J_{9,8}$ = 7.7 Hz), 4.16 (dt, 1 H, H-13, $^3J_{13,14} \approx$ 6.6 Hz), 3.66 (s, 3 H, OCH_3), 2.30 (t, 2 H, H-2, $^3J_{2,3}$ = 7.5 Hz), 2.17 (dt, 2 H, H-8, $^3J_{8,7} \approx$ 7.4 Hz), 1.60 (quintet, 2 H, H-3, $^3J_{3,4}$ = 7.3 Hz), 1.57 (m, 1 H, H-14), 1.51 (m, 1 H, H-14′), 1.45–1.35 (m, 2 H, H-7), 1.35–1.25 (m, 12 H, H-4, H-5, H-6, H-15, H-16, H-17), 0.89 (t, 3 H, H-18, $^3J_{18,17}$ = 6.9 Hz). The exact position of the hydroxy proton signal is undetermined, but it appears between δ_H 1.65 and 1.45.

The ^{13}C NMR spectrum of the 13-hydroxy isomer in Figure 5.7 is run with broadband proton decoupling; apparently, it is a routine spectrum without quantitative integral because the signals are of markedly different intensity yet each corresponding to a single chemically distinct carbon atom. Easily recognizable groups based on chemical shifts are the carbonyl (C-1), olefinic (C-9 to C-12), allylic methine (C-13), methoxy (OCH_3), C-2, allylic methylene (C-8), and C-18 carbon atoms. The remaining methyl-

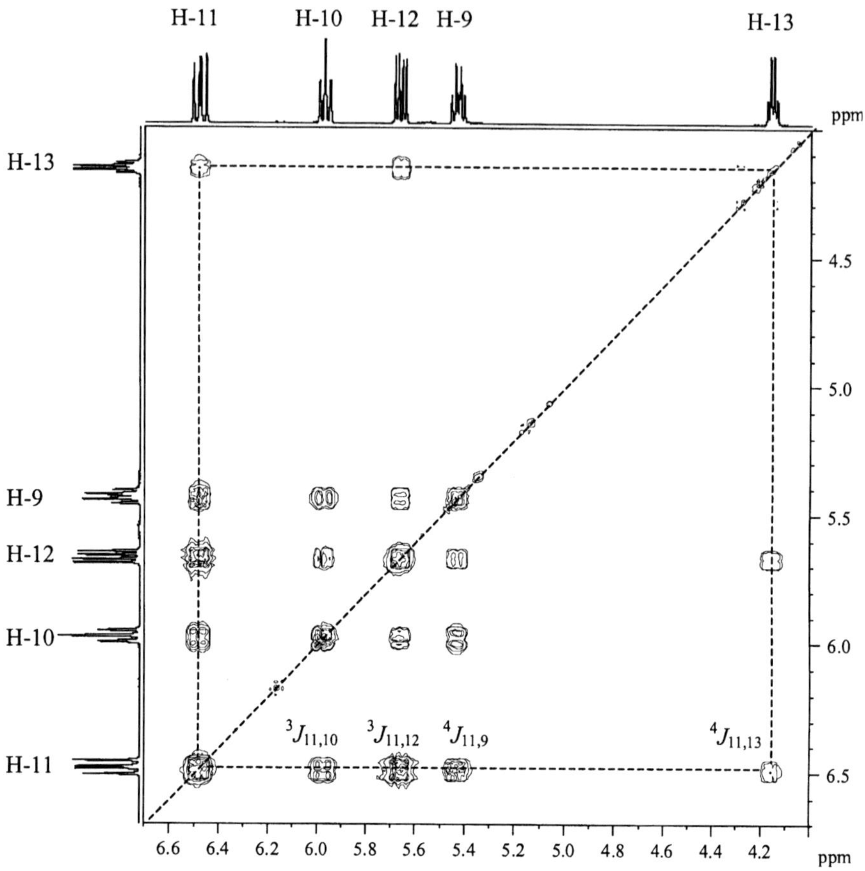

Fig. 5.5. 500 MHz ^{1}H-^{1}H long-range correlated spectroscopy (COSYLR) spectrum of Me 13-OH-9*c*,11*t*. The diagonal and cross peaks joined by dashed lines indicate the proton-to-proton chemical shift correlations of proton H-11.

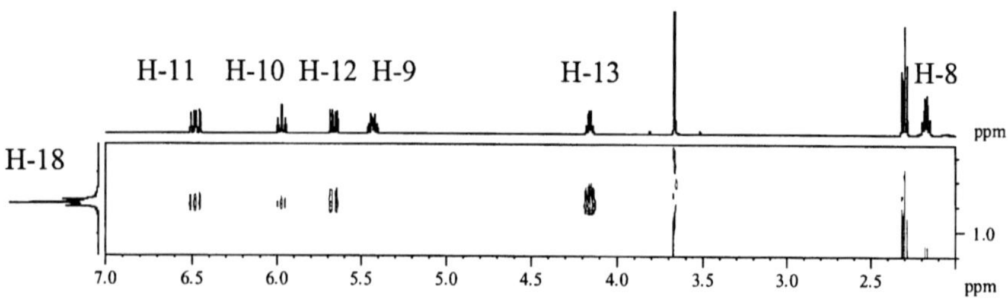

Fig. 5.6. Expanded trace of a 500 MHz total correlation spectroscopy (TOCSY) spectrum of Me 13-OH-9*c*,11*t*.

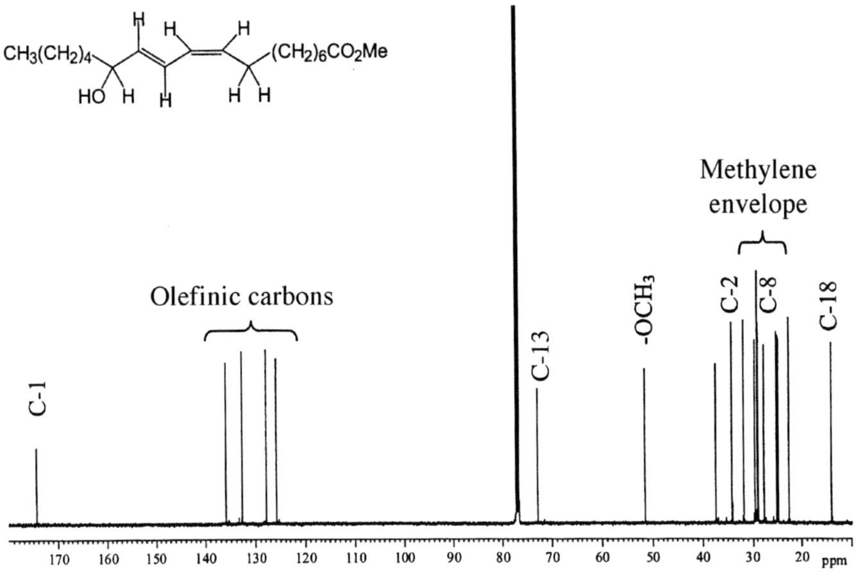

Fig. 5.7. 125 MHz ^{13}C NMR spectrum of Me 13-OH-9*c*,11*t*.

ene ^{13}C chemical shifts are more difficult to identify because they are clustered within a small region of the spectrum often referred to as the "methylene envelope." The *cis,trans* geometry of the conjugated double bond system is evident from the characteristic chemical shifts of allylic methine and methylene carbons supporting the previous stereochemical analysis from the ^{1}H NMR spectrum. The chemical shift at δ_C 72.90 (C-13) is characteristic of an allylic methine carbon atom adjacent to a *trans* double bond (Frankel *et al.* 1990) and the chemical shift at δ_C 27.68 (C-8) confirms that the allylic methylene carbon is adjacent to a *cis* double bond (Bus *et al.* 1976, Gunstone *et al.* 1977).

The previously identified protons can be correlated with their directly attached carbons in an HSQC spectrum (Fig. 5.8) allowing the identification of olefinic (C-9 to C-12), allylic methine (C-13), allylic methylene (C-8), homoallylic methylene (C-7 and C-14), C-2, C-3, and C-18 carbons. These assignments can be reconfirmed and yet further ones made by observing the correlations between protons and carbons over two and three bonds in an HMBC spectrum (Fig. 5.9). In particular, the H-18 methyl group protons have two strong correlation cross peaks, which allow the distinction of carbon C-16 from C-17 on chemical shift grounds. Carbon C-15 can be identified on the basis of correlations to previously assigned protons H-14 and H-13. More importantly, the position of the double bond system is confirmed by the correlation of C-16 to H-14, which in turn is correlated to the allylic methine carbon C-13. The fully assigned ^{13}C NMR data of Me 13-OH-9*c*,11*t* and that of other allylic hydroxy conjugated diene isomers are collected in Table 5.8. The assignment of the ^{13}C NMR spectrum in the

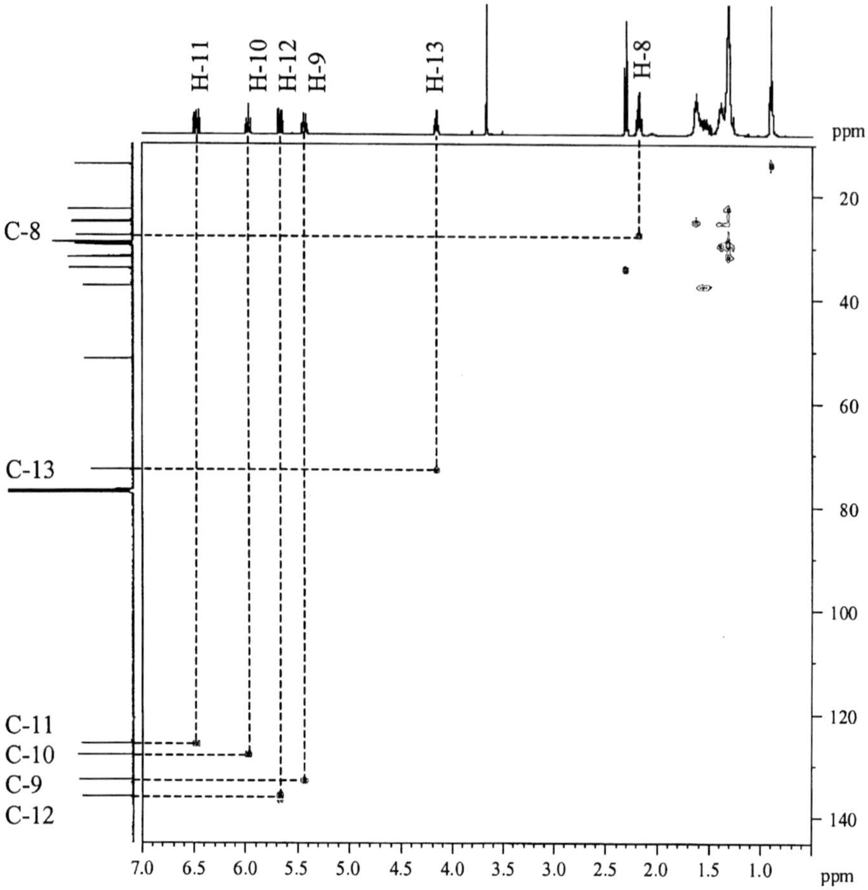

Fig. 5.8. Heteronuclear single quantum coherence (HSQC) spectrum of Me 13-OH-9*c*,11*t* with the olefinic, and the allylic carbon-to-proton chemical shift correlations indicated with dashed lines.

above example could have been established without prior knowledge of the position of the hydroxy group using NMR techniques alone. However, the determination of the position of hydroxy group in long chain fatty esters often requires either knowledge of the chemical shift values of other positional isomers or confirmation techniques other than NMR spectroscopy.

NMR Spectroscopy in the Analysis of Lipid Oxidation Products

NMR spectroscopy has established an important position in the analysis of lipids (excellent reviews by Gunstone 1993a, 1999, 2001, Lie Ken Jie and Mustafa

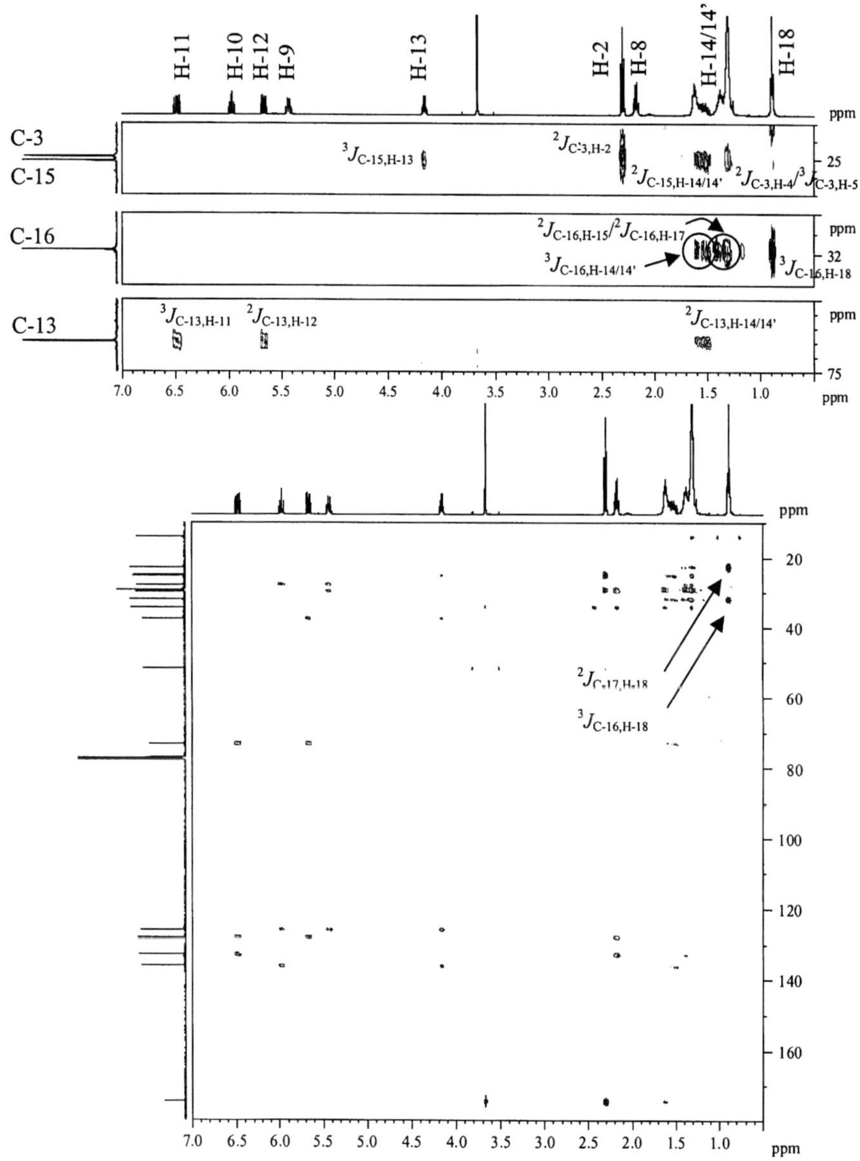

Fig. 5.9. Heteronuclear multiple bond correlation (HMBC) spectrum of Me 13-OH-9*c*,11*t* and expanded traces showing the chemical shift correlations of C-3, C-15, C-16, and C-13.

1997). The precise and systematic structural determinations of fatty acids by NMR spectroscopy laid the groundwork for the design of NMR applications for the analyses of the composition of margarines, butter, and natural fish, seed, and vegetable oils. Particularly ^{13}C NMR spectroscopy, with its wider chemical shift range, has proven useful in providing insight into the nature of lipid mixtures. For example, the presence or absence of *trans* fatty acids in hydrogenated oils can be established through the characteristic signals for olefinic and allylic carbon atoms also in quantitative mode (Miyake *et al.* 1998). However, NMR spectroscopy is still a relatively new technique for the study of lipid oxidation products. ^{1}H NMR spectroscopy and 2D NMR techniques (Claxson *et al.* 1994, Haywood *et al.* 1994, Lodge *et al.* 1995, Silwood and Grootveld 1999) as well as ^{13}C NMR spectroscopy (Medina *et al.* 1998) have been utilized to provide a general overview of classes of compounds formed during LDL peroxidation and thermal stressing of culinary oils and fish. The NMR data of pure oxidation products are, however, limited probably because the oxidation reactions produce complex mixtures and the separations can be tedious. When studying mixtures, the low sensitivity may be problematic especially in the case of minor compounds. Nevertheless, there are at least three approaches to the use of NMR spectroscopy in the analysis of lipid oxidation:

1. A compound specific analysis; characterization of the precise structure of an individual oxidation product (or its derivative) isolated from an oxidation mixture.
2. A bulk sample analysis; determination of the type of products in a chromatographic fraction isolated from an oxidation mixture.
3. A bulk sample analysis; analysis of oxidized oils and fats at different stages of oxidation.

The first approach is the most challenging and it produces crucial background knowledge for the interpretation of NMR spectra of mixtures of oxidation products. The other two can be done with simple sample preparation and without the need for derivatization (See review by Guillén and Ruiz 2001 and references therein).

Spectral Regions of Interest in the Analysis of Lipid Oxidation Products

As explained in the first part of this chapter, the NMR frequencies depend on the chemical environment of the particular nuclei; therefore the various functional groups in oxidized lipids resonate at slightly different frequencies. The characteristic chemical shift ranges of the protons and carbons of lipid oxidation products are depicted in Tables 5.1 and 5.2.

Chemical Shift and Solvent Effects

Chemical shifts may vary slightly with the experimental conditions, i.e., temperature, concentration of the solution, and even more with solvent. The temperature effects are most familiar for OH signals. Raising the temperature can cause coupling effects to disappear, for example, through inducing faster proton exchange or certain signals to move relative to others. When the same sample is run in different concentrations, the peak positions may show slight variations in different spectra, but the chemical shift differences between the signals within one spectrum are almost invariant. The solvent effects are an indication of molecular interaction between the dissolved molecules and the solvent. Effects of this sort are found mainly with polar compounds and become particularly important when the solvent molecules can arrange (with dipole-dipole or van der Waals interactions) in preferred orientations around the solute molecule or hydrogen bonding may occur. These effects are obtained when the spectrum is run in turn in solvents with different polarity and in some magnetically anisotropic solvents (e.g., benzene or other aromatic compounds).

In lipid chemistry, specific solvent effects have been systematically studied and used to advantage especially in steroid chemistry (Günther 1996a). However, to the best of our knowledge, there are no systematic studies on solvent effects on lipid oxidation products. The following example demonstrates the potential utility of the study of solvent effects in assigning the spectrum, i.e., in elucidation of structural, stereochemical, and conformational problems, and how this information can help in choosing the appropriate solvent for studying mixtures of compounds. The NMR spectrum of a mixture of two methyl linoleate hydroperoxides (Me 9-OOH-10*t*,12*c* and Me 13-OOH-9*c*,11*t*) was run in a relatively nonpolar solvent, $CDCl_3$, and a polar solvent, acetone-d_6 (Hämäläinen *et al.* 2001). In the ^{1}H NMR spectrum, the two hydroperoxide isomers differed only in the signals for the hydroperoxy protons in both solvents. In $CDCl_3$, these appeared as singlets at δ_H 8.05 for the 9-hydroperoxy isomer and at δ_H 7.97 for the 13-hydroperoxy isomer. In acetone-d_6, the hydroperoxy proton signals were shifted downfield, and appeared at δ_H 10.46 for the 9-hydroperoxy isomer and at δ_H 10.47 for the 13-hydroperoxy isomer. Interaction between the solvent molecules and the olefinic protons, although not as pronounced, was also evident. Moreover, the signals for the "outer" positioned olefinic protons in the conjugated diene system were partly overlapping in $CDCl_3$, whereas in acetone-d_6, the olefinic protons gave four separate resonance signals, which helps in the assigning of the spectrum. On the other hand, the ^{13}C NMR spectrum showed eight distinctive lines for the olefinic carbons in $CDCl_3$, whereas in acetone-d_6, the number of lines was only six. Thus, if olefinic signals are chosen as diagnostic signals (also known as "reporter" resonances), the appropriate choice for a ^{13}C NMR study of mixtures of hydroperoxides is $CDCl_3$.

Although it is not possible to make generalizations about solvent effects, the hydroperoxide proton signal, which is greatly influenced by the choice of solvent, deserves more attention. In general, hydroperoxy proton signals can be expected to be strongly dependent upon concentration, temperature, and the solvent employed in a manner similar to that of the resonances of the protons in OH, SH, and NH groups. The reason is that these protons can form hydrogen bonds and undergo exchange, and they have varying degrees of acidic character. However, their chem-

TABLE 5.1
Ranges of Chemical Shifts of ^1H Nuclei of Lipid Oxidation Products in $CDCl_3$[a]

Functional group		δ_H/ppm	n	References
-OO*H*	-CH_2OO*H*	7.9	3	Dussault and Sahli 1992
	-CH=CH-CH(OO*H*)-	7.70–7.92	5	Porter *et al.* 1994, Neff *et al.* 1990
	-CH=CH-CH=CH-CH(OO*H*)-	7.71–8.81	~31	Baba *et al.* 1992a, 1992b, 1994a; Claxon *et al.* 1994; Crombie *et al.* 1991; Dussault *et al.* 1993, Dussault and Lee 1995; Hämäläinen *et al.* 2001; Havrilla *et al.* 2000; Kenar *et al.* 1996; Nagata *et al.* 1989; Porter *et al.* 1990.
	dihydroperoxide -OO*H*s	7.74–8.25	40	Coxon *et al.* 1984; Neff *et al.* 1982, 1983
	hydroperoxy 1,2-dioxolanes -OO*H*	8.73–10.21	23	Coxon *et al.* 1981; Frankel *et al.* 1982a, 1982b; Michelich *et al.* 1980; Neff *et al.* 1981, 1982
-CH_2-OOH and -C*H*(OOH)-	-CH_2-OOH	4.0	3	Dussault and Sahli 1992
	-CH=CH-C*H*(OOH)-	*cis* 4.69 *trans* 4.25	4	Porter *et al.* 1994
	-CH=CH-CH*=CH-C*H*(OOH)-	**trans* 4.28–4.45	~24	Baba *et al.* 1992a, 1992b, 1994a; Crombie *et al.* 1991; Dussault *et al.* 1993, Dussault and Lee 1995; Hämäläinen *et al.* 2001; Kenar, *et al.* 1996; Onyango *et al.* 2001; Porter *et al.* 1990
	-CH=CH-C*H*(OOH)-CH=CH-	*cis,cis* 5.75 (in C_6D_6) *trans,trans* 4.67	2	Brash 2000; Corey and Nagata 1987
	dihydroperoxide –C*H*(OOH)	4.30–4.87	40	Coxon *et al.* 1984, Neff *et al.* 1982, 1983
	hydroperoxy 1,2-dioxolane -C*H*(OOH)-	3.85–4.20	27	Chan *et al.* 1980; Coxon *et al.* 1981; Frankel *et al.* 1982a, 1982b; Michelich *et al.* 1980; Neff *et al.* 1981, 1982
	hydroperoxy 1,2-dioxins -C*H*(OOH)-	4.08–4.22	6	Neff *et al.* 1983, Neff and Frankel 1984
-C*H*(OH)-	-CH=CH-C*H*(OH)-	*cis* 4.40–4.43 *trans* 4.00–4.03	10	Frankel *et al.* 1984, Kim *et al.* 2000, Porter *et al.* 1994

	-CH=CH-CH*=CH-C*H*(OH)-	**cis* 4.56–4.57 **trans* 4.10–4.21	49	Baba *et al.* 1992a; Chemin *et al.* 1992; Crilley *et al.* 1988; Dussault and Lee 1995; Gardner and Weislander 1970, 1972; Gueugnot *et al.* 1996; Havrilla *et al.* 2000; Henry *et al.* 1987; Hämäläinen *et al.* 2002; Just *et al.* 1983; Kann *et al.* 1990; Kato *et al.* 2001; Kenar *et al.* 1996; Kuklev *et al.* 1997; Lie Ken Jie *et al.* 1998; Martini *et al.* 1996a, 1996b; Moustakis *et al.* 1986; Nanda and Yada 2003, Porter *et al.* 1990; Reddy *et al.* 1994; Tranchepain *et al.* 1989; Zamboni and Rokach 1983
	-CH=CH-C*H*(OH)-CH=CH-	*cis,cis* 5.37 (in C_6D_6)	1	Brash 2000
	-CH=CH-C*H*(OH)-CH=CH-	*cis,trans* 4.95 (in C_6D_6)	1	Crilley *et al.* 1983
Double bond	-*CH*=*CH*-	4.5–8	NS	Friebolin 1998
Aldehyde	–*CH*=O	9–11	NS	Friebolin 1998
1,2-Dioxolane ring 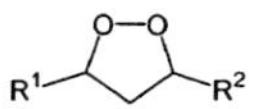	methine protons -*CH*-	4.44–4.84	40	Chan *et al.* 1980; Coxon *et al.* 1981; Frankel *et al.* 1982a, 1982b; Michelich *et al.* 1980; Neff *et al.* 1981, 1982
	methylene protons -CH_2-	2.13–2.88	43	Chan *et al.* 1980; Coxon *et al.* 1981; Frankel *et al.* 1982a, 1982b; Michelich *et al.* 1980; Neff *et al.* 1981, 1982
1,2-Dioxin ring 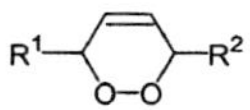	methine protons -*CH*-	4.48–4.66	~16	Bascetta *et al.* 1984b; Ideses *et al.* 1982; Neff *et al.* 1983; Neff and Frankel 1984
	olefinic protons -*CH*=*CH*-	5.90–6.01	14	Bascetta *et al.* 1984b; Ideses *et al.* 1982; Neff *et al.* 1983
Oxirane ring 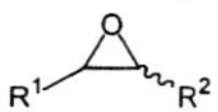	methylene and methine protons -CH_2-, -*CH*- (R^1 = alkyl, H)	2.34–3.72	~30	Bascetta *et al.* 1984a; Falck *et al.* 2001b; Friebolin *et al.* 1998; Günther 1996a; Lie Ken Jie and Lam 1995; Lie Ken Jie *et al.* 1997, 1999, 2003; Mosset *et al.* 1986; Tassignon *et al.* 1995
Furan ring 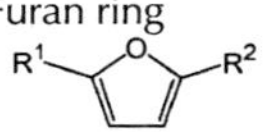	aromatic protons (R^1 = alkyl, H)	5.70–7.18	28	Lie Ken Jie *et al.* 1986

[a]n = number of entries; NS = not specified.

TABLE 5.2
Ranges of Chemical Shifts of ^{13}C Nuclei in Lipid Oxidation Products in $CDCl_3$[a]

Functional group		δ_C/ppm	n	References
$-CH_2$-OOH and -CH(OOH)-	-CH2OOH	76.89–77.08	9	Bascetta and Gunstone 1985, Dussault and Sahli 1992
	-CH=CH-CH(OOH)-	*cis* 81.1 *trans* 86.9–87.0	~10	Frankel *et al.* 1984, Garwood *et al.* 1977, Neff *et al.* 1990, Porter *et al.* 1994
	-CH=CH-CH*=CH(OO*H*)-	**trans* 85.9–86.8	~13	Dussault *et al.* 1993, Hämäläinen *et al.* 2001; 1995; Havrilla *et al.* 2000
	hydroperoxy 1,2-dioxolanes -CH-OOH	85.8–87.4	9	Frankel *et al.* 1982a, Michelich *et al.* 1980, Neff *et al.* 1981
$-CH_2$-OH and -CH(OH)-	$-CH_2$-OH	61.95–62.81	9	Bascetta and Gunstone 1985
	-CH(OH)-	63.02–73.21	17	Tulloch 1978
	-CH=CH-CH(OH)-	*cis* 67.5–67.8 *trans* 73.1–73.3	~7	Garwood *et al.* 1977, Frankel *et al.* 1984, Kim *et al.* 2000
	-CH=CH-CH*=CH-CH(OH)-	**cis* 67.94 **trans* 72.10–74.10	25	Chemin *et al.* 1992, Crilley *et al.* 1988, Frankel *et al.* 1990; Gueugnot *et al.* 1996, Henry *et al.* 1987, Hämäläinen *et al.* 2002, Kato *et al.* 2001, Kann *et al.* 1990, Martini *et al.* 1996a, 1996b, Nanda and Yada 2003, Tassignon *et al.* 1995
Double bond	-C=C-	100–150	NS	Friebolin *et al.* 1998
Allylic methylene	$-C=C-CH_2$-	*cis* 27.6–27.8 *trans* 32.01–32.85		Crilley *et al.* 1988; Frankel *et al.* 1984, 1990; Hämäläinen *et al.* 2002; Henry *et al.* 1987, Kann *et al.* 1990, Kato *et al.* 2001, Kim *et al.* 2000; Martini *et al.* 1996a; Tassignon *et al.* 1995

Carbonyl carbon	>C=O, and –CH=O	190–220	NS	Friebolin 1998
>C=0	-CO_2R, R = alkyl, H	160–180	NS	Friebolin. 1998
1,2-Dioxolane ring	methine carbons	82.6–85.8	18	Frankel *et al.* 1982a, Michelich 1980, Neff *et al.* 1981
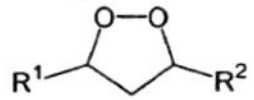	methylene carbons	40.8–43.5	13	Frankel *et al.* 1982a; Michelich *et al.* 1980; Neff *et al.* 1981, 1982
1,2-Dioxin ring	methine carbons	78.35	1	Bascetta *et al.* 1984b
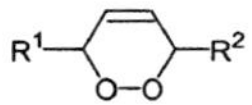	olefinic carbons	126.65	2	Bascetta *et al.* 1984b
Oxirane ring 	methine and methylene carbons	*cis* 46.71–58.08 *trans* 53.70–59.58	52	Bascetta and Gunstone 1985
Furan ring	carbons at 2- and 5-positions	140.70–156.72	24	Lie Ken Jie *et al.* 1986
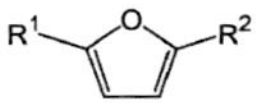	carbons at 3- and 4-positions	104.57–110.06	24	Lie Ken Jie *et al.* 1986

[a]n = number of entries; NS = not specified.

ical shifts are influenced not only by the above-mentioned experimental conditions, but also by impurities such as traces of acids (HCl is a common impurity in $CDCl_3$) and water. This means that the measured δ_H values are reproducible only under well-defined experimental conditions; consequently, the signal assignment of the hydroperoxy protons of different isomers in a mixture can be done only with great care, preferably using reference compounds. Chen *et al.* (1992) the effect of the presence of water on the chemical shifts of hydroperoxy protons was demonstrated in a work that studied the decomposition of methyl linoleate hydroperoxides demonstrated. The addition of water to the sample resulted in a downfield shift of both the hydroperoxide and the water proton resonances, and the extent of the shift was dependent on the water concentration. This observation was explained by hydrogen bonding, which naturally alters the shielding of these protons, i.e., they became less shielded and thus shift downfield. Furthermore, not only the position but also the shape of the resonance signal may be influenced by the presence of traces of acids or water. These impurities promote exchange processes and can cause coupling effects to disappear. The resonance signals of OH groups, for example, are usually singlets and broad, and the exact chemical shift values cannot be specified. However, if the ^3J values of H-C-O-H groups are of interest, the property of dimethyl sulfoxide to slow down the proton exchange can be used to advantage (Günther 1996a). The exchange of hydrogen atoms with deuterium is of great practical importance. After the H-D exchange, the corresponding signal disappears from the ^{1}H NMR spectrum; this can be used as a direct proof of hydroperoxide formation in an autoxidation study.

NMR Spectroscopy of Individual Compounds

The collection of data that follows is intended to serve as an aid for the interpretation of ^{1}H and ^{13}C NMR spectra of lipid oxidation products; it could prove helpful in designing new applications for analyzing mixtures of oxidation products (e.g., in a search for suitable "reporter" resonances). Because the assignment of the NMR spectra has not been the goal of much of the reviewed research, we note that the NMR data reported are often unassigned or only partially assigned. Furthermore, the assignments made on the basis of published data for similar carbons in other unsaturated long-chain fatty acids might not be correct, which serves as a warning not to overinterpret the NMR spectra. Several techniques are available for the confirmation of the assignments; as was shown in the above example, this can be a relatively straightforward task with the aid of 2D NMR spectroscopy.

Unless otherwise stated, the NMR spectra were run as pairs of enantiomers in $CDCl_3$. To avoid repetition, signals of functional groups common to all of the compounds such as olefinic, ester, or terminal methyl are not included in the lists of characteristic chemical shifts. Oxidations of unsaturated fatty acids result in mixtures of stereoisomeric products. Questions of stereochemistry are raised when there are examples in the literature to illustrate a specific point. For oxidation products that contain a cyclic structure (oxiranes, dioxolanes, dioxins, and furans), a numbering system in which the (two, three, four, and four) carbon atoms of the ring are considered to be part of "the fatty acid chain" and the esters carbonyl carbon as C-1 is often utilized instead of the IUPAC naming system. For example, a C_{18} furan fatty acid (FFA) can be considered to be a 2,5-disubstituted furan or a fatty acid with a furan ring (denotation $F_{8,11}$ refers to FFA where the four furan carbon atoms are located at C-8 to C-11 positions of the fatty acids carbon chain). The ^{13}C NMR spectroscopy of oxygenated fatty acids derivatives was reviewed previously by Gunstone (1993a, 2001), and by Lie Ken Jie and Mustafa (1997). Knothe and Nelsen (1998) presented the mathematical evaluation of the ^{13}C NMR signals of the saturated carbons in certain oxygenated fatty acids.

Mono- and Dihydroperoxides

Monohydroperoxides are formed as primary oxidation products in the autoxidation and photoxidation of monounsaturated and polyunsaturated fatty acids. The reaction with triplet oxygen proceeds by an autocatalytic free-radical chain reaction, whereas the reaction with singlet oxygen involves a concerted "ene" addition mechanism. Both autoxidation and photoxidation of polyunsaturated fatty acids can proceed to yield dihydroperoxides. Monohydroperoxides are also formed by lipoxygenase enzymes present in plants and animals. Certain enzymes, such as soybean lipoxygenase (Baba *et al.* 1992a, 1993, 1994a, 1994b) and potato lipoxygenase (Crombie *et al.* 1991), are highly regio- and stereospecific, and are thus utilized in the enantiospecific synthesis of optically active hydroperoxides and their derivatives.

The NMR data on lipid hydroperoxides are scarce, probably because these compounds are often analyzed as their hydroxy derivatives; these are not only more stable but they are also easier to separate. Thus, it is not a surprise that no systematic studies are available on the NMR of lipid hydroperoxides. Moreover, usually only the ^{1}H NMR data and those for mixtures of hydroperoxides are provided. The NMR spectra of hydroperoxides are run primarily in $CDCl_3$, and sometimes in $(CD_3)_2CO$ and CCl_4. Unfortunately, solvent effects cannot be determined reliably because sample concentrations or temperatures are not reported. Many studies also fail to report the hydroperoxy proton NMR signal, perhaps due to impurities that increase the proton exchange and thus make the signal broad and harder to detect. The ^{13}C NMR data of hydroperoxides are seldom provided, particularly in the older literature, perhaps because of the instability of these compounds and because at that time the measurement required a large amount of sample due to the insensitivity of the equipment. In more recent studies, the NMR data are sometimes included only as supplementary material.

Autoxidation produces racemic mixtures of hydroperoxides from achiral fatty acids. The autoxidation of oleic acid, for example, produces eight enantiomeric pairs of oleate hydroperoxides: 8-, 9-, 10-, and 11- hydroperoxy allylic *cis* and *trans* octadecenoates. The NMR spectra of a pair of enantiomers are identical, whereas those of the diastereoisomers usually differ. Stereoisomerism in a form of geometric isomerism arises from unsaturation. As explained earlier, the ^3J values of the olefinic protons reveal the geometry of the double bond: for protons in a *cis* double bond the $^3J_{cis}$ is 5–14 Hz and the typical value is 10 Hz, and for protons in a *trans* double bond the $^3J_{trans}$ is 11–19 Hz and the typical value is 16 Hz (Friebolin 1998). Alternatively, the geometry of the double bonds can be determined from the ^{13}C NMR spectrum because the ^{13}C chemical shifts of allylic methine and methylene carbons are indicative of their environment. The allylic carbon is shielded when adjacent to a *cis* double bond by ~5–6 ppm over that of the *trans* configuration. The allylic methine carbons of oleic acid hydroperoxides, for example, resonate when adjacent to a *cis* double bond at δ_C 81.1 (Frankel *et al.* 1984, Garwood *et al.* 1977) and when adjacent to a *trans* double bond at δ_C 86.9–87.0 (Frankel *et al.* 1984, Garwood *et al.* 1977, Porter *et al.* 1994).

Primary Monohydroperoxides. Dussault and Sahli (1992) reported the synthesis of primary C_{10}, C_{12}, and C_{16} alkyl hydroperoxides and their ^1H and ^{13}C NMR spectra. The hydroperoxy protons resonate in these compounds at δ_H 7.9 and the hydroperoxy-bearing methylene group at δ_H 4.0. In addition, Bascetta and Gunstone (1985) fully assigned the ^{13}C chemical shifts for a number of long-chain primary hydroperoxides. The hydroperoxy-bearing methylene carbon resonates at δ_C 76.89–77.08. The influence of the hydroperoxy group on the chemical shift of the α and β methylene carbon atoms is –1.89 and –3.58 ppm, respectively.

Secondary Allylic Monohydroperoxides from Monounsaturated Fatty Acids. NMR studies on allylic monounsaturated hydroperoxides such as those formed in the autoxidation of oleic acid are performed mainly on mixtures of isomers. However, Porter *et al.* (1994) presented the ^1H NMR spectra for pure *cis* and *trans* methyl 8-hydroperoxide-9-octadecenoates as well as the ^{13}C NMR spectrum for the *trans* isomer. The proton signals of special interest of these isomers are depicted in Figure 5.10 and the carbon data are listed in Table 5.3. The methine protons and the hydroperoxy-bearing methine carbons resonate in oleic acid hydroperoxides at a lower field (δ_H 4.25–4.69 and δ_C 81.1–87.0) than those of the primary hydroperoxides, because the former are also allylic protons and carbons. Furthermore, the allylic methine proton is less shielded (by ~0.4 ppm) when adjacent to a *cis* double bond than when adjacent to a *trans* double bond.

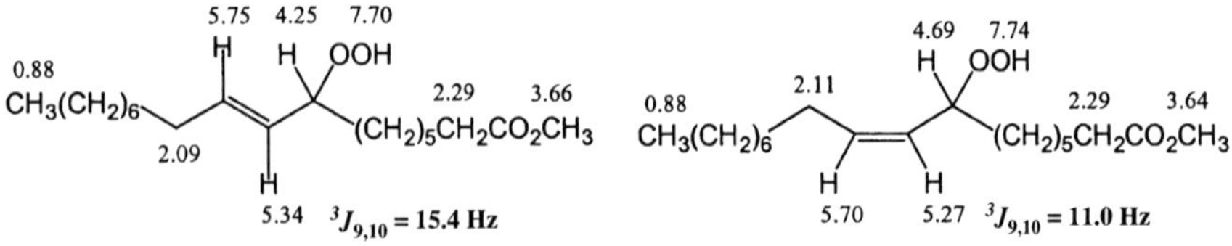

Fig. 5.10. The characteristic ^1H chemical shifts of methyl oleate hydroperoxides with ^3J values for the olefinic protons. *Source:* Porter et al. 1994.

TABLE 5.3
^{13}C Chemical Shifts (ppm) for Hydroperoxides of Methyl Oleate, Methyl Ricinoleate, and Methyl Linoleate[a]

Carbon nucleus	[26194-98-9] Me 8-OOH-9*t*[b]	[95502-85-5] Me 9-OOH-12-OH-10*t*[c]	[95502-84-4] Me 10-OOH-12-OH-8*t*[c]	[60900-57-4] Me 9-OOH-10*t*,12*c*[d]	[60900-56-3] Me 13-OOH-9*c*,11*t*[d]	[96192-73-3] Me 13*S*-OOH-9*c*,11*t*[e]*
C-1	174.28	174.31	174.21	174.24	174.35	174.4
C-2	34.01*	33.94	33.88	34.09	34.09	34.1
C-3	24.80*	24.74	24.69	24.91	24.90	24.9
C-4	*b*	28.86/29.06/29.57	28.53/29.09	29.02–29.30	28.92–29.40	28.9/2x29.0/29.4
C-5	*b*	28.86/29.06/29.57	28.53/29.09	29.02–29.30	28.92–29.40	28.9/2x29.0/29.4
C-6	*b*	28.86/29.06/29.57	28.53/29.09	29.02–29.30	28.92–29.40	28.9/2x29.0/29.4
C-7	*b*	25.07/25.21	25.30/25.40	25.23	28.92–29.40	28.9/2x29.0/29.4
C-8	86.98	32.29	128.31/135.67	32.55	27.73	27.7
C-9	128.37/137.25	86.02	128.31/135.67	86.76	133.82	133.9
C-10	128.37/137.25	129.85/137.89	83.33	131.00	127.47	127.6
C-11	32.32*	129.85/137.89	39.98	130.11	129.96	130.0
C-12	*b*	72.38	68.30	127.30	131.21	131.3
C-13	*b*	37.04	37.39	134.16	86.78	86.8
C-14	*b*	25.07/25.21	25.30/25.40	27.85	32.55	32.5
C-15	*b*	28.86/29.06/29.57	28.53/29.09	29.02–29.30	25.02	25.0
C-16	*b*	31.67	31.65	31.48	31.76	31.7
C-17	22.65*	22.46	22.40	22.57	22.54	22.5
C-18	14.09*	13.91	13.86	14.10	14.06	14.0
OCH_3	51.46*	51.31	51.29	51.50	51.52	51.5

[a]The complete nomenclature is as follows: Methyl 8-hydroperoxy-9-*trans*-octadecenoate (Me 8-OOH-9*t*); Methyl 9-hydroperoxy-12-hydroxy-10-*trans*-octadecenoate (Me 9-OOH-12-OH-10*t*); Methyl 10-hydroperoxy-12-hydroxy-8-*trans*-octadecenoate (Me 10-OOH-12-OH-8*t*); Methyl 9-hydroperoxy-10-*trans*,12-*cis*-octadecadienoate (Me 9-OOH-10*t*,12*c*); Methyl 13-hydroperoxy-9-*cis*,11-*trans*-octadecadienoate (Me 13-OOH-9*c*,11*t*).

[b]*Source:* Porter *et al.* 1994 (unassigned δ_C 25.09, 28.94, 29.02, 29.09, 29.16, 29.25, 29 39, 31.57, 31.85).

[c]*Source:* Bascetta *et al.* 1984a.

[d]*Source:* Hämäläinen *et al.* 2001 (original article is corrected here).

[e]*Source:* Dussault *et al.* 1993.*

*Assignments made by the reviewer.

Bascetta *et al.* (1984a) partly assigned the ^1H and ^{13}C (Table 5.3) NMR spectra of two methyl ricinoleate (methyl 12-hydroxyoleate) hydroperoxides. Interestingly, in one of the products in which the hydroperoxy-bearing methine group is allylic to a *trans* double bond, the methine carbon resonance shifts upfield from the usual value of such an allylic group to δ_C 83.33 by the influence of the β-hydroxy group. In addition, the hydroperoxy protons in these compounds resonate at a considerably lower field (δ_H 9.2–9.4) compared with those of methyl oleate hydroperoxides, suggesting that hydrogen bonding is occurring.

Secondary Allylic Monohydroperoxides from Polyunsaturated Fatty Acids. The characteristic ^1H and ^{13}C NMR signals of allylic hydroperoxy conjugated dienes are essentially the same as those in the corresponding hydroxy derivatives introduced in the first part of this chapter. Moreover, the ^1H NMR spectrum of a hydroperoxide is almost identical to the corresponding hydroxy derivative. The most notable differences are the presence of a hydroperoxy instead of a hydroxy proton signal, and the shift of the allylic methine proton resonance of the hydroperoxide downfield (by ~0.2 ppm) from that of the corresponding alcohol.

In methyl linoleate hydroperoxides with an allylic monohydroperoxide conjugated diene structure, the hydroperoxy proton resonates at δ_H 7.71–8.81 in $CDCl_3$ (for references, see Table 5.1), at δ_H 8.8–9.0 in CCl_4 (Hall and Roberts 1966), and at δ_H 10.40–10.65 in $(CD_3)_2CO$ (Chen *et al.* 1992, Hämäläinen *et al.* 2001). The solvent effects are not so apparent at the allylic methine proton resonance, which appears at δ_H 4.2–4.4. In a methyl linoleate hydroperoxide with a bis-allylic monohydroperoxide nonconjugated diene structure, namely in methyl 11-hydroperoxy-9*c*,12-*c*-octadecadienoate, the hydroperoxy proton resonates at δ_H 7.55, the bis-allylic methine proton at δ_H 5.75, and the olefinic protons at δ_H 5.55–5.70 in C_6D_6 (Brash 2000).

The ^1H NMR data are available for pure methyl linoleate hydroperoxides from autoxidation (Porter *et al.* 1990) and from enzymatic oxidation reactions (Crombie *et al.* 1991, Dussault *et al.* 1993, Dussault and Lee 1995). Kenar *et al.* (1996) provided the ^1H NMR data for two pure hydroperoxide isomers from cholesterol linoleate autoxidation, and Baba *et al.* (1992a, 1992b, 1994a) for one synthetic cholesterol fatty acid ester hydroperoxide and for several synthetic triacylglycerol hydroperoxides derived from linoleic acid. Hämäläinen *et al.* (2001) assigned the ^{13}C NMR spectra of the two *cis,trans* methyl linoleate hydroperoxides (Table 5.3) with the aid of 2D NMR spectroscopy. Dussault *et al.* (1993) and Dussault and Lee (1995) reported the ^{13}C NMR spectra for optically active enzymatic oxidation products of linoleic acid and of a C_{18} diacid.

Hydroperoxyeicosatetraenoic acids (HPETEs) are formed in the peroxidation of arachidonic acid. The ^1H NMR data were presented for synthetic 8- and 12-HPETE isomers (Nagata *et al.* 1989), the ^1H and ^{13}C NMR spectra unassigned for enantiomerically pure synthetic 5*S*-HPETE and 15*S*-HPETE and their methyl esters (Dussault and Lee 1995), and for 15*S*-HPETE cholesterol ester (Havrilla *et* *al.* 2000). Baba *et al.* (1993 and 1994b) provided ^{1}H NMR spectra for an optically active diacylglycerophospholipid hydroperoxide and for a diacylglycerophosphatidyl-L-serine hydroperoxide derived from arachidonic acid. In addition, Corey and Nagata (1987) reported the ^{1}H NMR spectrum for a 15-hydroperoxypentaenoic acid (15-HPEPE). In 15-HPEPE, the methine proton is bis-allylic with respect to two *trans* double bonds and it resonates at δ_H 4.67.

Dihydroperoxides. In methyl linoleate and methyl linolenate dihydroperoxides, the hydroperoxy protons resonate at δ_H 7.74–8.25 and the methine protons at δ_H 4.30–4.87. Neff *et al.* (1982, 1983) presented the ^{1}H NMR data for two pure 8,13- and one 9,14-dihydroperoxyoctadecadienoate isomers and for 12 isomeric dihydroperoxyoctadecatrienoates. Coxon *et al.* (1984) provided the ^{1}H NMR data for two diastereoisomeric all *cis* isomers and for two diastereoisomeric *trans,cis,trans* isomers of conjugated 9,16-dihydroperoxyoctadecatrienoate.

Hydroperoxy Epidioxides and Hydroperoxy Bis-Epidioxides

Oxidation of unsaturated fatty acids, particularly those containing more than two double bonds, can proceed readily to incorporate two or even three molecules of oxygen to yield hydroperoxy epidioxides or hydroperoxy bis-epidioxides (Chan *et al.* 1980, Frankel *et al.* 1984). Formation of hydroperoxy epidioxides, or more precisely hydroperoxy 1,2-dioxolanes, is an important process in the photoxidation of methyl linoleate (Frankel *et al.* 1982a, Michelich 1980, Neff *et al.* 1982) and in the autoxidation of methyl linolenate (Chan *et al.* 1980, Coxon *et al.* 1981, Neff *et al.* 1981). These oxidation reactions produce homoallylic peroxy radicals that are required for 1,3-cyclization into the 5-membered epidioxides (Scheme 5.1A). This ring closure is highly selective, affording a *cis*-1,2-dioxolane ring (Michelich 1980, O'Connor *et al.* 1981).

Polyunsaturated fatty acids (Bascetta *et al.* 1984b) or fatty acid hydroperoxides with a conjugated diene structure react with singlet oxygen to give 6-membered epidioxides, i.e., 1,2-dioxins. The reaction is a Diels-Alder reaction with singlet oxygen acting as a dienophile (Scheme 5.1B). For the reaction to occur, the diene must be capable of achieving the s-*cis* conformation. Hence, the *trans,trans* isomer is more reactive than the *cis,trans* isomer because the substituent R^2 in the latter suffers from steric crowding in the s-*cis* conformation. The photoxidation of methyl linoleate hydroperoxides (Neff *et al.* 1983) and methyl linolenate hydroperoxides (Neff and Frankel 1984) are thought to occur after the primary *cis,trans* hydroperoxides isomerize to *trans,trans* isomers.

There are no systematic studies available on the NMR spectroscopy of hydroperoxide epidioxides or on hydroperoxy bis-epidioxides, probably because the oxidation reactions of polyunsaturated fatty acids result in complex mixtures of different positional and stereoisomers of these cyclic peroxides in small amounts. Stereoisomerism arises from unsaturation (geometric isomers) and from chiral carbon atoms. The number of possible stereoisomers of saturated hydroperoxy 1,2-

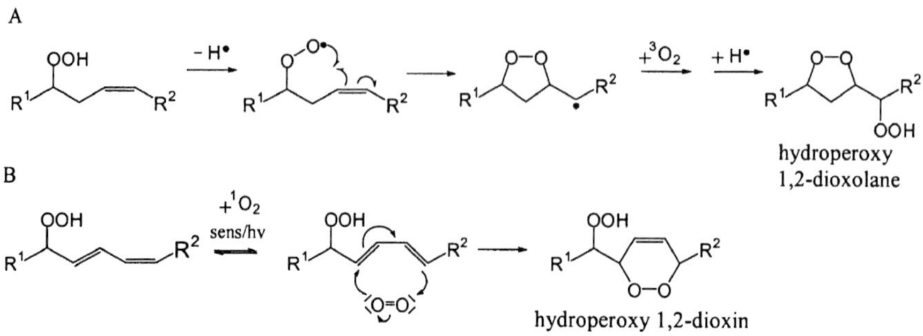

Scheme 5.1. Formation of (A) hydroperoxy 1,2-dioxolane and (B) hydroperoxy 1,2-dioxin.

dioxolane lipid model compounds, which have three chiral carbon atoms, is already eight. Frankel *et al.* (1981) reported the synthesis and the characteristic ^{1}H and ^{13}C chemical shifts (Table 5.4) for two such model compounds.

Unsaturated Hydroperoxy 1,2-Dioxolanes. The signals of interest in the ^{1}H NMR spectra of unsaturated hydroperoxy 1,2-dioxolanes include the hydroperoxy proton appearing at δ_H 8.73–10.21, the ring methine protons at δ_H 4.44–4.84, the ring methylene protons at δ_H 2.13–2.88, and the methine proton on the hydroperoxy-bearing carbon at δ_H 3.85–4.20. Characteristic ^{13}C chemical shifts of unsaturated hydroperoxy 1,2-dioxolanes include the ring methine carbons at δ_C 82.6–85.75, the ring methylene carbon at δ_C 40.8–43.5, and the methine carbon bearing the hydroperoxy group at δ_C 85.75–87.4.

^{1}H NMR spectroscopy is an extremely valuable tool for determining the structure and stereochemistry of the isolated unsaturated hydroperoxy 1,2-dioxolanes. Chemical shift and coupling constant considerations support not only a five-membered ring structure (Coxon *et al.* 1981, Neff *et al.* 1982) but can also provide proof for the positional configuration of the isomers. The diagnostic chemical shifts used for the latter purpose include those of the allylic methylene (Coxon *et al.* 1981) and the terminal methyl groups (Neff *et al.* 1982). The resonance of the terminal methyl group at δ_H 1.78, for example, reveals that this methyl group is attached to an olefinic carbon. In addition to the double bond geometry determination, ^{1}H NMR spectroscopy provides means to distinguish the *erythro* and *threo* diastereoisomers of hydroperoxy 1,2-dioxolanes; the chemical shifts of the methine proton on the hydroperoxy-bearing carbon and of one of the methylene protons in a ring reveal the stereochemical relation between the hydroperoxy group relative to the adjacent 1,2-dioxolane ring. In the *erythro* isomer, these signals appear at δ_H 3.85–4.11 and at δ_H 2.13–2.24, and in the *threo* isomer at δ_H 4.10–4.20 and at δ_H 2.40–2.47, respectively. In addition, the ^3J values between the ring methine proton and the methine proton on the hydroperoxy-bearing carbon are significantly different in the two diastereoisomers (Frankel *et al.* 1982a). In ^{13}C NMR spectra, the *erythro* and *threo* isomers can be distinguished by the ring methylene carbon resonance.

The hydroperoxy protons of unsaturated hydroperoxy 1,2-dioxolanes resonate due to intramolecular hydrogen bonding at a considerably lower field than those in acyclic monohydroperoxides. Moreover, Coxon *et al.* (1981) discovered that the downfield shift is significantly larger in the *threo* isomer than in the *erythro* isomer, suggesting that intramolecular hydrogen bonding is stronger in the former. They expect a "quasi-chair" conformation for the chelated species of the 1,2-dioxolane (Fig. 5.11), which means that in the *threo* isomer, the R^1 substituent at the hydroperoxy bearing methine carbon is "quasi-equatorial," whereas in the *erythro* isomer, it occupies the less favorable "quasi-axial" position. This energetically less favorable chelated species arising from the *erythro* isomer may account for the less strong hydrogen bonding observed.

Frankel *et al.* (1982a) examined the ^{1}H and ^{13}C (Table 5.4) NMR data of four isomeric hydroperoxy epidioxides (all as enantiomeric pairs) from the photoxidation of methyl linoleate. The four isomers were identified as diastereoisomeric pairs of two positional isomers, i.e., 9-hydroperoxy-10,12-epidioxy-13*t*-octadecenoate and 13-hydroperoxy-10,12-epidioxy-8*t*-octadecenoate. The different positional isomers with the same relative stereochemistry (e.g., the 9*S*-OOH-10*R*,12*S*-isomer and the 13*S*-OOH-10*S*,12*R*-isomer) could not be distinguished by ^{1}H NMR spectroscopy, but epimers (e.g., the 9*S*-OOH-10*R*,12*S*-isomer and the 9*R*-OOH-10*R*,12*S*-isomer) differed at the resonances of the methine proton on the hydroperoxy-bearing carbon atom (H-9), and of one of the ring methylene protons (H-11a); protons H-9 and H-11a resonate at δ_H 3.93 and 2.13 for the 9*S*,10*R*,12*S*-isomer (*erythro* isomer), and at δ_H 4.20 and 2.40 for the 9*R*,10*R*,12*S*-isomer (*threo* isomer). In the ^{13}C NMR spectra, the two positional isomers with the same relative stereochemistry differed at the methylene carbon adjacent to the hydroperoxy-bearing methine carbon (C-8 at δ_C 30.9 and C14 at δ_C 31.8), and the epimers differed significantly only at the resonance of the methylene ring carbon (at δ_C 40.8 for the 9*S*,10*R*,12*S*-isomer and at δ_C 43.3 for the 9*R*,10*R*,12*S*-isomer).

Chan *et al.* (1980) reported the characteristic ^{1}H chemical shifts for one and Coxon *et al.* (1981) for six isomers of conjugated diene hydroperoxy 1,2-dioxolanes from autoxidation of individual methyl linolenate hydroperoxides. An example of these compounds together with characteristic chemical shifts is depicted in Figure 5.12. Neff *et al.* (1981, 1982) provided the ^{1}H and ^{13}C (Table 5.4) NMR data for three conjugated diene hydroperoxy 1,2-dioxolanes from autoxidation of methyl linolenate, and some ^{1}H and ^{13}C chemical shifts for six unsaturated

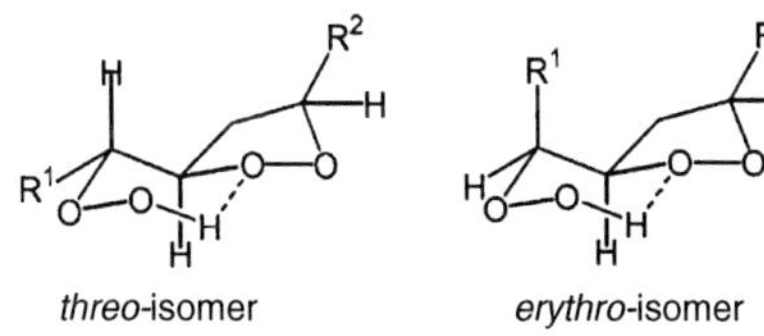

Fig. 5.11. Intramolecular H-bonding in diastereoisomeric hydroperoxy 1,2-dioxolanes.

TABLE 5.4
^{13}C Chemical Shifts (ppm) of Selected Hydroperoxy Epidioxides[a]

Carbon nucleus	[79760-42-2] Me 9*S*-OOH-10*R*,12*R*-epidioxide[b]	[79813-61-9] Me 9*R*-OOH-10*R*,12*R*-epidioxide[b]	[81445-28-5] Me 9*R*-OOH-10*R*,12*S*-epidioxide-13*t*[c]	[81445-26-3] Me 9*S*-OOH-10*R*,12*S*-epidioxide-13*t*[c]	[81445-56-9] Me 13*R*-OOH-10*S*,12*R*-epidioxide-8*t*[c]
C-1			174.4	174.4	174.4
C-2			34.1	34.1	34.1
C-3			24.9	24.9	24.9
C-4			28.7/29.1/29.3	29.0/29.2/29.6	28.5/28.8
C-5			28.7/29.1/29.3	29.0/29.2/29.6	28.5/28.8
C-6			28.7/29.1/29.3	29.0/29.2/29.6	28.5/28.8
C-7			25.6	25.7	32.3
C-8	33.5	32.0	30.9	30.9	124.4/139.0
C-9	83.3	84.1	85.9	86.1	124.4/139.0
C-10	81.3	82.0	82.9	83.3	82.9
C-11	42.4	42.5	43.3	40.9	43.3
C-12	72.5	73.0	84.0	83.8	84.1
C-13	33.9	32.9	124.2/139.3	124.0/139.5	85.9
C-14			124.2/139.3	124.0/139.5	31.8
C-15			32.1	32.1	25.3
C-16			32.1	32.1	31.8
C-17			22.2	22.5	22.6
C-18			13.9	13.8	14.0
OCH_3			51.5	51.8	51.5

[a]The complete nomenclature is as follows: Methyl 9-hydroperoxy-10,12-epidioxy-13-*trans*,15-*cis*-octadecanoate epidioxide-13*t*); Methyl 13-hydroperoxy-10,12-epidioxy-8-*trans*-octadecenoate (Me 13-OOH-10,12-epidioxide-8*t*); 16-hydroperoxy-13,15-epidioxy-9-*cis*,11-*trans*-octadecadienoate (Me 16-OOH-13,15-epidioxide-9c,11*t*);
[b]*Source:* Frankel *et al.* 1981.
[c]*Source:* Frankel *et al.* 1982a.
[d]*Source:* Neff *et al.* 1981.

(both conjugated and nonconjugated) hydroperoxy 1,2-dioxolanes from photoxidation of methyl linolenate. In the ^{13}C NMR spectra, the diastereoisomeric pairs differed only at the resonance for the ring methylene carbon.

Hydroperoxy Bi-1,2-Dioxolanes. The 1,3-cyclization of methyl linolenate 10- and 15-monohydroperoxides provides a pathway not only for hydroperoxy 1,2-dioxolanes but also for hydroperoxy bi-1,2-dioxolanes. Frankel and co-workers provided the ^{1}H NMR data for two positional isomers of hydroperoxy bi-1,2-dioxolanes as pairs of diastereoisomers (Frankel *et al.* 1982b, Neff *et al.* 1982). An example of these compounds together with the critical chemical shifts is presented in Figure 5.12. The positional isomers could be distinguished from each other on the basis of the chemical shifts of the terminal methyl groups, and the diastereoisomers based on the chemical shifts of the hydroperoxy proton and of the methine proton on the hydroperoxy-bearing carbon.

[81445-27-4] Me 13*S*-OOH-10*S*,12*R*-epidioxide-8*t*[c]	[78544-87-3] Me 9-OOH-10,12-epidioxide-13*t*,15*c*[d]	[74879-03-1] Me 16-OOH-13,15-epidioxide-9*c*,11*t*[d]	[74879-03-1] Me 16-OOH-13,15-epidioxide-9*t*,11*t*[d]
174.3	174.3	174.3	—
34.1	34.1	34.1	34.1
24.9	24.9	25.0	25.0
28.5/28.7/28.9	29.0/29.6	29.1/29.4	29.1
28.5/28.7/28.9	29.0/29.6	29.1/29.4	29.1
28.5/28.7/28.9	29.0/29.6	29.1/29.4	29.1
32.3	29.0/29.6	29.1/29.4	29.1
124.2/139.3	29.0/29.6	27.8	—
124.2/139.3	86.0	126.3/127.3/131.8/135.2	126.6/127.3
83.3	83.0/83.8	126.3/127.3/131.8/135.2	126.6/127.3
40.8	41.3	126.3/127.3/131.8/135.2	126.6/127.3
84.0	83.0/83.8	126.3/127.3/131.8/135.2	126.6/127.3
86.1	126.2/126.6/131.8/136.8	82.9/83.5	82.6/83.8
31.7	126.2/126.6/131.8/136.8	41.3	43.7
25.5	126.2/126.6/131.8/136.8	82.9/83.5	82.6/83.8
31.7	126.2/126.6/131.8/136.8	87.4	87.2
22.5	25.6	22.8	22.1
14.0	14.0	10.2	10.2
51.4	51.3	51.4	51.4

(Me 9-OOH-10,12-epidioxide); Methyl 9-hydroperoxy-10,12-epidioxy-13-*trans*-octadecenoate (Me 9-OOH-10,12-
Methy 9-hydroperoxy-10,12-epidioxy-13*t*,15*c*-octadecadienoate (Me 9-OOH-10,12-epidioxide-13*t*,15*c*); Methyl
Methyl 16-hydroperoxy-13,15-epidioxy-9-*trans*,11-*trans*-octadecadienoate (Me 16-OOH-13,15-epidioxide-9*t*,11*t*).

1,2-Dioxins and Hydroperoxy 1,2-Dioxins. Bascetta *et al.* (1984b) reported the characteristic ^{1}H (Fig. 5.12) and ^{13}C chemical shifts for 9,12-epidioxy-10-octadecenoate formed in the photoxidation of methyl 9*t*,11*t*-octadecadienoate. The methine carbon atoms of the 1,2-dioxin ring resonate at considerably higher field (δ_C 78.35) than those of the 1,2-dioxolane ring.

Neff *et al.* (1983) provided the ^{1}H NMR data for four (two positional isomers as pairs of diastereoisomers) hydroperoxy 1,2-dioxins formed in the photoxidation of methyl linoleate (see Fig. 5.12). Unlike in the 1,2-dioxolanes, there is no apparent difference in the chemical shifts between the diastereoisomers. The ^{1}H NMR spectra contained characteristic signals at δ_H 4.12–4.17 (the C*H*-OOH group), at δ_H 4.61–4.66 (the ring methine proton adjacent to the hydroperoxy-bearing carbon), at δ_H 4.48–4.53 (the other ring methine proton), and at δ_H 5.90–6.01 (the olefinic ring protons). In addition, the spin-spin coupling constants provided support for the six-membered ring structure as well as for the *cis* geometry of the double bond.

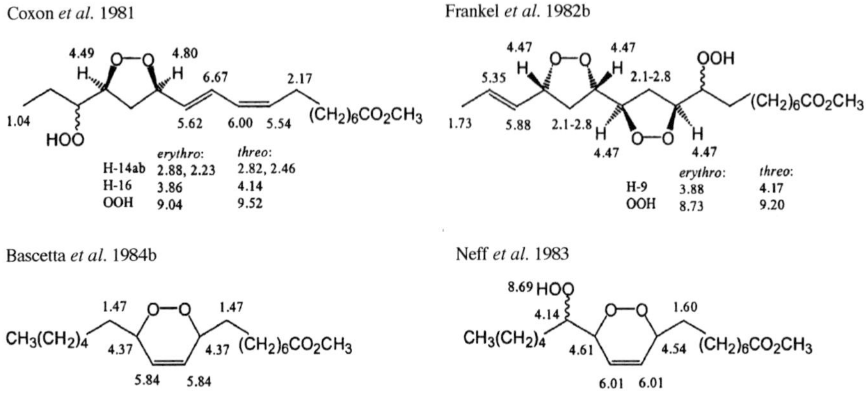

Fig. 5.12. Characteristic ^{1}H chemical shifts of selected lipid epidioxides.

Hydroxy Compounds

Alcohols can be formed by several mechanisms in the oxidation of polyunsaturated fatty acids. Lipid oxy radicals derived from lipid hydroperoxides by homolytic cleavage of the peroxide bond, for instance, provide several pathways to different hydroxy compounds (Scheme 5.2). In addition, the reactions of lipid hydroperoxides (Scheme 5.3B) and those of peroxyl radicals (e.g., the termination reaction by the Russell mechanism) can lead to hydroxy compounds.

Lipid hydroperoxides are often studied as their hydroxy derivatives or as fully saturated hydroxy derivatives. The former are obtained by chemoselective reduc-

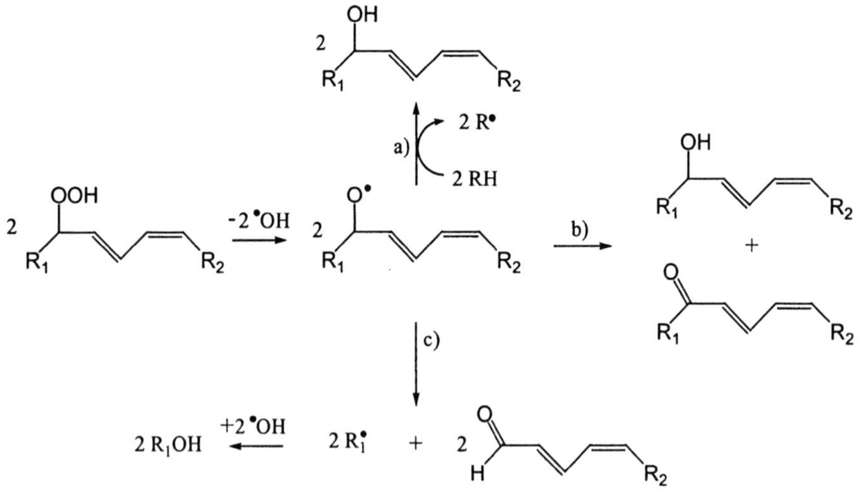

Scheme 5.2. Formation of alcohols from lipid oxy radicals by a) hydrogen atom abstraction, b) disproportionation, c) β-scission followed by radical-radical coupling reaction.

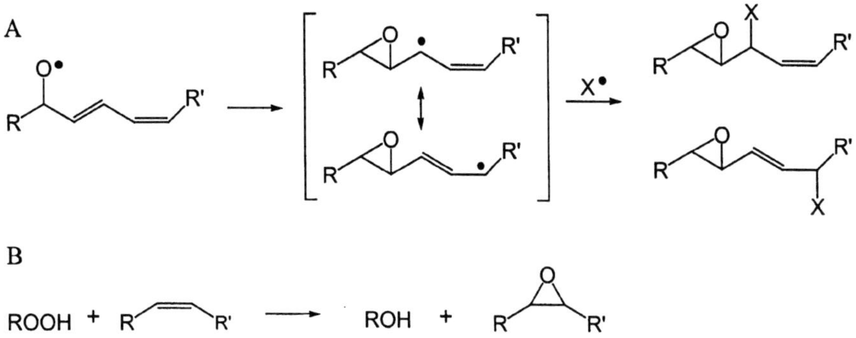

Scheme 5.3. Possible pathways for oxirane formation in the autoxidation of polyunsaturated fatty acids.

tion of the hydroperoxy group into a hydroxy group, e.g., with $NaBH_4$, and the latter, e.g., by palladium-catalyzed hydrogenation. In nature, hydroxy fatty acids are biologically important fatty acid metabolites produced mainly by plant systems. For example, three different enzymatic routes convert arachidonic acid to hydroxyeicosatetraenoic acids (HETEs).

As mentioned earlier, the NMR data of hydroxy compounds are more abundant than data concerning the corresponding hydroperoxides. Tulloch (1978) performed a systematic study of hydroxyoctadecanoates (hydroxystearic acids) and examined their solvent effects (Tulloch 1966). Pfeffer *et al.* (1992 and 1994) examined the effects of homoallylic and bis-homoallylic substitution on the olefinic ^{13}C chemical shifts of fatty acid methyl esters and Lie Ken Jie and Cheng (1995) extended that work. Hämäläinen *et al.* (2002) studied the NMR spectra of the hydroxy derivatives of CLA methyl ester hydroperoxides. Knothe synthesized several allylic mono- and dihydroxy compounds and their saturated analogs. This work was reviewed previously; thus, it is not included in this chapter (Knothe 1997).

Primary Alcohols. Bascetta and Gunstone (1985) assigned the ^{13}C chemical shifts for several long-chain saturated and unsaturated primary alcohols. The hydroxy-bearing methylene carbon resonates at δ_C 61.95–62.81. The influence of the primary hydroxy groups on the chemical shifts of the α and β methylene carbon atoms is +3.34 and –3.69 ppm, respectively.

Saturated Secondary Alcohols. Tulloch (1978) assigned unambiguously most of ^{13}C chemical shifts of 17 regioisomeric methyl hydroxyoctadecanoates and their acetate derivatives with the aid of deuterated model compounds. The effects of the hydroxy and of the acetate group, as calculated by comparison with methyl octadecanoate, are summarized in Table 5.5. A typical value for the methine carbon in a mid-chain hydroxy ester that is not much affected by the carboxylate or the terminal methyl group is 71.8 ppm (7 through 13-hydroxy isomers) and for the carbons α and

TABLE 5.5
Effects of the Hydroxy Group and the Acetate Group on ^{13}C Chemical Shifts of Nearby Carbon Atoms[a]

X	CHX	α	β	γ	δ	ε	ζ	η	θ
OH	+42.20[b]	+7.80[c]	–4.00[c]	+0.01	–0.09	–0.11	–0.06	–0.05	–0.04
OAc	+44.60[b]	+4.40[b]	–4.40[b]	–0.20	–0.20	–0.16	–0.11	–0.08	–0.07

[a]*Source:* Tulloch 1978.
[b]The average value for the effect in the 4 to 15-hydroxy isomers.
[c]The average value for the effect in the 5 to 16-hydroxy isomers.

β to the hydroxy-bearing carbon, 37.5 and 25.6 ppm, respectively. According to Tulloch, the oxygenated aliphatic esters up to 18 carbon atoms differ sufficiently in their spectra to allow identification of the isomers.

Monounsaturated Secondary (Allylic, Homoallylic, and Bis-Homoallylic) Alcohols. The ^{1}H NMR data for several pure hydroxy derivatives of methyl oleate hydroperoxides are available (Frankel *et al.* 1984, Kim *et al.* 2000, Porter *et al.* 1994). The ^{13}C NMR data for a few of these hydroxy derivatives are listed in Table 5.6. Using 2D NMR spectroscopy, Knothe *et al.* (1994) correlated the downfield resonance of the two olefinic carbon resonances of several allylic hydroxy *trans* monoene compounds to olefinic protons adjacent to the hydroxy-bearing carbon atom, which is the reverse of the previous assignment (Frankel *et al.* 1984). Furthermore, according to Knothe, two monohydroxy compounds with the hydroxy group in allylic positions on the opposite sides of the double bond can be distinguished by the differences in the olefinic carbon signals. He also discovered that when the hydroxy group is located between the double bond and the terminal methyl group, both the hydroxy group and the C-1 influence the olefinic chemical shifts, whereas the influence of C-1 is not apparent when the hydroxy group is located between the double bond and the C-1 group (Knothe *et al.* 1994).

Signals of particular interest in the ^{13}C NMR spectra of homoallylic and bis-homoallylic fatty acids alcohols are those most affected by the hydroxy group (i.e., the carbons α and β to the hydroxy-bearing carbon atom). Consider, for example, the unusual resonances for methylene carbons allylic to *cis* double bonds. A homoallylic hydroxy group shifts the allylic methylene carbon resonance downfield and it appears at δ_C 35.1–35.4, whereas a bis-homoallylic hydroxy group shifts the resonance upfield and it appears at δ_C 23.5–23.6. The ^{13}C NMR data of selected homoallylic and bis-homoallylic fatty acids and esters are summarized in Table 5.7. Lanser (1998) and Lanser and Manthey (1999) presented the ^{1}H and ^{13}C NMR data for keto hydroxy acids (bioconversion products of oleic acid) and Hou *et al.* (1994) for hydroxy acid (bioconversion products of linoleic acid), in which the hydroxy group is at the homoallylic position. Pfeffer *et al.* (1992) fully assigned the ^{13}C NMR data for three bis-homoallylic hydroxyoctadecenoic acid methyl esters. Lie Ken Jie and Cheng (1995) clarified the conflicting assignments of the

TABLE 5.6
^{13}C Chemical Shifts (ppm) of Methyl Hydroxyoctadecenoates[a]

Carbon nucleus	[64713-64-0] Me 9-OH-10*c*[b]	[10075-09-9] Me 9-OH-10*t*[b]	[118745-44-1] Me 10-OH-8*t*[c]	[95080-13-0] Me 12-OH-10*c*[b]	[95080-12-9] Me 12-OH-10*t*[b]
C-1	174.3	174.2	174.20	174.2	174.3
C-2	34.1	34.1	34.01	34.0	34.1
C-3	25.0	25.0	24.62	24.9	25.0
C-4			28.57[d]		
C-5			28.78[d]		
C-6			29.26[d]		
C-7			32.01		
C-8	37.6	37.4	131.97		
C-9	67.8	73.2	133.14	27.6	32.1
C-10	132.1[e]	133.2[f]	73.27	132.8[e]	131.9[f]
C-11	132.8[e]	132.1[f]	37.28	131.9[e]	133.3[f]
C-12	27.7	32.2	25.48	67.5	73.1
C-13			28.81[d]	37.6	37.4
C-14			29.55[d]		
C-15			29.55[d]		
C-16	31.8	31.8	31.86	31.8	31.9
C-17	22.6	22.7	22.66	22.6	22.6
C-18	14.0	14.0	14.10	14.0	14.1
OCH_3	51.4	51.4	?	51.3	51.4

[a]The complete nomenclature is as follows: Methyl 9-hydroxy-10-*cis*-octadecenoate (Me 9-OH-10*c*); Methyl 9-hydroxy-10-*trans*-octadecenoate (Me 9-OH-10t); Methyl 10-hydroxy-8-*trans*-octadecenoate (Me 10-OH-8*t*); Methyl 12-hydroxy-10-*cis*-octadecenoate (Me 12-OH-10*c*); Methyl 12-hydroxy-10-*trans*-octadecenoate (Me 12-OH-10*t*).
[b]*Source:* Frankel *et al.* 1984.
[c]*Source:* Kim *et al.* 2000 (the text refers to 10(*S*)-OH-8*t* [263023-34-3], but judging from the carbonyl resonance, this is an ester).
[d]Assignments considered doubtful by the reviewer.
[e]Values should be interchanged?
[f]Values interchanged by the reviewer based on Knothe *et al.* 1994.
? = Not reported.

^{13}C chemical shifts for the olefinic carbon atoms of homoallylic substituted unsaturated fatty esters in their NMR study of homoallylic and bis-homoallylic substituted fatty ester derivatives using 2D NMR techniques. They concluded that "for a C_n=C_{n+1} bond, the carbon chemical shift of the C_n is always greater than that of the C_{n+1} in a homoallylic [C_n=C_{n+1}-CH_2-CH(X)-, where X = OH, N_3, OAc, Cl or oxo] or bis-homoallylic [C_n=C_{n+1}-CH_2-CH_2-CH(X)-] system."

Diunsaturated Secondary Allylic Alcohols. The characteristic ^1H and ^{13}C NMR signals of allylic hydroxy conjugated octadecadienoic acids were introduced in the first part of this chapter. The critical ^1H chemical shifts for an unusual allylic hydroxy *cis,trans* conjugated octadecadienoate (in which the hydroxy group is allylic to a *cis* double bond) from autoxidation of CLA methyl ester (Hämäläinen *et al.* 2002), and

TABLE 5.7
^{13}C Chemical Shifts (ppm) of Selected Octadecenoic Acids and Methyl Esters with Homoallylic- and Bis-Homoallylic Hydroxy Groups[a]

Carbon nucleus	[221302-77-8] 7-OH-16-oxo-9c[b]	Unknown* Me 7-OH-17-oxo-9c[c]	[141724-85-8] Me 9-OH-5c[d]	[141724-83-6] Me 9-OH-12c[e]	[141724-84-7] Me 10-OH-6c[d]	[34932-12-2] 10-OH-12c[f]	[41989-07-5] Me 12-OH-9c[e]
C-1	178.9	174.2	174.28	174.355	174.15	179.4	174.343
C-2	35.4	34.0	33.45	34.089	33.89	34.0	34.093
C-3	24.7	24.9	24.89	24.918	24.49	24.6	24.937
C-4	29.4	29.1	26.56	28.829–29.463	29.06	28.9–29.5	28.829–29.463
C-5	25.3	25.4	130.55	28.829–29.463	26.75	28.9–29.5	28.829–29.463
C-6	36.5	36.6	129.09	28.829–29.463	129.85	28.9–29.5	28.829–29.463
C-7	71.3	71.3	23.55	25.571	129.65	28.9–29.5	28.829–29.463
C-8	35.4	35.4	37.75	37.325	23.53	25.6	27.389
C-9	125.3	125.1	71.51	71.678	37.54	36.6	133.217
C-10	133.2	133.4	37.24	37.432	71.43	71.6	125.280
C-11	27.2	27.3	25.70	23.604	37.22	35.1	35.377
C-12	29.0	29.4	29.73	129.184	25.64	125.0	71.514
C-13	28.6	29.0	29.67	130.654	29.70	133.4	36.874
C-14	23.7	29.0	29.61	27.210	29.58	27.3	25.743
C-15	42.3	23.7	29.35	28.829–29.463	29.24	28.9–29.5	28.829–29.463
C-16	212.1	43.7	31.92	31.538	31.85	31.5	31.866
C-17	35.9	209.3	22.70	22.588	22.62	22.5	22.641
C-18	7.8	29.9	14.12	14.080	14.04	14.0	14.100
OCH_3	—	51.5	?	51.466	?	—	51.463

[a]The complete nomenclature is as follows: 7-Hydroxy-16-oxo-9-*cis*-octadecenoic acid (7-OH-16-oxo-9*c*); Methyl 7-hydroxy-17-oxo-9-*cis*-octadecenoate (Me 7-OH-17-oxo-9*c*); Methyl 9-hydroxy-5-*cis*-octadecenoate (Me 9-OH-5*c*); Methyl 9-hydroxy-12-*cis*-octadecenoate (Me 9-OH-12*c*); Methyl 10-hydroxy-6-*cis*-octadecenoate (Me 10-OH-6*c*); 10-Hydroxy-12-*cis*-octadecenoic acid (10-OH-12*c*); Methyl 12-hydroxy-9-*cis*-octadecenoate (Me 12-OH-9*c*).

[b]*Source:* Lanser 1998.

[c]*Source:* Lanser and Manthey 1999.

[d]*Source:* Pfeffer *et al.* 1992 (? = not reported).

[e]*Source:* Lie Ken Jie and Cheng 1995.

[f]*Source:* Hou 1994.

*CA number for the corresponding acid is [248242-95-7].

for the nonconjugated hydroxy derivatives of methyl linoleate hydroperoxides (Brash 2000, Haslbeck *et al.* 1983) are illustrated in Figure 5.13. In addition, Kobayashi *et al.* (1987) presented the ^{1}H NMR data for synthetic 13-hydroxy-9*t*,11*c*-octadecadienoic acid in CCl_4, and Crilley *et al.* (1988) the ^{1}H and ^{13}C NMR data for a synthetic bis-allylic hydroxy octadecadienoic acid methyl ester in C_6D_6 (Fig. 5.13).

The ^{13}C NMR data of allylic hydroxy conjugated octadecadienoic acids and methyl esters are collected in Table 5.8. In all but one of these compounds, the hydroxy-bearing methine carbon is adjacent to a *trans* double bond. This is evident from the chemical shifts of the allylic methine carbons; those adjacent to a *trans* double bond appear in the range 72.85–72.95 ppm and the one adjacent to a *cis* double bond appears at δ_C 67.94.

The ^{1}H and ^{13}C NMR data are available for all four hydroxy compounds corresponding to the methyl linoleate hydroperoxides, i.e., *cis,trans* and *trans,trans* isomers of methyl 9-hydroxy-10,12-octadecadienoate (Me 9-OH-10,12) and methyl 13-hydroxy-9,11-octadecadienoate (13-OH-9,11) (Crilley *et al.* 1988, Frankel *et al.* 1990, Gardner and Weisleder 1970 and 1972, Hämäläinen *et al.* 2002, Kann *et al.* 1990, Kato *et al.* 2001, Kuklev *et al.* 1997, Martini *et al.* 1996a, Piazza *et al.* 1997, Porter *et al.* 1990, Tassignon *et al.* 1995, Tranchepain *et al.* 1989). These compounds are obtained either from autoxidation reactions or from synthetic procedures. The latter include optically active hydroxy compounds. NMR data were reported also for the corresponding acids (Henry *et al.* 1987, Kann *et al.* 1990, Kobayashi *et al.* 1987, Moustakis *et al.* 1986, Tranchepain *et al.* 1989). In addition, the ^{1}H NMR data for all hydroxy derivatives of cholesterol linoleate hydroperoxides (Baba *et al.* 1992a, Kenar *et al.* 1996), and ^{1}H and ^{13}C NMR spectra unassigned for the acetate and benzoate derivatives of Me 9-OH-10*c*,12*t* and Me 13-OH-9*c*,11*t* (Kato *et al.*

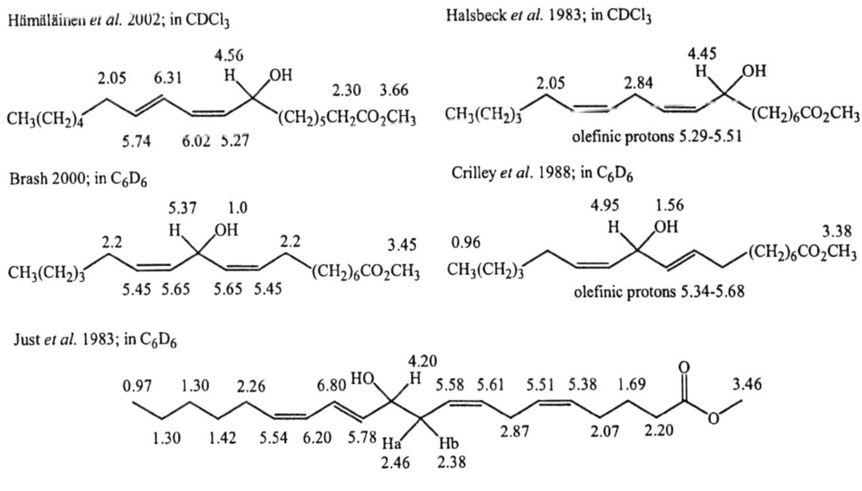

Fig. 5.13. Characteristic ^{1}H chemical shifts of unusual hydroxy diene esters and 11-HETE.

TABLE 5.8

^{13}C Chemical Shifts (ppm) of Hydroxyoctadecadienoic Acid and Methyl Esters[a]

Carbon nucleus	[215378-75-9] Me 8-OH-9*c*,11*t*[b]	[116595-30-3] Me 8-OH-9*t*,11*t*[b]	[6084-82-8] Me 9-OH-10*t*,12*c*[b]	[69257-16-5]/[10075-07-7] Me 9*R*/*S*-OH-10*t*,12*c*[c*]	[6157-02-4] Me 9-OH-10*t*,12*t*[b]	[464914-50-9] Me 12-OH-8*t*,10*t*[b]	[109837-85-6] Me 13-OH-9*c*,11*t*[b]	[10219-70-2]/[24058-13-7] Me 13*R*/*S*-OH-9*c*,11*t*[c*]	[32819-31-1] 13-OH-9*t*,11*t*[d*]	[28392-55-4] Me 13-OH-9*t*,11*t*[d*]
C-1	174.28	174.28	174.31	174.3	174.31	174.27	174.34	174.3	179.51	174.30
C-2	34.08	34.08	34.09	34.0	34.09	34.08	34.08	34.1	33.95	34.07
C-3	24.88	24.88	24.92	24.9	24.92	24.88	24.90	24.9	24.62	24.89
C-4	*b*	*b*	*b*	*c*	*b*	*b*	*b*	*c*	*d*	*d*
C-5	*b*	*b*	*b*	*c*	*b*	*b*	*b*	*c*	*d*	*d*
C-6	25.16	25.25	*b*	*c*	*b*	*b*	*b*	*c*	*d*	*d*
C-7	37.40	37.23	25.35	25.3	25.36	32.53	29.48	29.5	*d*	*d*
C-8	67.94	72.85	37.31	37.3	37.29	135.26	27.68	27.7	32.55	32.56
C-9	131.34	133.50	72.91	72.8	72.87	129.59	132.79	132.8	135.37	135.88
C-10	130.62	131.06	135.76	135.7	133.57	130.88	127.84	127.9	129.52	129.49
C-11	125.01	129.38	125.89	125.8	131.02	133.76	125.74	125.7	130.97	130.90
C-12	137.53	135.70	127.67	127.6	129.41	72.90	135.97	136.0	133.57	133.69
C-13	32.85	32.66	133.10	133.0	135.65	37.36	72.90	72.9	72.95	72.89
C-14	*b*	*b*	27.76	27.7	32.62	25.41	37.34	37.3	37.23	37.28
C-15	*b*	*b*	*b*	*c*	*b*	29.24	25.13	25.1	25.10	25.12
C-16	31.71	31.73	31.47	31.4	31.42	31.80	31.79	31.8	31.76	31.76
C-17	22.62	22.62	22.55	22.5	22.53	22.61	22.61	22.6	22.59	22.59
C-18	14.09	14.09	14.05	14.0	14.04	14.08	14.05	14.1	14.03	14.03
OCH_3	51.45	51.46	51.45	51.4	51.45	51.46	51.47	51.5	—	51.45

[a] The complete nomenclature is as follows: Methyl 8-hydroxy-9-*cis*,11-*trans*-octadecadienoate (Me 8-OH-9*c*,11*t*); Methyl 8-hydroxy-9-*trans*,11-*trans*-octadecadienoate (Me 8-OH-9*t*,11*t*); Methyl 9-hydroxy-10-*trans*,12-*cis*-octadecadienoate (Me 9-OH-10*t*,12*c*); Methyl 9-hydroxy-10-*trans*,12-*trans*-octadecadienoate (Me 9-OH-10*t*,12*t*); Methyl 12-hydroxy-8-*trans*,10-*trans*-octadecadienoate (Me 12-OH-8*t*,10*t*); Methyl 13-hydroxy-9-*cis*,11-*trans*-octadecadienoate (Me 13-OH-9*c*,11*t*); 13-Hydroxy-9-*trans*,11-*trans*-octadecadienoic acid (13-OH-9*t*,11*t*); Methyl 13-hydroxy-9-*trans*,11-*trans*-octadecadienoate (Me 13-OH-9*t*,11*t*).

[b] *Source:* Hämäläinen *et al.* 2002 (unassigned: Me 8-OH-9*c*,11*t* δ_C 28.90, 29.08, 29.16, 29.20; Me 8-OH-9*t*,11*t* δ_C 28.89, 29.07, 29.18, 29.21; Me 9-OH-10*t*,12*c* δ_C 29.07, 29.17, 29.31, 29.35; Me 9-OH-10*t*,12*t* δ_C 28.92, 29.07, 29.11, 29.35; Me 12-OH-8*t*,10*t* δ_C 28.78, 28.97, 29.00; Me 13-OH-9*c*,11*t* δ_C 28.96, 29.05, 29.06).

[c] *Source:* Kato *et al.* 2001* [unassigned: Me 9*R*/*S*-OH-10*t*,12*c* δ_C 29.0, 29.1 (2 C), 29.3; Me 13*R*/*S*-OH-9*c*,11*t* δ_C 29.0, 29.1 (2 C)].

[d] *Source:* Kann *et al.* 1990* [unassigned: 13-OH-9*t*,11*t* δ_C 28.88, 28.94, 29.00, 29.06; Me 13-OH-9*t*,11*t* δ_C 28.93, 29.05 (2 C), 29.10].

*Assignment made by the reviewer.

2001) and assigned for methyl 9*S*-methoxy-10*t*,12*t*-octadecadienoate (Tassignon *et al.* 1995) are available. Yamauchi *et al.* (1990) assigned the ^{1}H and ^{13}C NMR spectra for four diastereoisomers of α-tocopherone derivatives of both *cis,trans* positional isomers of methyl linoleate hydroperoxides. In addition, Hämäläinen *et al.* (2002) assigned the ^{1}H and ^{13}C NMR spectra of six hydroxy derivatives of 9*c*,11*t*-CLA methyl ester hydroperoxides with the aid of 2D NMR spectroscopy.

Other Polyunsaturated Secondary Allylic Alcohols. The structural features of polyunsaturated hydroxy fatty acids having more than two double bonds make the NMR spectroscopy of these compounds interesting. They usually have both conjugated and isolated double bonds, and the difference between these types of olefinic protons can be seen in ^{1}H NMR spectrum (for an example, see Fig. 5.13). In addition, the position of the hydroxy group varies, i.e., the hydroxy group can be attached to a bis-allylic position as in 10*R*-HETE, or to an allylic (to a *trans* double bond) position that is at the same time a homoallylic position as in 11-HETE (Fig. 5.13). The full assignment of the ^{1}H NMR spectrum of 11-HETE was performed with the aid of a 2D *J* resolved spectrum and confirmed by spectral simulation (Just *et al.* 1983). As can be expected, the diastereotopic protons (Ha and Hb) at the carbon atom adjacent to the hydroxy-bearing carbon atom resonate at different frequencies.

Polyunsaturated C_{18} hydroxy fatty acids, which have more than two double bonds, have interested researchers as targets of stereoselective synthesis because of their biological activity. Despite this, the availability of NMR data is limited. For example, Rao and Reddy (1986) reported the stereoselective synthesis of two positional isomers of hydroxy octadecatrienoic fatty acids (HOTE), but with no NMR data. However, the ^{1}H and ^{13}C NMR data are available for the synthetic 9*S*-HOTE methyl ester (Martini *et al.* 1996a), for 13-HOTE (Gardner *et al.* 1972, Nanda and Yadav 2003, Reddy *et al.* 1994), and for 16*R*-HOTE methyl ester and its benzoate derivative (Kato *et al.* 2001). In addition, ^{1}H NMR data were reported for C_{16} hydroxy trienoic fatty acid (Lumin *et al.* 1991) and ^{1}H and ^{13}C NMR data for C_{20} hydroxy trienoic fatty acid (Nanda and Yadav 2003).

NMR data are also available for several hydroxy derivatives of arachidonic acid hydroperoxides, HETEs. The ^{1}H NMR data were presented for synthetic 5*R*-HETE methyl ester (Zamboni and Rokach 1983), for 11-HETE (Corey *et al.* 1980), for 11*R*- and 11*S*-HETE methyl esters (Just *et al.* 1983), and for 12-HETE (Corey *et al.* 1980). Both the ^{1}H and ^{13}C (unassigned) NMR data were reported for 5*S*-HETE methyl ester (Dussault and Lee 1995, Gueugnot *et al.* 1996), for bis-allylic 10*R*-HETE and its methyl ester in CD_3CN (Yeola *et al.* 1996), for 12*S*- and 12*R*-HETE methyl esters (Chemin *et al.* 1992, Nicolaou *et al.* 1986, Yadagiri *et al.* 1986), and for 15*S*-HETE methyl ester and 5*S*,15*S*-diHETE methyl ester (Martini *et al.* 1996b). In addition, Havrilla *et al.* (2000) provided the ^{1}H NMR data for all positional hydroxy derivatives of cholesterol arachidonate hydroperoxides (Ch-HETEs), and Baba *et al.* (1993 and 1994b) for optically active diacylglycerophospholipid and diacylglycerophosphatidyl-L-serine HETE derivatives.

Oxo compounds

A wide range of saturated and unsaturated aldehydes and ketones is formed as volatile secondary products in the autoxidation and photoxidation of unsaturated fatty acids (Grosch 1987) (see Scheme 5.2). Many of the simple aldehydes and ketones are easily synthesized or commercially available and thus are not reviewed here in detail. Takeoka *et al.* (1995), for example, reported the synthesis together with ^{1}H NMR data for several oxoaldehydes. The aldehyde function is readily identified from the proton resonance at δ_H 9.71–9.81. In ^{13}C NMR spectra, the aldehyde carbonyl carbon appears usually in the range 190 to 220 ppm (Friebolin 1998). In addition, Tulloch and Mazurek (1976) and Tulloch (1977) synthesized all positional isomers of ketostearic acids and fully assigned their ^{13}C NMR spectra with the aid of deuterated model compounds. The presence of a keto group in the alkyl chain of the fatty ester is evident from the chemical shift of the carbonyl carbon at ~δ_C 210. The influence of a keto group on the chemical shift of the neighboring carbon atoms was determined and the shift parameters reported. The effect is strongest on the α and β methylene carbons that resonate in mid-chain ketostearic acids at ~δ_C 43 and 24, respectively. In addition, ^{1}H and ^{13}C NMR data are available among others for keto-enoic acid methyl esters (Bascetta *et al.* 1984a, Lie Ken Jie and Lam 1995 and 1996), keto-dienoic acid methyl esters (Kuklev *et al.* 1997, Tokita *et al.* 1999, Tokita and Morita 2000), and keto-trienoic acids (Koch *et al.* 2002), and ^{1}H NMR data for a keto-tetraenoic acid (Kerdesky *et al.* 1987). The ^{13}C NMR data of selected oxo (keto) acids were collected and presented by Gunstone (2001).

Oxiranes

Oxiranes (epoxides) are formed in the autoxidation of unsaturated fatty acids by intramolecular radical rearrangement of allylic oxy radicals (Scheme 5.3A). These oxiranes are generally *trans* 1,2-epoxides and have a substituent at the α or γ carbon atom. For example, methyl 12,13-epoxy-9-hydroperoxy-10-octadecanoate was isolated from the autoxidation of methyl linoleate (Imagawa *et al.* 1982). Moreover, this rearrangement was of major importance in the homolysis of linoleic acid hydroperoxides (Gardner *et al.* 1978, Gardner and Kleiman 1981, Hamberg *et al.* 1975). Alternatively, oxiranes are formed in the autoxidation of unsaturated fatty acids by intermolecular addition of the hydroperoxide oxygen across a double bond. This usually results in a *cis* oxirane, because most naturally occurring fatty acid have *cis* double bonds and the geometry of the double bond is retained in the reaction (Scheme 5.3B). Epoxidation of a secondary lipid oxidation product, 5-hydroxy-2-nonenal, by fatty acid hydroperoxides to a mutagenic oxirane product (Chen and Chung 1996) serves as an example of this type of reaction.

Epoxidation is an important biological process. Cytochrome P-450 reductase, for example, catalyzes one of the metabolic pathways of arachidonic acid, which leads to the formation of four regioisomeric *cis* epoxyeicosatrienoic acids (5,6-; 8,9-; 11,12-; and 14,15-EET) (Falck and Manna 1982, Falck *et al.* 1984b). Another enzyme, peroxygenase, catalyzes the intra- and intermolecular transfer of oxygen from fatty acid hydroperoxides to form epoxides (Piazza *et al.* 2003). Epoxy fatty acids, such as vernolic acid (*cis*-12,13-epoxyoleic acid), are found naturally in seed oils (Gunstone 1954).

Epoxidation is also an important industrial process. Polyunsaturated fatty acids have been subjected to chemical epoxidation by various epoxidizing agents, partly because these compounds are useful precursors for further synthesis. One of the most commonly utilized alkene epoxidizing agents is *m*-chloroperoxybenzoic acid (MCPBA). Epoxidized oils can also be used directly as plasticizers and plastic stabilizers (Gunstone 1993b).

Although there are a number of investigations on oxiranes (Jackman and Sternhell 1969 and references therein) and also on the solvent effects induced, for example, by benzene and pyridine (Williams *et al.* 1968), the NMR data of the lipid oxidation products, particularly ^{1}H NMR data, are limited. However, Bascetta and Gunstone (1985) performed a systematic study on ^{13}C NMR spectroscopy of saturated epoxy esters.

^{1}H NMR spectroscopy provides a way to determine the stereochemistry of oxiranes. Oxiranes have a rigid ring structure and thus stereochemically distinct vicinal coupling constants. According to the Karplus relation, which shows the dependence of ^3J values on dihedral angle, $^3J_{cis}$ is always > $^3J_{trans}$ for any given pair. Based on the literature reviewed in this chapter, the $^3J_{cis}$ values of oxiranes vary between 2.2 to 5.0 Hz and $^3J_{trans}$ values between 1.4 and 3.1 Hz. The ^3J values of oxiranes are thus displaced toward the lower values from the ranges of those of cyclopropane for which $^3J_{cis}$ values are usually between 6 and 10 Hz and $^3J_{trans}$ values between 3 and 6 Hz (Friebolin 1998, Jackman and Sternhell 1969). A typical ^{1}H chemical shift for oxirane methylene and methine (substituted oxiranes) protons is between 2.3 to 3.7 ppm. The coupling constants and the ^{1}H chemical shift of oxirane protons are depicted in Figure 5.14. In addition, $^3J_{CH}$ coupling constants of substituted oxiranes can be used as an aid for stereochemical determinations (Kingsbury *et al.* 1978).

When studying substituted oxiranes, such as fatty acid oxidation products, additional questions of stereochemistry that arise from chiral carbon atoms have to be considered. Monosubstituted oxiranes have one chiral carbon atom; they can

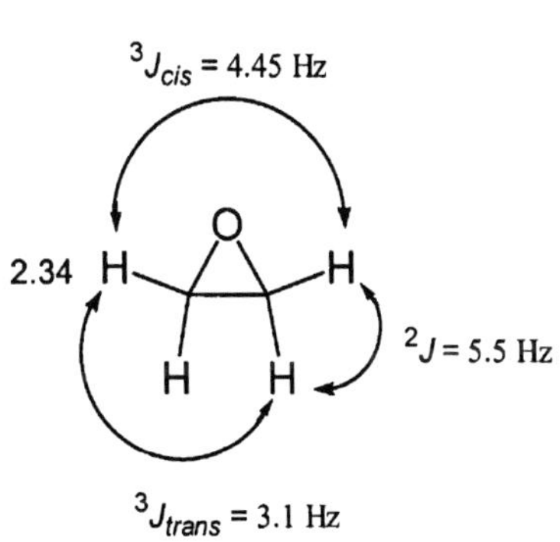

Fig. 5.14. The ^{1}H chemical shift and coupling constants of oxirane [75-21-8]. *Sources:* Friebolin 1998, Günther 1996a.

exist in two enantiomeric forms having identical NMR spectra. Saturated oxirane fatty acid esters (except the terminal oxiranes) have two chiral oxirane methine carbon atoms and can therefore exist in four different stereoisomeric forms. It should be noted that the coupling constants reveal only the relative stereochemistry of oxiranes; the saturated *cis* oxirane esters are *R***S** isomers (i.e., *RS* or *SR* isomers), whereas the *trans* oxirane esters are *R***R** isomers. In hydroperoxy oxirane and hydroxy oxirane fatty acid esters, there are three chiral carbon atoms; thus, the number of possible stereoisomers is eight. In α-hydroperoxy or α-hydroxy oxiranes, the $^3J_{HH}$ values reveal the relative stereochemistry between the α group and the oxirane ring. In α-hydroxy oxiranes, the ^3J value between the oxirane methine proton and the α-methine proton is ~5 Hz for a *threo* and 3 Hz for the *erythro* isomer (Omar *et al.* 2003, Piazza *et al.* 1997, Schieberle *et al.* 1985 and 1986, Tassignon *et al.* 1995). Furthermore, the ^{1}H NMR spectra of several *trans* oxirane alcohols showed that the α-methine protons of the *erythro* isomers resonate (at δ_H 3.8 ± 0.1) consistently downfield from those of the corresponding *threo* isomers (at δ_H 3.5 ± 0.1) (Mihelich 1979).

If a selective epoxidation of unsaturated alcohol is performed with a pure enantiomer with known stereochemistry, NMR spectroscopy can be utilized to determine the absolute stereochemistry of the oxirane alcohol formed. For example, the Sharpless epoxidation of 13*S*-hydroxy-9*c*,11*t*-octadecadienoic acid in the presence of the L(+)-diisopropyl tartrate produced diastereoselectively the *erythro* isomer of an α-hydroxy *trans* oxirane. This was evident from the ^{1}H NMR spectrum of the product because the ^3J value between the oxirane methine proton and the α-methine proton was 3.21 Hz and the α-methine proton resonated at δ_H 4.01 (Omar *et al.* 2003). The absolute stereochemistry of the oxirane must then be 11*S*,12*S*, because that of the hydroxy-bearing carbon (C-13) is *S*. As can be expected, the corresponding ^3J value in the *threo* isomer is larger (^3J = 4.80 Hz), and the α-methine proton resonates upfield at δ_H 3.62 from the resonance of the *erythro* isomer.

Saturated Oxirane Esters. The signals of special interest in the ^{13}C NMR spectrum of 17,18-epoxyoctadecanoate, a saturated terminal oxirane fatty acid ester, are the oxirane methylene carbon resonating at δ_C 46.71 and the oxirane methine carbon at δ_C 51.08. Similar shifts are reported for monosubstituted alkyl oxiranes (Bascetta and Gunstone 1985). Bascetta and Gunstone (1985) reported the ^{13}C chemical shifts of almost complete series of methyl *cis* and *trans* epoxyoctadecanoates. The two oxirane methine carbon atoms give one resonance signal in mid-chain (8,9- through 13,14-isomers) *cis* saturated oxirane esters at ~δ_C 56.85, whereas in mid-chain *trans* saturated oxirane esters at ~δ_C 58.51. When the oxirane function is closer to either end of the oxirane fatty acid ester, the influence of the terminal methyl group (14,15- through 16,17-isomers) or that of the ester group (3,4- through 7,8-isomers) can be detected, i.e., the oxirane methine carbon atoms give different signals, and the chemical shift values fall in the range 52.29 to 58.08 ppm for the *cis* and 53.70 to 59.58 ppm for the *trans* isomers. The largest diastereoisomeric differences observed in the series of epoxyoctadecanoates are those of the carbon atoms α to the oxirane ring; the α carbon atoms of the *cis* isomers resonate ~4.05 ppm upfield from those of the *trans* isomer. The effect of the oxirane function on the chemical shifts of nearby carbon atoms, as calculated by Bascetta and Gunstone, is presented in Table 5.9. As Gunstone pointed out, the values for β carbon atoms are significantly larger than the corresponding values for olefinic compounds and thus could help to identify the oxirane function (Gunstone 2001).

Unsaturated Oxirane Esters. Bascetta and Gunstone (1985) fully assigned the ^{13}C chemical shifts of seven unsaturated oxirane esters, which have 1–5 methylene groups between the oxirane ring and the double bond. The ^{13}C NMR spectrum of methyl *cis*-12,13-epoxy oleate, where the oxirane ring is at the homoallylic position, is particularly interesting because it can be envisaged to be formed in the autoxidation of methyl linoleate by intermolecular addition of the methyl linoleate hydroperoxide oxygen across one of the double bonds. Some characteristic ^{13}C chemical shifts of this unsaturated oxirane ester are depicted in Figure 5.15. In addition, Alaiz *et al.* (1989) reported some chemical shifts for the three monoepoxides produced from ethyl linoleate, and Piazza *et al.* (2003) the ^{1}H and ^{13}C NMR data for monoepoxides (regioisomers not determined) from linolenic acid and methyl linoleate in C_6D_6.

The influence of α,β-unsaturation on the ring methine ^{13}C chemical shifts can be clearly detected, when unsaturated oxirane esters in which the oxirane ring is at the allylic position to an alkene (Lie Ken Jie *et al.* 2003) or to an acetylene (Lie Ken Jie *et al.* 1999) group, are compared with saturated analogs. For example, when a *trans* oxirane is allylic to a *cis* double bond, the ring methine carbons give two signals, and the resonance of the ring methine carbon adjacent to the double bond shifts upfield by 4.07 ppm and that of the other ring methine carbon downfield by 1.69 ppm. A triple bond in the place of the double bond shifts the oxirane methine carbon resonances in the same directions by (–)12.79 and (+)2.12 ppm, respectively (Fig. 5.15).

Despite considerable synthetic efforts, the NMR data on EET are very limited. Enantiospecific synthesis of the *cis* stereoisomers of the 5,6-, 8,9- and 11,12-EET regioisomers were reported (Corey *et al.* 1980, Falck *et al.* 1984a, Frykman *et al.* 1997, Han *et al.* 2000, Mosset *et al.* 1986, Moustakis *et al.* 1985), but only incompletely assigned or unassigned NMR data are provided for 5,6- and 11,12-EET (Frykman *et al.* 1997, Mosset *et al.* 1986). In addition, all of the diastereoisomers

TABLE 5.9
Effects of *cis* and *trans* Oxirane Rings on ^{13}C Chemical Shifts of Nearby Carbon Atoms[a]

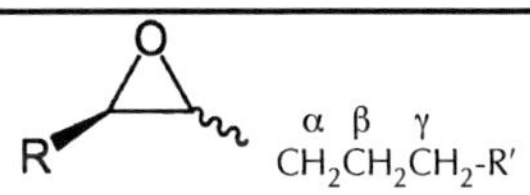

	α	β	γ
cis	−1.71	−2.93	−0.38
trans	+2.57	−3.50	−0.23

[a]*Source:* Bascetta and Gunstone 1985 (based on values from 6,7- through 16,17-epoxy esters).

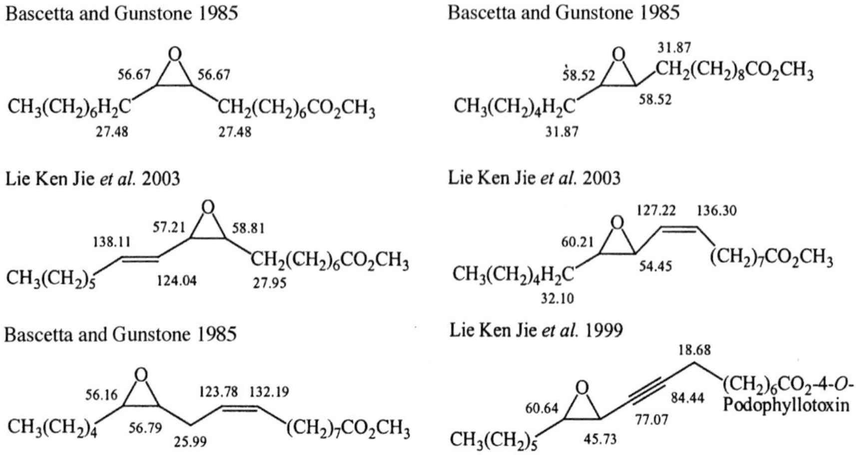

Fig. 5.15. Characteristic ^{13}C chemical shifts of selected epoxy fatty acids.

of 14,15-EET were synthesized (Corey *et al.* 1979 and 1980, Ennis and Baze 1986, Falck *et al.* 1984a and 2001a, Moustakis *et al.* 1985) as well as two 14,15-EET metabolites, but only unassigned ^1H NMR data are reported for 14*R**,15*S**-EET (Falck *et al.* 1984a) and unassigned ^1H and ^{13}C NMR data for one of the metabolites, i.e., methyl 9,10-epoxyoctadec-6*c*,12*c*-dienoate (Falck *et al.* 2001b).

Furans

Alkylfurans, ranging from 2-butylfuran to 2-octylfuran (Takeoka *et al.* 1995), and monosubstituted alkylcarboxylate furans were identified as volatile decomposition products of used frying oils. More specifically, 2-pentyl furan and methyl 8-(2-furyl)-octanoate are formed in the autoxidation of methyl linoleate and methyl linolenate (Chang *et al.* 1966, Gallasch and Spiteller 2000, Grosch 1987).

FFAs are found naturally, for example, in seed oils and in fish lipids. The FFAs of fish are tri- or tetrasubstituted; they have 1-2 methyl groups attached at the 3- or/and 4-position(s) of the furan ring and may vary in the chain lengths of the alkyl and alkylcarboxyl substituents (mainly C_{18} and C_{20} fatty acids). In addition, FFAs were established as secondary oxidation products in the autoxidation of CLA (Yurawecz *et al.* 1995). Moreover, FFA esters have been utilized as substrates for oxidation reactions. Dimethyl FFA esters are markedly more subject to autoxidation and polymerization than monomethyl FFA esters (Rahn *et al.* 1979). The autoxidation of one of the most common FFA esters, methyl 9,12-epoxyoctadeca-9,11-dienoate, results in the formation of oxo-furyl compounds (Sehat *et al.* 1998), whereas the ultrasonically stimulated oxidation reactions of a C_{18} FFA ester lead to non-furan oxidation products (Lie Ken Jie *et al.* 1997). Disubstituted furans are also formed in the epoxidation of *Biota* seed oil (Lie Ken Jie *et al.* 1988).

The NMR data of FFA are abundant. Lie Ken Jie and Lam (1978) and Lie Ken Jie *et al.* (1986) performed systematic ^{13}C NMR studies on isomeric C_{18} FFAs and FFA esters. In addition, Lie Ken Jie *et al.* (1983) presented ^{1}H and ^{13}C NMR data for several chemical transformation products of C_{18} FFAs and monomethyl substituted FFAs. However, no solvent effects studies involving FFAs are reported.

^{1}H NMR spectroscopy of FFAs provides means to determine experimentally not only the substituent and steric effects but also the ring current effect, i.e., the heteroaromatic character of furans. The aromaticity of furans arises from the aromatic sextet: 4 π-electrons of the two double bonds and an unshared electron pair of oxygen. These six electrons are delocalized and form a closed ring of electrons. When an external magnetic field is imposed upon a furan ring, as in the NMR measurement, the closed loop of aromatic electrons circulates in a diamagnetic ring current, which, according to Maxwell's law, sends out a field of its own. This induced field is parallel to the external field in the area of the furan protons in the molecular plane and outside the ring; thus, the protons experience a local field that is greater than the external magnetic field. It follows that the proton signals are deshielded, i.e., shifted downfield compared with where they would have been in the absence of the diamagnetic ring current. Hence, the olefinic protons in nonaromatic 4,5-dihydrofuran are found at δ_H 6.22 and 4.82, whereas the aromatic protons of furan are located at δ_H 7.40 and 6.60 (Fig. 5.16). The proton H-2 is shifted downfield by this magnetic anisotropy of the furan ring by ~1.2 ppm, and the H-3 proton by ~1.8 ppm. As the distance from the center of the ring increases, the deshielding influence decreases, as can be detected from the methyl group signal resonances in 2-methylfuran and 3-methylfuran. These groups are shifted downfield by ~1 ppm.

The ability to sustain an induced ring current is currently taken as a qualitative criterion for aromatic character, and compounds with this ability are termed *diatropic*. More about ring currents and π-electron effects in ^{1}H NMR spectra of number of hetero-aromatics are readily available in the literature (e.g., Abraham and Reed 2002, Page *et al.* 1965).

The ring current effect is apparent in alkylfurans. The ^{1}H chemical shifts of 2-pentyl furan are depicted in Figure 5.17. Multiplicity of the furan proton signals confirms the site of substitution; 2-monosubstituted furans display doublets for protons H-3 and H-5 and a doublet of doublets for proton H-4. In isomeric C_{18}

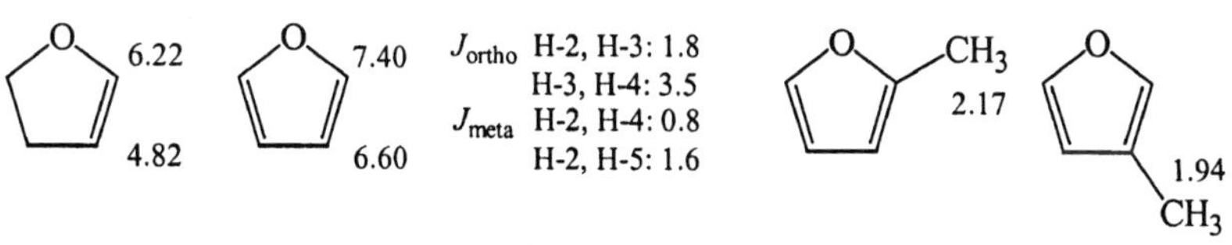

Fig. 5.16. The ^{1}H chemical shifts of special interest of 4,5-dihydrofuran [1191-99-7], furan [110-00-9], 2-methylfuran [534-22-5], and 3-methylfuran [930-27-8], and the *J* values for furan. *Sources:* Günther 1996a, Jackman and Sternhell 1969.

FFAs and FFA esters, the heteroaromaticity of the furan ring influences the chemical shifts of the methylene and methyl protons to some distance along the alkyl chain. The effects of the furan ring on ^{1}H chemical shifts of nearby methylene and methyl groups, as determined by Lie Ken Jie *et al.* (1986), are summarized in Table 5.10. The effect is always positive, hence indicating a downfield shift.

Not only does the heteroaromaticity of the furan ring affect the ^{1}H chemical shifts of the substituents, but the substituents themselves also influence the ^{1}H chemical shifts of the furan ring. The mid-chain FFA esters have a characteristic two-proton signal at ~δ_H 5.7 (Fig. 5.18). However, the furan protons give different signals when under the influence of the ester group ($F_{2,5}$ through $F_{7,10}$ isomers). This influence is at its strongest when the ester group is directly attached to the furan ring, i.e., when the methoxycarbonyl group is conjugated with the furan ring. The ester group is an electron withdrawing group with respect to the furan ring and hence, the chemical shifts of the furan protons shift downfield. In the $F_{2,5}$ isomer, for example, the furan protons resonate at δ_H 6.97 (H-3) and at δ_H 6.05 (H-4). Moreover, the conjugation results in a significant downfield shift of the methoxy proton signal (Lie Ken Jie *et al.* 1986). Similar effects on the furan ring protons are observed when the furan ring is in conjugation with an aldehyde group as can be seen when 5-hexyl-2-furaldehyde (an autoxidation product of $F_{9,12}$; Sehat *et al.* 1998) is compared with a C_{18} furan aldehyde in which the aldehyde group is further away from the furan ring (a chemical transformation product of C_{18} FFA; Lie Ken Jie *et al.* 1983) (Fig. 5.18). Interestingly, despite the ring current effect, the aldehyde proton is more shielded and shifted upfield.

In mono- and dimethyl substituted FFAs, the ring current effect on the additional methyl substituent(s) is clear (Fig. 5.19). The methyl group(s) is (are) deshielded by ~1 ppm. The synthesis of methyl (Lie Ken Jie *et al.* 1983) and dimethyl C_{18} FFA (Lie Ken Jie and Ahmad 1981, Lie Ken Jie *et al.* 1983) as well as methyl (Lie Ken Jie and Sinha 1980, Rahn *et al.* 1979) and dimethyl C_{20} FFAs (Rahn *et al.* 1979) and their ^{1}H NMR data are available.

In ^{13}C NMR spectroscopy, the ring current effect is less important. Some characteristic ^{13}C chemical shifts of an alkylfuran, of FFA ester and an FFA oxidation product, and of two methyl substituted FFA esters are depicted in Figures 5.17, 5.18, and 5.19, respectively. Lie Ken Jie *et al.* (1986) reported the synthesis of all isomeric C_{18} FFA methyl esters with full ^{13}C chemical shift assignments for most of the isomers. They concluded that the differences in the spectra are sufficient to allow the identification of all positional C_{18} FFA esters by ^{13}C NMR spectroscopy. In C_{18} FFA ester isomers ($F_{4,7}$ through $F_{15,18}$), the chemical shifts for the carbons at the 2- and 5-position fall in the range 140.70 to 156.72 ppm and those for the

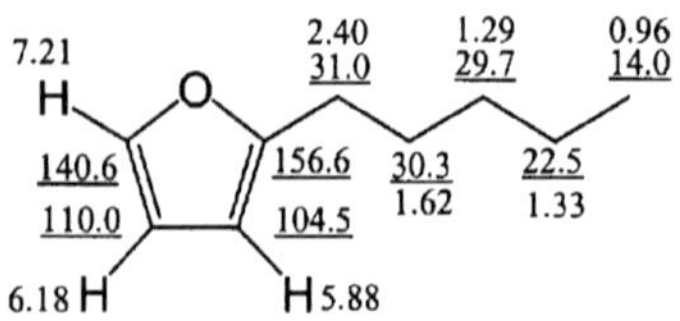

Fig. 5.17. The ^{1}H and ^{13}C (underlined) chemical shifts of 2-pentylfuran [3777-69-3]. *Sources:* Pouchert 1993, and Rahn *et al.* 1979.

TABLE 5.10
Effects of the Furan Ring on ^1H Chemical Shifts of Nearby Methylene and Methyl Groups[a]

	 R_2 = long-chain alkyl group (for CH_2 groups) or $-(CH_2)_nCH_3$, n = 0–5 (for CH_3 groups)					
	α	β	γ	δ	ε	ζ
CH_2	1.266	0.338	0.074	—	—	—
CH_2, R^1 = H	1.324	0.368	—	—	—	—
CH_3	1.337	0.323	0.073	0.053	0.028	0.012

[a]*Source:* Lie Ken Jie *et al.* 1986.

carbons at the 3- and 4-position in the range 104.57 to 110.06 ppm. Values "unaffected" by the proximity of the terminal methyl group or the ester group are δ_C 154.6 and δ_C 105.0, respectively.

A study of furan and its 2-methyl and 2,5-dimethyl derivatives demonstrates that the introduction of a methyl group into a furan ring will shift the resonance of the directly bonded ^{13}C nucleus downfield by ~9 ppm (Page *et al.* 1965). This is true also for methyl and dimethyl FFA esters. The α-deshielding effect of a methyl group on the furan ring ^{13}C chemical shift can be detected by comparing the FFA ester in Figure 5.18 to a monomethyl FFA ester in Figure 5.19. Moreover, the β effects of a methyl group attached to the C-4 or C-3 position of the furan ring seem to be similar in value but have opposite sign. The deshielding β effect (i.e., positive β effect) suggests according to Rahn *et al.* (1979) that the methyl group causes deformation of the ring. The ^{13}C NMR data are available for methyl and dimethyl C_{18} FFAs (Lie Ken Jie *et al.* 1983) and methyl and dimethyl fish C_{20} FFAs (Rahn *et al.* 1979).

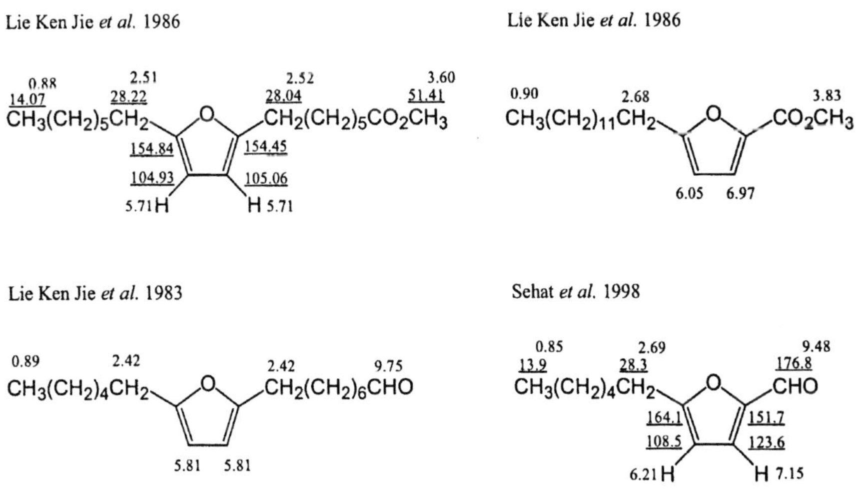

Fig. 5.18. Characteristic ^1H and ^{13}C (underlined) chemical shifts of furan fatty acids and furan aldehydes.

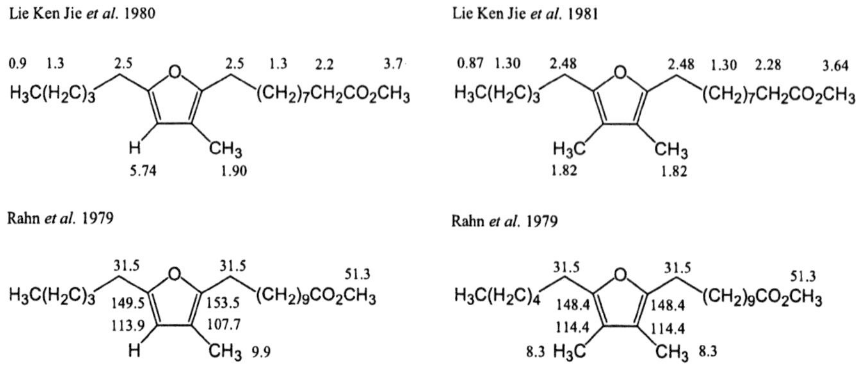

Fig. 5.19. Characteristic ^1H (C_{18} isomers) and ^{13}C (C_{20} isomers) chemical shifts of methyl and dimethyl furan fatty acid esters.

Concluding Remarks

The importance of the systematic study of pure compounds should be stressed here because it lays the foundations for the study of mixtures. In using NMR data for the development of NMR applications for the analysis of lipid oxidation, it is necessary to first study some pure compounds and to assign their NMR spectra as fully and reliably as possible. This offers a way to determine a group of resolvable and identifiable peaks (i.e., "reporter" resonances), which could then be used for regiospecific analysis of the NMR spectrum of a mixture. In lipid chemistry, this approach has proven to be useful, for instance, in the study of mixtures of triacylglycerols (Gunstone 1993a, Lie Ken Jie *et al.* 1997b and references therein) and of mixtures of CLA isomers (Davis *et al.* 1999). Furthermore, understanding the effects of concentration, temperature, and solvent makes the interpretation of mixtures more reliable and helps in the choice of the ideal solvent for a particular purpose. ^1H NMR spectroscopy can be used for studying classes of compounds, and ^{13}C NMR spectroscopy is the technique of choice when studying structurally closely related compounds because of the larger chemical shift dispersion.

Acknowledgment

T.I. Hämäläinen wishes to acknowledge funding support from The Finnish Cultural Foundation and Professor T. Hase for proofreading the manuscript.

References

Abraham, R.J., and Reid, M. (2002) ^1H Chemical Shifts in NMR. Part 18 Ring Currents and π-Electron Effects in Hetero-Aromatics, *J. Chem. Soc. Perkin Trans. 2*, 1081–1091.

Alaiz, M., Maza, M.P., Zamora, R., Hidalgo, F.J., Millan, F., and Vioque, E. (1989) Epoxidation of (Z)-9-(Z)-12-(Z)-15-Octadecatrienoate with *m*-Chloroperbenzoic Acid, *Chem. Phys. Lipids 49*, 221–224.

Baba, N., Tahara, S., Nakajima, S., Iwasa, J., Kaneko, T., and Matsuo, M. (1992a) Synthesis of Cholesteryl 13-Hydroperoxyoctadecadienoate and Its Derivative with Lipoxygenase, *Biosci. Biotechnol. Biochem. 56*, 540.

Baba, N., Hirota, N., Tahara, S., Nakajima, S., Iwasa, J., Kaneko, T., and Matsuo, M. (1992b) Chemoenzymatic Synthesis of Triacylglyceride Hydroperoxides, *Biosci. Biotechnol. Biochem. 56*, 1694–1695.

Baba, N., Shigeta, Y., Kishida, Y., Nakajima, S., Tahara, S., Kaneko, T., Matsuo, M., and Shimizu, S. (1993) Chemoenzymatic Synthesis of Optically Active Diacylglycerophospholipid Hydroperoxides Derived from Arachidonic Acid, *Biosci. Biotechnol. Biochem. 57*, 1615–1616.

Baba, N., Hirota, N., Umino, H., Matsuo, K., Nakajima, S., Tahara, S., Kaneko, T., and Matsuo, M. (1994a) Synthesis of Triacylglyceride Hydroperoxides Derived from Linoleic Acid, *Biosci. Biotechnol. Biochem. 58*, 1547–1548.

Baba, N., Aoishi, A., Shigeta, Y., Nakajima, S., Kaneko, T., Matsuo, M., and Shimizu, S. (1994b) Chemoenzymatic Synthesis of Phosphatidyl-L-serine Hydroperoxide, *Biosci. Biotechnol. Biochem. 58*, 1927–1928.

Bascetta, E., and Gunstone, F.D. (1985) ^{13}C Chemical Shifts of Long-Chain Epoxides, Alcohols and Hydroperoxides, *Chem. Phys. Lipids 36*, 253–261.

Bascetta, E., Gunstone, F.D., and Scrimgeour, C.M. (1984a) Photosensitized Oxidation of 12,13-Epoxy-, 12-Hydroxy-, 12-Oxo- and 12-Bromooleate, *Chem. Phys. Lipids 35*, 349–362.

Bascetta, E., Gunstone, F.D., and Scrimgeour, C.M. (1984b) Synthesis, Characterisation, and Transformations of a Lipid Cyclic Peroxide, *J. Chem. Soc. Perkin Trans. 1*, 2199–2205.

Brash, A.R. (2000) Autoxidation of Methyl Linoleate: Identification of the Bis-Allylic 11-Hydroperoxide, *Lipids 35*, 947–952.

Bus, J., Sies, I., and Lie Ken Jie, M.S.F. (1976) ^{13}C-NMR of Methyl, Methylene and Carbonyl Carbon Atoms of Methyl Alkenoates and Alkynoates, *Chem. Phys. Lipids 17*, 501–518.

Chan, H.W.-S., Matthew, J.A., and Coxon, D.T. (1980) A Hydroperoxy-Epidioxide from the Autoxidation of a Hydroperoxide of Methyl Linolenate, *J. Chem. Soc. Chem. Commun.*, 235–236.

Chang, S.S., Smouse, T.H., Krishnamurthy, R.G., Mookherjee, B.D., and Reddy, R.B. (1966) Isolation and Identification of 2-Pentylfuran as Contributing to the Reversion Flavor of Soybean Oil, *Chem. Ind.*, 1926–1927.

Chauret, D.C., Durst, T., Arnason, J.T., Sanchez-Vindas, P., San Roman, L., Poveda, L., and Keifer, P.A. (1996) Novel Steroids from Trichilia Hirta as Identified by Nanoprobe INADEQUATE 2D-NMR Spectroscopy, *Tetrahedron Lett. 37*, 7875–7878.

Chemin, D., Gueugnot, S., and Linstrumelle, G. (1992) An Efficient Stereocontrolled Synthesis of 12(*R*)-HETE and 12(*S*)-HETE, *Tetrahedron 48*, 4369–4378.

Chen, H.-J.C., and Chung, F.-L. (1996) Epoxidation of *trans*-4-Hydroxy-2-nonenal by Fatty Acid Hydroperoxides and Hydrogen Peroxide, *Chem. Res. Toxicol. 9*, 306–312.

Chen, H., Lee, D.J., and Schanus, E.G. (1992) The Inhibitory Effect of Water on the Co^{2+} and Cu^{2+} Catalyzed Decomposition of Methyl Linoleate Hydroperoxides, *Lipids 27*, 234–239.

Claxson, A.W.D., Hawkes, G.E., Richardson, D.P., Naughton, D.P., Haywood, R.M., Chander, C.L., Atherton, M., Lynch, E.J., and Grootveld, M.C. (1994) Generation of Lipid Peroxidation Products in Culinary Oils and Fats During Episodes of Thermal Stressing: A High Field ^{1}H NMR Study, *FEBS Lett. 335*, 81–90.

Corey, E.J., and Nagata, R. (1987) Synthesis of Three New Dehydroarachidonic Acid Derivatives and Their Oxidation by Soybean Lipoxygenase, *Tetrahedron Lett. 28*, 5391–5394.

Corey, E.J., Niwa, H., and Falck, J.R. (1979) Selective Epoxidation of Eicosa-*cis*-5,8,11,14-tetraenoic (Arachidonic) Acid and Eicosa-*cis*-8,11,14-trienoic Acid, *J. Am. Chem. Soc. 101*, 1586–1587.

Corey, E.J., Marfat, A., Falck, J.R., and Albright, J.O. (1980) Controlled Chemical Synthesis of the Enzymically Produced Eicosanoids 11-, 12-, and 15-HETE from Arachidonic Acid and Conversion into the Corresponding Hydroperoxides (HPETE), *J. Am. Chem. Soc. 102*, 1433–1435.

Coxon, D.T., Price, K.R., and Chan, H.W.-S. (1981) Formation, Isolation and Structure Determination of Methyl Linolenate Diperoxides, *Chem. Phys. Lipids 28*, 365–378.

Coxon, D.T., Peers, K.E., and Rigby, N.M. (1984) Selective Formation of Dihydroperoxides in the α-Tocopherol Inhibited Autoxidation of Methyl Linolenate, *J. Chem. Soc. Chem. Commun.*, 67–68.

Crilley, M.M.L., Golding, B.T., and Pierpoint, C. (1988) Palladium(II)-Catalyzed Rearrangements of Allylic Acetates in the Syntheses of Methyl (10*E*,12*Z*)-9-Hydroxyoctadeca-10,12-dienoate (α-Dimorphecolate) and (2*E*,4*Z*)-Deca-2,4-dienal, *J. Chem. Soc. Perkin Trans. 1*, 2061–2067.

Crombie, L., Morgan, D.O., and Smith E.H. (1991) An Isotopic Study (^{2}H and ^{18}O) of the Enzymic Conversion of Linoleic Acid into Colneleic Acid with Carbon Chain Fracture: The Origin of Shorter Chain Aldehydes, *J. Chem. Soc. Perkin Trans. 1*, 567–575.

Davis, A.L., McNeill, G.P., and Caswell, D.C. (1999) Identification and Quantification of Conjugated Linoleic Acid Isomers in Fatty Acid Mixtures by ^{13}C NMR Spectroscopy, in *Advances in Conjugated Linoleic Acid Research, Volume 1* (Yurawecz, M.P., Mossoba, M.M., Kramer, J.K.G., Pariza, M.W., and Nelson, G.J., eds.), pp. 164–179, AOCS Press, Champaign, IL.

Dussault, P., and Sahli, A. (1992) 2-Methoxy-2-propyl Hydroperoxide: A Convenient Reagent for the Synthesis of Hydroperoxides and Peracids, *J. Org. Chem. 57*, 1009–1012.

Dussault, P., and Lee, Q. (1995) A Chemoenzymatic Approach to Hydroperoxyeicosatetraenoic Acids. Total Synthesis of 5(*S*)-HPETE, *J. Org. Chem. 60*, 218–226.

Dussault, P., Sahli, A., and Westermeyer, T. (1993) An Organometallic Approach to Peroxyketals, *J. Org. Chem. 58*, 5469–5474.

Ennis, M.D., and Baze, M.E. (1986) Asymmetric Total Synthesis of 14(*R*),15(*S*)-, 14(*S*),15(*R*)-, 14(*R*),15(*R*)-, and 14(*S*),15(*S*)-Epoxyeicosatrienoic Acids, *Tetrahedron Lett. 27*, 6031–6034.

Falck, J.R., and Manna, S. (1982) 8,9-Epoxyarachidonic Acid: A Cytochrome P-450 Metabolite, *Tetrahedron Lett. 23*, 1755–1756.

Falck, J.R., Manna, S., and Capdevila, J. (1984a) Enantiospecific Synthesis of Methyl 11,12- and 14,15-Epoxyeicosatrienoate, *Tetrahedron Lett. 25*, 2443–2446.

Falck, J.R., Manna, S., Jacobson, H.R., Estabrook, R.W., Chacos, N., and Capdevila, J. (1984b) Absolute Configuration of Epoxyeicosatrienoic Acids (EETs) Formed During Catalytic Oxygenation of Arachidonic Acid by Purified Rat Liver Microsomal Cytochrome P-450, *J. Am. Chem. Soc. 106*, 3334–3336.

Falck, J.R., Reddy, Y.K., Haines, D.C., Reddy, K.M., Krishna, U.M., Graham, S., Murry, B., and Peterson, J.A. (2001a) Practical, Enantiospecific Syntheses of 14,15-EET and Leukotoxin B (Vernolic Acid), *Tetrahedron Lett. 42*, 4131–4133.

Falck, J.R., Kumar, P.S., Reddy, Y.K., Zou, G., and Capdevila, J.H. (2001b) Stereospecific Synthesis of EET Metabolites *via* Suzuki-Miyaura Coupling, *Tetrahedron Lett. 42*, 7211–7212.

Frankel, E.N., Weisleder, D., and Neff, W.E. (1981) Synthesis of a Saturated Lipid Hydroperoxy Cyclic Peroxide, *J. Chem. Soc. Chem. Commun.*, 766–767.

Frankel, E.N., Neff, W.E., Selke, E., and Weisleder, D. (1982a) Photosensitized Oxidation of Methyl Linoleate: Secondary and Volatile Thermal Decomposition Products, *Lipids 17*, 11–18.

Frankel, E.N., Neff, W.E., and Weisleder, D. (1982b) Formation of Hydroperoxy Bis-Epidioxides in Sensitized Photo-Oxidized Methyl Linoleate, *J. Chem. Soc. Chem. Commun.*, 599–600.

Frankel, E.N., Garwood, R.F., Khambay, B.P.S., Moss, G.P., and Weedon, B.C.L. (1984) Stereochemisty of Olefin and Fatty Acid Oxidation. Part 3. The Allylic Hydroperoxides from the Autoxidation of Methyl Oleate, *J. Chem. Soc. Perkin Trans. 1*, 2233–2240.

Frankel, E.N., Neff, W.E., and Weisleder, D. (1990) Determination of Methyl Linoleate Hydroperoxides by ^{13}C Nuclear Magnetic Resonance Spectroscopy, in *Methods in Enzymology* (Parker, I., and Glazer, A.N., eds.), Vol. 186, pp. 380–387, Academic Press, New York.

Friebolin, H. (1998) *Basic One- and Two-Dimensional NMR Spectroscopy*, 3rd ed., Wiley-VCH, Darmstadt.

Frykman, H.B., and Isbell, T.A. (1997) Synthesis of 6-Hydroxy δ-Lactones and 5,6-Dihydroxy Eicosanoic/Docosanoic Acids from Meadowfoam Fatty Acids *via* a Lipase-Mediated Self-Epoxidation, *J. Am. Oil Chem. Soc. 74*, 719-722.

Gallasch, B.A.W., and Spiteller, G. (2000) Synthesis of 9,12-Dioxo-10(*Z*)-dodecenoic Acid, a New Fatty Acid Metabolite Derived from 9 Hydroperoxy-10,12-octadecadienoic Acid in Lentil Seed (*Lens culinaris* Medik.), *Lipids 35*, 953–960.

Gardner, H.W., and Weisleder, D. (1970) Lipoxygenase from *Zea mays*: 9-*D*-Hydroperoxy-*trans*-10, *cis*-12-octadecadienoic Acid from Linoleic Acid, *Lipids 5*, 678–683.

Gardner, H.W., and Weisleder, D. (1972) Hydroperoxides from Oxidation of Linoleic and Linolenic Acids by Soybean Lipoxygenase. Proof of the *trans*-11 Double Bond, *Lipids 7*, 191–193.

Gardner, H.W., and Kleiman, R. (1981) Degradation of Linoleic Acid Hydroperoxides by a Cysteine Ferric Chloride Catalyst as a Model for Similar Biochemical Reactions. II. Specificity in Formation of Fatty Acid Epoxides, *Biochim. Biophys. Acta 665*, 113–125.

Gardner, H.W., Weisleder, D., and Kleiman, R. (1978) Formation of *trans*-12,13-Epoxy-9-hydroperoxy-*trans*-10-octadecadienoic Acid from 13-*L*-Hydroperoxy-*cis*-9,*trans*-11-octadecadienoic Acid Catalyzed by Either a Soybean Extract or Cysteine-$FeCl_3$, *Lipids 13*, 246–252.

Garwood, R.F., Khambay, B.P.S., Weedon, B.C.L., and Frankel, E.N. (1977) Allylic Hydroperoxides from the Autoxidation of Methyl Oleate, *J. Chem. Soc. Chem. Commun.*, 364–365.

Griffiths, L. (2000) Towards the Automatic Analysis of ^{1}H NMR Spectra, *Magn. Reson. Chem. 38*, 444–451.

Grosch, W. (1987) Reactions of Hydroperoxides—Products of Low Molecular Weight, in *Autoxidation of Unsaturated Lipids* (Chan, H.W.-S., ed.), pp. 95–139, Academic Press, London.

Gueugnot, S., Alami, M., Linstrumelle, G., Mambu, L., Petit, Y., and Larcheveque, M. (1996) An Efficient Total Synthesis of 5-(*S*)-HETE, *Tetrahedron 52*, 6635–6646.

Guillén, M.D., and Ruiz, A. (2001) High Resolution ^{1}H Nuclear Magnetic Resonance in the Study of Edible Oils and Fats, *Trends Food Sci. Technol. 12*, 328–338.

Gunstone, F.D. (1954) Fatty Acids. Part II. The Nature of the Oxygenated Acid Present in *Vernonia Anthelmintica* (Willd.) Seed Oil, *J. Chem. Soc.*, 1611–1616.

Gunstone, F.D. (1993a) High Resolution ^{13}C NMR Spectroscopy of Lipids, in *Advances in Lipid Methodology—Two* (Christie, W.W., ed.), pp. 1–68, The Oily Press, Dundee, Scotland.

Gunstone, F.D. (1993b) The Study of Natural Epoxy Oils and Epoxidized Vegetable Oils by Carbon-13 Nuclear Magnetic Resonance Spectroscopy, *J. Am. Oil Chem. Soc. 70*, 1139–1144.

Gunstone, F.D. (1999) High-Resolution NMR Spectroscopy as an Analytical Tool, *Lipid Technol. 11*, 39–41.

Gunstone, F.D. (last updated 2001) ^{13}C NMR Chemical Shifts for Fatty Acids and Their Derivatives, in http://www.lipid.co.uk/infores/index.htm, accessed 12. 13. 2004.

Gunstone, F.D., Pollard, M.R., Scrimgeour, C.M., and Vedanayagam, H.S. (1977) Fatty Acids. Part 50. ^{13}C Nuclear Magnetic Resonance Studies of Olefinic Fatty Acids and Esters, *Chem. Phys. Lipids 18*, 115-129.

Günther, H. (1996a) *NMR Spectroscopy*, 2nd ed., John Wiley and Sons, Chichester, UK.

Günther, H. (1996b) Standard Definitions of Terms, Symbols, Conventions, and References Relating to High-Resolution Nuclear Magnetic Resonance (NMR) Spectroscopy, in *NMR Spectroscopy*, 2nd ed., pp. 540–548, John Wiley and Sons, Chichester, UK.

Hall, G.E., and Roberts, D.G. (1966) A Study by Infrared and Proton Magnetic Resonance Spectroscopy of the Monohydroperoxides of Oleate and Linoleate Esters, *J. Chem. Soc.*, 1109–1112.

Hämäläinen, T.I., Sundberg, S., Mäkinen, M., Hase, T., Kaltia, S., and Hopia, A. (2001) Hydroperoxide Formation During Autoxidation of Conjugated Linoleic Acid Methyl Ester, *Eur. J. Lipid Sci. Technol. 103*, 588–593.

Hämäläinen, T.I., Sundberg, S., Hase, T., and Hopia, A. (2002) Stereochemistry of the Hydroperoxides Formed During Autoxidation of CLA Methyl Ester in the Presence of α-Tocopherol, *Lipids 37*, 533–540.

Han, X., Crane, S.N., and Corey, E.J. (2000) A Short Catalytic Enantioselective Synthesis of the Vascular Antiinflammatory Eicosanoid (11*R*,12*S*)-Oxidoarachidonic Acid, *Org. Lett. 2*, 3437–3438.

Harris, R.K., Kowalewsksi, J., and Cabral de Menezes, S. (1997) Parameters and Symbols for Use in Nuclear Magnetic Resonance, *Pure Appl. Chem. 69*, 2489–2495.

Haslbeck, F., Grosch, W., and Firl, J. (1983) Formation of Hydroperoxides with Unconjugated Diene Systems During Autoxidation and Enzymic Oxygenation of Linoleic Acid, *Biochim. Biophys. Acta 750*, 185–193.

Havrilla, C.M., Hachey, D.L, and Porter, N.A. (2000) Coordination (Ag+) Ion Spray-Mass Spectrometry of Peroxidation Products of Cholesterol Linoleate and Cholesterol Arachidonate: High-Performance Liquid Chromatography-Mass Spectrometry Analysis of Peroxide Products from Polyunsaturated Lipid Autoxidation, *J. Am. Chem. Soc. 122*, 8042–8055.

Haywood, R.M., Claxson, A.W.D., Hawkes, G.E., Richardson, D.P., Naughton, D.P., Coumbarides, G., Hawkes, J., Lynch, E.J., and Grootveld, M.C. (1994) Detection of Aldehydes and Their Conjugated Hydroperoxydiene Precursors in Thermally-Stressed Culinary Oils and Fats: Investigations Using High Resolution Proton NMR Spectroscopy, *Free Rad. Res. 22*, 441–482.

Henry, D.Y., Gueritte-Voegelein, F., Insel, P.A., Ferry, N., Bouguet, J., Potier, P., Sevenet, T., and Hanoune, J. (1987) Isolation and Characterization of 9-Hydroxy-10-*trans*,12-*cis*-octadecadienoic Acid, a Novel Regulator of Platelet Adenylate Cyclase from *Glechoma hederacea* L. Labiatae, *Eur. J. Biochem. 170*, 389–394.

Hou, C.T. (1994) Conversion of Linoleic Acid to 10-Hydroxy-12(*Z*)-octadecenoic Acid by *Flavobacterium* Sp. (NRRL B-14859), *J. Am. Oil Chem. Soc. 71*, 975–978.

Ideses, R.S., Arnon, K., and Jacob, T. (1982) Cyclic Peroxide: An Isolable Intermediate in Singlet Oxygen Oxidation of Pheromones to the Furan System, *Chem. Ind.*, 409–410.

Imagawa, T., Kasai, S., Matsui, K., and Nakamura, T. (1982) Methyl Hydroperoxy-Epoxy-Octadecenoate as an Autoxidation Product of Methyl Linoleate: A New Inhibitor-Uncoupler of Mitochondrial Respiration, *J. Biochem. 92*, 1109–1121.

Jackman, L.M., and Sternhell, S. (1969) *Applications of Nuclear Magnetic Resonance Spectroscopy in Organic Chemistry*, 2nd ed., Pergamon Press, Oxford.

Just, G., Luthe, C., and Viet, M.T.P. (1983) The Synthesis of 11*R*- and 11*S*-HETE [(5Z, 8Z, 12E, 14Z)-11-Hydroxyeicosatetraenoic Acid] and of 11-R,S-HPETE [(5Z, 8Z, 12E, 14Z)-11-Hydroperoxyeicosatetraenoic acid] Methyl Esters, *Can. J. Chem. 61*, 712–717.

Kann, N., Rein, T., Aakermark, B., and Helquist, P. (1990) New Functionalized Horner-Wadsworth-Emmons Reagents: Useful Building Blocks in the Synthesis of Polyunsaturated Aldehydes. A Short Synthesis of (±)-(*E,E*)-Coriolic Acid, *J. Org. Chem. 55*, 5312–5323.

Kato, T., Nakai, T., Ishikawa, R., Karasawa, A., and Namai, T. (2001) Preparation of the Enantiomers of Hydroxy-C18 Fatty Acids and Their Anti-Rice Blast Fungus Activities, *Tetrahedron Asymmetry 12*, 2695–2701.

Kenar, J.A., Havrilla, C.M., Porter, N.A., Guyton, J.R., Brown, S.A., Klemp, K.F., and Selinger, E. (1996) Identification and Quantification of the Regioisomeric Cholesteryl Linoleate Hydroperoxides in Oxidized Human Low Density Lipoprotein and High Density Lipoprotein, *Chem. Res. Toxicol. 9*, 737–744.

Kerdesky, F.A.J., Schmidt, S.P., Holms, J.H., Dyer, R.D., Carter, G.W., and Brooks, D.W. (1987) Synthesis and 5-Lipoxygenase Inhibitory Activity of 5-Hydroperoxy-6,8,11,14-eicosatetraenoic Acid Analogs, *J. Med. Chem. 30*, 1177–1186.

Kim, H., Gardner, H.W., and Hou, C.T. (2000) 10(*S*)-Hydroxy-8(*E*)-octadecenoic Acid, an Intermediate in the Conversion of Oleic Acid to 7,10-Dihydroxy-8(*E*)-octadecenoic Acid, *J. Am. Oil Chem. Soc. 77*, 95–99.

Kingsbury, C.A., Durham, D.L., and Hutton, R. (1978) $^3J_{CH}$ Coupling Constants in Oxiranes, Thiiranes, and Cyclopropanes, *J. Org. Chem. 43*, 4696–4700.

Knothe, G. (1997) NMR Characterization of Fatty Compounds Obtained *via* Selenium Dioxide-Based Oxidations, in *New Techniques and Applications in Lipid Analysis* (McDonald, R.E., and Mossoba, M.M., eds.), pp. 121–138, AOCS Press, Champaign, IL.

Knothe, G., and Nelsen, T.C. (1998) Evaluation of the ^{13}C NMR Signals of Saturated Carbons in Some Long-Chain Compounds, *J. Chem. Soc. Perkin Trans. 2*, 2019–2026.

Knothe, G., Bagby, M.O., Weisleder, D., and Peterson, R.E. (1994) Allylic Mono- and Dihydroxylation of Isolated Double Bonds with Selenium Dioxide-*tert*-butyl Hydroperoxide. NMR Characterization of Long-Chain Enols, Allylic and Saturated 1,4-Diols, and Enones, *J. Chem. Soc. Perkin Trans. 2*, 1661–1669.

Kobayashi, Y., Okamoto, S., Shimazaki, T., Ochiai, Y., and Sato, F. (1987) Synthesis and Physiological Activities of Both Enantiomers of Coriolic Acid and Their Geometric Isomers, *Tetrahedron Lett. 28*, 3959–3962.

Koch, T., Hoskovec, M., and Boland, W. (2002) Efficient Syntheses of (10*E*,12*Z*,15*Z*)-9-Oxo- and (9*Z*,11*E*,15*Z*)-13-Oxo-octadecatrienoic Acids; Two Stress Metabolites of Wounded Plants, *Tetrahedron 58*, 3271–3274.

Krucker, M., Lienau, A., Putzbach, K., Grynbaum, M.D., Schuler, P., and Albert, K. (2004) Hyphenation of Capillary HPLC to Microcoil ^{1}H NMR Spectroscopy for the Determination of Tocopherol Homologues, *Anal. Chem. 76*, 2623–2628.

Kuklev, D.V., Christie, W.W., Durand, T., Rossi, J.C., Vidal, J.P., Kasyanov, S.P., Akulin, V.N., and Bezuglov, V.V. (1997) Synthesis of Keto- and Hydroxydienoic Compounds from Linoleic Acid, *Chem. Phys. Lipids 85*, 125–134.

Lanser, A.C. (1998) Bioconversion of Oleic Acid to a Series of Keto-Hydroxy and Dihydroxy Acids by *Bacillus* Species NRRL BD-447: Identification of 7-Hydroxy-16-oxo-9-*cis*-octadecenoic Acid, *J. Am. Oil Chem. Soc. 75*, 1809–1813.

Lanser, A.C., and Manthey, L.K. (1999) Bioconversion of Oleic Acid by *Bacillus* Strain NRRL BD-447: Identification of 7-Hydroxy-17-oxo-9-*cis*-octadecenoic Acid, *J. Am. Oil Chem. Soc. 76*, 1023–1026.

Lie Ken Jie, M.S.F., and Lam, C.H. (1978) Fatty Acids. Part XVI. The Synthesis of All Isomeric C_{18} Furan-Containing Fatty Acids, *Chem. Phys. Lipids 21*, 275–287.

Lie Ken Jie, M.S.F., and Sinha, S. (1980) Synthesis of a Fish C_{20} Furanoid Fatty Acid from the Lipid Extract of the Latex of the Rubber Plant (*Hevea brasiliensis*), *J. Chem. Soc. Chem. Commun.*, 1002–1003.

Lie Ken Jie, M.S.F., and Ahmad, F. (1981) Conversion of Linoleic and Latex Furanoid Acid to Fish C_{18} Dimethyl Furanoid Isomers, *J. Chem. Soc. Chem. Commun.*, 1110–1111.

Lie Ken Jie, M.S.F., and Ahmad, F. (1983) Fatty Acids: Part 26. Partial Synthesis of C_{18} Mono- and Dimethylfuranoid Fatty Esters, *J. Am. Oil Chem. Soc. 60*, 1783–1785.

Lie Ken Jie, M.S.F., and Cheng, K.I. (1995) Nuclear Magnetic Resonance Spectroscopic Analysis of Homoallylic and Bis Homoallylic Substituted Methyl Fatty Acid Esters Derivatives, *Lipids 30*, 115–120.

Lie Ken Jie, M.S.F., and Lam, C.K. (1995) Ultrasound-Assisted Epoxidation Reaction of Long-Chain Unsaturated Fatty Esters, *Ultrasonics Sonochem. 2*, S11–S14.

Lie Ken Jie, M.S.F., and Lam, C.K. (1996) Regiospecific Oxidation of Unsaturated Fatty Esters with Palladium(II) Chloride/*p*-Benzoquinone: A Sonochemical Approach, *Chem. Phys. Lipids 81*, 55–61.

Lie Ken Jie, M.S.F., and Mustafa, J. (1997) High-Resolution Nuclear Magnetic Resonance Spectroscopy—Applications to Fatty Acids and Triacylglycerols, *Lipids 32*, 1019–1034.

Lie Ken Jie, M.S.F., Sinha, S., and Ahmad, F. (1983) Fatty Acids: Part 25. Chemical Transformations of C_{18} Furanoid Fatty Esters, *J. Am. Oil Chem. Soc. 60*, 1777–1782.

Lie Ken Jie, M.S.F., Bus. J., Groenewegen, A., and Sies, I. (1986) Fatty Acids, Part 28. ^{1}H- and ^{13}C-Nuclear Magnetic Resonance Studies of 2,5-Disubstituted C_{18} Furanoid Ester Isomers, *J. Chem. Soc. Perkin Trans. 2*, 1275–1278.

Lie Ken Jie, M.S.F., Lao, H.B., and Zheng, Y.F. (1988) Lipids in Chinese Medicine. Characterization of All *cis* 5,11,14,17-Eicosatetraenoic Acid in *Biota orientalis* Seed Oil and a Study of Oxo/Furanoid Ester Derived from *Biota* Oil, *J. Am. Oil Chem. Soc. 65*, 597–600.

Lie Ken Jie, M.S.F., Pasha, K.M., and Lam, C.K. (1997) Ultrasonically Stimulated Oxidation Reactions of 2,5-Disubstituted C_{18} Furanoid Fatty Ester, *Chem. Phys. Lipids 85*, 101–106.

Lie Ken Jie, M.S.F., Mustafa, J., and Pasha, K.M. (1998) An Efficient Ultrasound-Assisted Zinc Reduction of Fatty Esters Containing Conjugated Enynol and Conjugated Enynone Systems, *Lipids 33*, 941–945.

Lie Ken Jie, M.S.F., Mustafa, J., and Pasha, K.M. (1999) Synthesis and Spectral Characteristics of Some Unusual Fatty Esters of Podophyllotoxin, *Chem. Phys. Lipids 100*, 165–170.

Lie Ken Jie, M.S.F., Lam, C.N.W., Ho, J.C.M., and Lau, M.M.L. (2003) Epoxidation of a Conjugated Linoleic Acid Isomer, *Eur. J. Lipid Sci. Technol. 105*, 391–396.

Lodge, J.K., Sadler, P.J., Kus, M.L., and Winyard, P.G. (1995) Copper-Induced LDL Peroxidation Investigated by ^{1}H-NMR Spectroscopy, *Biochim. Biophys. Acta 1256*, 130–140.

Louden, D., Handley, A., Taylor, S., Lenz, E., Miller, S., Wilson, I.D., Sage, A., and Lafont, R. (2001) Spectroscopic Characterization and Identification of Ecdysteroids Using High-Performance Liquid Chromatography Combined with On-Line UV-Diode Array, FT-Infrared and ^{1}H-Nuclear Magnetic Resonance Spectroscopy and Time of Flight Mass Spectrometry, *J. Chromatogr. A 910*, 237–246.

Lumin, S., Falck, J.R., and Schwartzman, M.L. (1991) A Concise Synthesis of (*R*)-Hydroxy-*E,Z*-Diene Fatty Acids: Preparation of 12(*R*)-HETE, Tetranor-12(*R*)-HETE, and 13(*R*)-HODE, *Tetrahedron Lett. 32*, 2315–2318.

Martini, D., Buono, G., Montillet, J.-L., and Iacazio, G. (1996a) Chemoenzymic Synthesis of Methyl 9(*S*)-HODE (Dimorphecolic Acid Methyl Ester) and Methyl 9(*S*)-HOTE Catalyzed by Barley Seed Lipoxygenase, *Tetrahedron Asymmetry 7*, 1489–1492.

Martini, D., Buono, G., and Iacazio, G. (1996b) Regiocontrol of Soybean Lipoxygenase Oxygenation. Application to the Chemoenzymic Synthesis of Methyl 15(*S*)-HETE and Methyl 5(*S*),15(*S*)-diHETE, *J. Org. Chem. 61*, 9062–9064.

Medina, I., Sacchi, R., Giudicianni, I., and Aubourg, S. (1998) Oxidation of Fish Lipids During Thermal Stress as Studied by ^{13}C Nuclear Magnetic Resonance Spectroscopy, *J. Am. Oil Chem. Soc. 75*, 147–154.

Mihelich, E.D. (1979) Vanadium-Catalyzed Epoxidations. I. A New Selectivity Pattern for Acyclic Allylic Alcohols, *Tetrahedron Lett. 49*, 4729–4732.

Mihelich, E.D. (1980) Structure and Stereochemistry of Novel Endoperoxides Isolated from the Sensitized Photooxidation of Methyl Linoleate. Implications for Prostaglandin Biosynthesis, *J. Am. Chem. Soc. 102*, 7141–7143.

Miyake, Y., and Yokomizo, K. (1998) Determination of *cis*- and *trans*-18:1 Fatty Acid Isomers in Hydrogenated Vegetable Oils by High-Resolution Carbon Nuclear Magnetic Resonance, *J. Am. Oil Chem. Soc. 75*, 801–805.

Mosset, P., Yadagiri, P., Lumin, S., Capdevila, J., and Falck, J.R. (1986) Arachidonate Epoxygenase. Total Synthesis of Both Enantiomers of 8,9- and 11,12-Epoxyeicosatrienoic Acid, *Tetrahedron Lett. 27*, 6035–6038.

Moustakis, C.A., Viala, J., Capdevila, J., and Falck, J.R. (1985) Total Synthesis of the Cytochrome P-450 Epoxygenase Metabolites 5(*R*),6(S)-, 5(*S*),6(*R*)-, and 14(*R*),15(*S*)-Epoxyeicosatrienoic Acid (EET) and Hydration Products 5(*R*),6(*R*)- and 14(*R*),15(*R*)-Dihydroxyeicosatrienoic acid (DHET), *J. Am. Chem. Soc. 107,* 5283–5285.

Moustakis, C.A., Weerasinghe, D.K., Mosset, P., Falck, J.R., and Mioskowski, C. (1986) Synthesis of 12(*R*),13(*S*)-Oxido-9Z-octadecenoic (Vernolic) and 13(*S*)-Hydroxy-9Z,11*E*-octadecadienoic (Coriolic) Acids, *Tetrahedron Lett. 27*, 303–304.

Nagata, R., Kawakami, M., Matsuura, T., and Saito, I. (1989) Synthesis of 12-Hydroperoxyeicosatetraenoic Acid (12-HPETE). On the Stereochemistry in the Conversion of 12(*S*)-HETE to 12-HPETE, *Tetrahedron Lett. 30*, 2817–2920.

Nanda, S., and Yadav, J.S. (2003) Asymmetric Synthesis of Unnatural (*Z,Z,E*)-Octadecatrienoid and Eicosatrienoid by Lipoxygenase-Catalyzed Oxygenation, *Tetrahedron Asymmetry 14*, 1799–1806.

Neff, W.E., and Frankel, E.N. (1984) Photosensitized Oxidation of Methyl Linoleate Monohydroperoxides: Hydroperoxy Cyclic Peroxides, Dihydroperoxides and Hydroperoxy Bis-Cyclic Peroxides, *Lipids 19*, 952–957.

Neff, W.E., Frankel, E.N., and Weisleder, D. (1981) High-Pressure Liquid Chromatography of Autoxidized Lipids: II. Hydroperoxy-Cyclic Peroxides and Other Secondary Products from Methyl Linolenate, *Lipids 16*, 439–448.

Neff, W.E., Frankel, E.N., and Weisleder, D. (1982) Photosensitized Oxidation of Methyl Linolenate. Secondary Products, *Lipids 17*, 780–790.

Neff, W.E., Frankel, E.N., Selke, E., and Weisleder, D. (1983) Photosensitized Oxidation of Methyl Linoleate Monohydroperoxides: Hydroperoxy Cyclic Peroxides, Dihydroperoxides, Keto Esters and Volatile Thermal Decomposition Products, *Lipids 18*, 868–876.

Neff, W.E., Frankel, E.N., and Miyashita, K. (1990) Autoxidation of Polyunsaturated Triacylglycerols. I. Trilinoleoylglycerol, *Lipids 25*, 33–39.

Nicolaou, K.C., Ladduwahetty, T., Taffer, I.M., and Zipkin, R.E. (1986) A General Strategy for the Synthesis of Monohydroxyeicosatetraenoic Acids. Total Synthesis of 5(*S*)-Hydroxy-6(*E*),8,11,14(*Z*)-eicosatetraenoic Acid (5-HETE) and 12(*S*)-Hydroxy-5,8,14(*Z*),10(*E*)-eicosatetraenoic Acid (12-HETE), *Synthesis*, 344–347.

O'Connor, D.E., Mihelich, E.D., and Coleman, M.C. (1981) Isolation and Characterization of Bicycloendoperoxides Derived from Methyl Linolenate, *J. Am. Chem. Soc. 103*, 223–224.

Omar, M.N., Moynihan, H.A., and Hamilton, R.J. (2003) Asymmetric Sharpless Epoxidation of 13*S*-Hydroxy-9*Z*,11*E*-octadecadienoic Acid (13*S*-HODE), *Eur. J. Lipid Sci. Technol. 105*, 43–44.

Onyango, A.N., Inoue, T., Nakajima, S., Baba, N., Kaneko, T., Matsuo, M., and Shimizu, S. (2001) Synthesis and Stability of Phosphatidylcholines Bearing Polyenoic Acid Hydroperoxides at the *sn*-2 Position, *Angew. Chem. Int. Ed. Engl. 40*, 1755–1757.

Page, T.F., Jr., Alger, T., and Grant, D.M. (1965) Carbon-13 Nuclear Magnetic Resonance Spectra of Furan, Pyrrole, Thiophene, and Some of Their Methyl Derivatives, *J. Am. Chem. Soc. 87*, 5333–5339.

Pfeffer, P.E., Sonnet, P.E., Schwartz, D.P., Osman, S.F., and Weisleder, D. (1992) Effects of Bis Homoallylic and Homoallylic Hydroxyl Substitution on the Olefinic ^{13}C Resonance Shifts in Fatty Acid Methyl Esters, *Lipids 27*, 285–288.

Pfeffer, P.E., Sonnet, P.E., Schwartz, D.P., Osman, S.F., and Weisleder, D. (1994) Effects of Bis Homoallylic and Homoallylic Hydroxyl Substitution on the Olefinic ^{13}C Resonance Shifts in Fatty Acid Methyl Esters, *Lipids 29*, 913.

Piazza, G.J., Foglia, T.A., and Nuiiez, A. (1997) Enantioselective Formation of an α,β-Epoxy Alcohol by Reaction of Methyl 13(*S*)-Hydroperoxy-9(*Z*),11(*E*)-octadecadienoate with Titanium Isopropoxide, *J. Am. Oil Chem. Soc. 74*, 1385–1390.

Piazza, G.J., Nuñez, A., and Foglia, T.A. (2003) Epoxidation of Fatty Acids, Fatty Methyl Esters, and Alkenes by Immobilized Oat Seed Peroxygenase, *J. Mol. Catal. B: Enzymatic 21*, 143–151.

Porter, N.A., Dussault, P., Breyer, R.A., Kaplan, J., Morelli, J. (1990) The Resolution of Racemic Hydroperoxides: A Chromatography-Based Separation of Perketals Derived from Arachidonic, Linoleic, and Oleic Acid Hydroperoxides, *Chem. Res. Toxicol. 3*, 236–243.

Porter, N.A., Mills, K.A., and Carter, R.L. (1994) A Mechanistic Study of Oleate Autoxidation: Competing Peroxyl H-Atom Abstraction and Rearrangement, *J. Am. Chem. Soc. 116*, 6690–6696.

Pouchert, C.J. (1993) *The Aldrich Library of ^{13}C and ^{1}H FT NMR Spectra* (Pouchert, C.J., and Behnke, J., eds.), Aldrich Chemical Co., Milwaukee, Wis.

Rahn, C.H., Sand, D.M., Wedmid, Y., Schlenk, H., Krick, T.P., and Glass, R.L. (1979) Synthesis of Naturally Occurring Furan Fatty Acids, *J. Org. Chem. 44*, 3420–3424.

Rao, A.V.R., and Reddy, E.R. (1986) Stereoselective Synthesis of Hydroxyoctadecatrienoic Acids. The Self Defensive Substances in Rice Plant, *Tetrahedron Lett. 27*, 2279–2282.

Rapp, E., Jakob, A., Schefer, A.B., Bayer, E., and Albert, K. (2003) Splitless On-Line Coupling of Capillary High-Performance Liquid Chromatography, Capillary Electrochromatography and Pressurized Capillary Electrochromatography with Nuclear Magnetic Resonance Spectroscopy, *Anal. Bioanal. Chem. 376*, 1053–1061.

Reddy, K., Lakshmi, N., Reddy, S.P., and Sharma, G.V.M. (1994) Synthesis of 13-Hydroxy-9Z,11*E*,15Z-octadecatrienoic Acid, *Synth. Commun. 24*, 1395–1401.

Schieberle, P., Trebert, Y., Firl, J., and Grosch, W. (1985) Photolysis of Unsaturated Fatty Acid Hydroperoxides. 2. Products from the Anaerobic Decomposition of Methyl 13(*S*)-Hydroperoxy-9(*Z*),11(*E*)-octadecadienoate Dissolved in Methanol, *Chem. Phys. Lipids 37*, 99–114.

Schieberle, P., Trebert, Y., Firl, J., and Grosch, W. (1986) Photolysis of Unsaturated Fatty Acid Hydroperoxides. 3. Products from the Aerobic Decomposition of Methyl 13(*S*)-Hydroperoxy-9(*Z*),11(*E*)-octadecadienoate Dissolved in Methanol, *Chem. Phys. Lipids 41*, 101–116.

Sehat, N., Yurawecz, M.P., Roach, J.A.G., Mossoba, M.M., Eulitz, K., Mazzola, E.P., and Ku, Y. (1998) Autoxidation of Furan Fatty Acid Ester, Methyl 9,12-Epoxyoctadeca-9,11-dienoate, *J. Am. Oil Chem. Soc. 75*, 1313–1319.

Shapira, B., Karton, A., Aronzon, D., and Frydman, L. (2004) Real-Time 2D NMR Identification of Analytes Undergoing Continuous Chromatographic Separation, *J. Am. Chem. Soc. 126*, 1262–1265.

Silwood, C.J.L., and Grootveld, M. (1999) Application of High-Resolution, Two-Dimensional ^{1}H and ^{13}C Nuclear Magnetic Resonance Techniques to the Characterization of Lipid Oxidation Products in Autoxidized Linoleoyl/Linolenoylglycerols, *Lipids 34*, 741–756.

Takeoka, G.R, Buttery, R.G., and Perrino, C.T., Jr. (1995) Synthesis and Occurrence of Oxoaldehydes in Used Frying Oils, *J. Agric. Food Chem. 43*, 22–26.

Tassignon, P., de Waard, P., de Wit, D., and de Buyck, L. (1995) The Regio- and Stereoselectivity of the MCPBA-Epoxidation of Methyl Dimorphecolate and Derivatives, *Ind. Crops Products 3*, 273–280.

Tokita, M., and Morita, M. (2000) Identification of New Geometric Isomers of Methyl Linoleate Hydroperoxide and Their Chromatographic Behavior, *Biosci. Biotechnol. Biochem. 64*, 1044–1046.

Tokita, M., Iwahara, J., and Morita, M. (1999) New Geometric Isomers of Oxooctadecadienoate in Copper-Catalyzed Decomposition Products of Linoleate Hydroperoxide, *Biosci. Biotechnol. Biochem. 63*, 993–997.

Tranchepain, I., Le Berre, F., Dureault, A., Le Merrer, Y., and Depezay, J.C. (1989) Total Enantiospecific Syntheses of 13(*S*)-Hydroxy-9*Z*,11*E*-octadecadienoic (Coriolic) Acid and Its 13(*S*)-*N*-Tosylamino Analog, *Tetrahedron 45*, 2057–2065.

Tulloch, A.P. (1966) Solvent Effects on the Nuclear Magnetic Resonance Spectra of Methyl Hydroxystearates, *J. Am. Oil Chem. Soc. 43*, 670–674.

Tulloch, A.P. (1977) Deuterium Isotope Effects and Assignment of Carbon-13 Chemical Shifts in Spectra of Methyl Octadecanoate and the Sixteen Isomeric Oxooctadecanoates, *Can. J. Chem. 55*, 1135–1142.

Tulloch, A.P. (1978) Carbon-13 NMR Spectra of All the Isomeric Methyl Hydroxy- and Acetoxyoctadecanoates. Determination of Chemical Shifts by Deuterium Isotope Effects, *Org. Magn. Reson. 11*, 109–115.

Tulloch, A.P., and Mazurek, M. (1976) Carbon-13 Nuclear Magnetic Resonance Spectroscopy of Saturated, Unsaturated, and Oxygenated Fatty Acid Methyl Esters, *Lipids 11*, 228–234.

Williams, D.H., Ronayne, J., Moore, H.W., and Shelden, H.R. (1968) Solvent Shifts of Proton Resonances Induced by Benzene and Pyridine in Epoxides and Ethers. An Aid to Structure Elucidation, *J. Org. Chem. 33*, 998–1002.

Wilson, I.D. (2000) Multiple Hyphenation of Liquid Chromatography with Nuclear Magnetic Resonance Spectroscopy, Mass Spectrometry and Beyond, *J. Chromatogr. A 892*, 315–327.

Yadagiri, P., Lumin, S., Mosset, P., Capdevila, J., and Falck, J.R. (1986) Enantiospecific Total Synthesis of 8- and 12-Hydroxyeicosatetraenoic Acid, *Tetrahedron Lett. 27*, 6039–6040.

Yamauchi, R., Matsui, T., Kato, K., and Ueno, Y. (1990) Reaction Products of α-Tocopherol with Methyl Linoleate-Peroxyl Radicals, *Lipids 25*, 152–158.

Yeola, S.N., Saleh, S.A., Brash, A.R., Prakash, C., Taber, D.F., and Blair, I.A. (1996) Synthesis of 10(*S*)-Hydroxyeicosatetraenoic Acid: A Novel Cytochrome P-450 Metabolite of Arachidonic Acid, *J. Org. Chem. 61*, 838–841.

Yurawecz, M.P., Hood, J.K., Mossoba, M.M., Roach, J.A.G., and Ku, Y. (1995) Furan Fatty Acids Determined as Oxidation Products of Conjugated Octadecadienoic Acid, *Lipids 30*, 595–598.

Zamboni, R., and Rokach, J. (1983) Stereospecific Synthesis of 5*S*-HETE, 5*R*-HETE, and Their Transformation to 5-(±)-HPETE, *Tetrahedron Lett. 24*, 999–1002.

Chapter 6

Analysis of Lipid Oxidation by ESR Spectroscopy

Mogens L. Andersen, Joaquin Velasco, and Leif H. Skibsted

Food Chemistry, Department of Food Science, The Royal Veterinary and Agricultural University, DK-1958 Frederiksberg C, Denmark

Introduction

Lipid oxidation has been studied largely by analysis of relatively stable compounds such as the hydroperoxides produced in the propagation step of the oxidative process and by analysis of secondary oxidation products produced by the cleavage of the hydroperoxides. The initiation of lipid oxidation involves the formation of lipid radicals reacting at diffusion-limited rates with oxygen to form peroxyl radicals. The peroxyl radicals propagate the reaction chain by abstraction of hydrogen radicals from other lipid molecules which turn into new radical-carrying species. The formation of secondary lipid oxidation products also depends on radical intermediates formed by homolytical cleavage of the hydroperoxides.

The key role of radicals in lipid oxidation makes it of interest to explore methods for the detection of radicals to follow oxidation under various conditions. Lipid oxidation is characterized by a lag phase in which antioxidants present in the system yield protection by scavenging radicals involved in the initiation and propagation steps. The end of the lag phase, when antioxidants become depleted, is characterized accordingly by a change in the balance between formation and scavenging of radicals in the system.

Radicals may be detected by their magnetic moment, and development of electron spin resonance (ESR) spectroscopy as a sensitive method has facilitated direct measurement of radicals in biological systems. A new generation of simple ESR spectrometers is now becoming available and will allow routine measurements in industrial laboratories also. The recent developments in ESR spectroscopy make detection and quantification of radical species involved in lipid oxidation of interest for shelf-life prediction and for mechanistic studies of lipid oxidation in relation to optimal protection of food and beverages by antioxidants. ESR spectroscopy is also finding increased use in medical and pharmacologic studies related to oxidative stress in living organisms (Utsumi and Yamada 2003).

ESR is based on the magnetic properties of species with unpaired electrons. In the presence of an external magnetic field, the magnetic moment of a radical with one unpaired electron takes two different orientations with respect to the direction of the field, i.e., in the same and in the opposite direction. As a result, two different states of energy are obtained and the difference in energy between these states is a

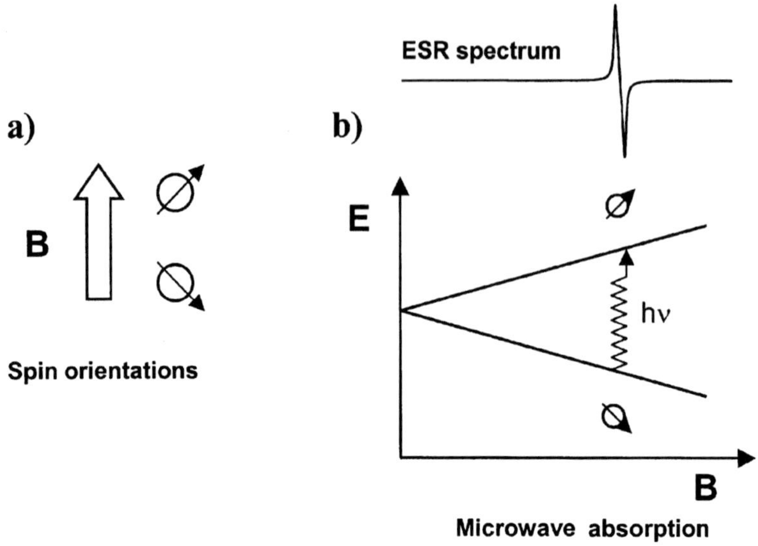

Fig. 6.1. (a) Representative scheme of the different spin orientations adopted by an unpaired electron in the presence of a magnetic field (B). (b) Dependence of the spin energy estates on the strength of the magnetic field; and resonance conditions. Each radical is characterized by a *g*-value according to $h\nu = g\,\beta_e\,\beta$, where β_e is the Bohr magneton.

function of the strength of the magnetic field (Fig. 6.1). Usually, the sample is irradiated with electromagnetic radiation at a constant frequency, typically in the microwave region, and the magnetic field is varied to achieve the resonance conditions. Absorption of electromagnetic radiation is observed when the difference in energy between the two spin states equals the energy of the irradiation. Most ESR spectrometers record the first derivative of the absorption as a function of the strength of the applied magnetic field (Fig. 6.1). The magnetic moment of the unpaired electron may couple with magnetic moments of nearby magnetic nuclei present in the molecule, leading to splitting of the energy levels of the unpaired electron and generating the known hyperfine splitting. As a result, the simple ESR absorption line splits into different lines according to the spin multiplicity $2I + 1$, where I stands for the spin of the magnetic nucleus. The hyperfine splitting is not dependent on the strength of the magnetic field, and provides useful information about the identity of the radical species. The hyperfine splitting is characterized by the coupling constant, which is the separation of the lines involved in the splitting and is expressed in units of magnetic field. There are excellent books and reviews available in the literature about the basic principles and applications of ESR, which the reader is referred to (Eaton and Eaton 1997, Rosen *et al.* 1999, Weil *et al.* 1994).

In addition to providing information about the concentration of radical species, further information about their chemical and steric structures and insights into the odd-electron distribution can be obtained from the ESR spectra (Fischer 1973). Information on the nature of reaction intermediates is contained in the absorption of electromagnetic radiation, the number and positions of the spectral lines, and in the line widths.

The application of ESR spectroscopy in the area of lipid oxidation is relatively recent, and only the brewing industry is currently using ESR spectroscopy as a standard method in quality control (Andersen and Skibsted 1998, Uchida and Ono 1996). Developments in the use of ESR spectroscopy in the food and feed areas are expected to be stimulated by the possibility of using direct measurements for a number of products without extraction steps and purification procedures.

Direct measurement of radicals formed in the product represents the most simple and straightforward method. However, for many products such as oils, direct measurement is not possible because the steady-state concentration of radicals is too low. The spin-trapping technique, based on stabilizing radicals through reactions with specially designed molecular traps to accumulate resonance-stabilized radicals, was developed for use in various biological systems. Another technique, known as spin scavenging, is based on opposite principles because relatively stable radicals are added to the system and their depletion through radical-radical reactions is followed. These different ESR techniques will be presented and discussed together with a short presentation of the spin-probing technique used for studies of molecular mobility and oxygen depletion in food systems. Some of the features of these techniques are summarized in Table 6.1.

Direct Detection of Radicals

Detection at Low Temperatures

The radicals involved in lipid oxidation have very short lifetimes because they are very reactive species that undergo bimolecular reactions at rates close to the diffusion-controlled limit. Their steady-state concentrations are therefore very low; in fluid systems, they fall below the detection limit of ESR (Chiba and Kaneda 1984, Schaich and Borg 1980). The lowest limit for the direct detection of organic radicals was reported to be in the range 10^{-8}–10^{-9} M under optimal conditions (Andersen and Skibsted 2002, McNamee 1984, Schaich and Borg 1980); however, higher concentrations are required for reasonable spectral resolution (Schaich and Borg 1980). Direct detection is possible only at very low temperatures, where bimolecular reactions are so slow that lipid-derived radicals become longer lived. Chiba and Kaneda (1984) reported detection of peroxyl radicals derived from unsaturated fatty acid methyl esters in cyclopropane under conditions in which the solvent remained in the liquid state (–113°C). Samples oxidized at 40°C in the dark did not show any ESR signal at –196°C. Use of the *tert*-butoxyl radical gener-

TABLE 6.1
General Features of Different ESR Techniques for Evaluation of Lipid Oxidation

Technique	Applications	Food systems	Advantages	Disadvantages
Direct measurement	Identification of radicals at low temperatures Quantification of radicals in low mobility systems	Oils and model lipid systems Dry and dehydrated foods	No introduction of foreign substances Radical detection on intact samples	Use of UV light or radical initiators Broad signals with little information about the nature of radicals
Spin trapping	Identification and quantification of radical species at ambient temperatures Fast determination of oxidative stability Evaluation of radical scavenging capacity of antioxidants	Oils and a variety of foods (e.g., meat, raw milk, mayonnaise, cream cheese)	Short-lived radicals may be studied Possibility for assignment of radicals	Dependence on kinetic factors Introduction of foreign substances interfering with the oxidation chain. Identification of radicals requires additional analytical methods
Spin scavenging	Measurement of radical quenching Evaluation of radical scavenging capacity of antioxidants	Oils and different food systems	Known initial concentration of radicals	Introduction of a foreign substance (a stable radical)
Spin probing	Determination of oxygen consumption Determination of oxygen permeation through solids	Foods with liquid phases Oil-encapsulating glassy solid systems	*In situ* determination at microscopic level	Introduction of a foreign substance (a stable radical) High oxygen concentrations are required

ated by photolysis of di-*tert*-butyl peroxide was necessary to increase the concentration of lipid peroxyl radicals to detectable levels.

Photolysis of lipid hydroperoxides in polyunsaturated lipids at –196°C allowed detection of carbon-centered radicals with the unpaired electron localized in the saturated part of the fatty acid chain (Yanez *et al.* 1987, Zhu and Sevilla 1990). It was suggested that these radicals are formed by abstraction of hydrogen atoms by alkoxyl and hydroxyl radicals released in the photolysis of hydroperoxides. Later, Geoffroy *et al.* (2000) pointed out that formation of these carbon-centered radicals may be due to the participation of only hydroxyl radicals because the alkoxyl radicals that are formed simultaneously during photolysis are much less mobile and reactive. When the temperature of the photolyzed lipids was slowly increased, the ESR spectra of the carbon-centered radicals were replaced by those of peroxyl radicals. The formation of peroxyl radicals can take place only when the temperature is high enough to allow oxygen to migrate through the frozen sample. When the mobility of peroxyl radicals further increases with temperature, these radicals are able to abstract hydrogen atoms from unsaturated lipids, yielding carbon-centered radicals stabilized by conjugation (Becker *et al.* 1987, Geoffroy *et al.* 2000, Yanez *et al.* 1987, Zhu and Sevilla 1990). Yanez *et al.* (1987) suggested that at certain temperatures, the new carbon-centered radicals predominate over peroxyl radicals because the rate of oxygen diffusion is slower than that of the reaction between peroxyl radicals and the lipid substrate to form carbon-centered radicals. The ESR signal of these radicals disappears at higher temperatures when the mobility of lipid radicals becomes high enough to allow their recombination.

A very weak ESR signal was detected by photolysis of trilinolein at –196°C when hydroperoxides were almost completely removed by adsorption chromatography or reduced with triphenylphosphine, indicating that the absence of hydroperoxides prevents the formation of primary carbon-centered radicals at –196°C. Nevertheless, the addition of oxygen and a photosensitizer to the purified lipid led to the formation of carbon-centered and peroxyl radicals upon photolysis at –196°C, suggesting that the initiation of oxidation was due to the action of singlet oxygen (Geoffroy *et al.* 2000).

Detection in Dehydrated or Dry Foods

The direct detection of radicals at room temperature by ESR is often possible in dry foods. The low mobility of the constituents in dry foods prevents bimolecular decay of radicals that are formed as a result of various oxidative processes. This allows the accumulation of detectable concentrations of physically trapped radicals under the normal conditions of processing and storage of foods.

The ESR spectrum of the stable radicals in dry foods consists in most cases of a single line with a peak-to-peak line width from 4 to 12 Gauss and a *g*-value in the range 2.003–2.006 (*cf.* Fig. 6.1). The identity of the stable radicals in dry foods is unknown; however, it is unlikely that they are lipid-derived radicals because of the high reactivity of the latter. It was suggested that the stable ESR signals can be assigned to radicals formed upon heat treatment of carbohydrates (Gonis *et al.* 1995), oxidized polyphenols (Gallez *et al.* 2000), protein radicals (Schaich 2002), or oxidized Maillard products (Hofmann *et al.* 2002).

The formation of stable radicals in dry foods has nevertheless been linked to lipid oxidation. Oxidizing methyl linoleate was shown to be able to generate stable radicals in wheat flours and starch (Schaich and Rebello 1999). The amount of stable radicals in dry foods has in various cases been shown to correlate with markers of lipid oxidation and rancidity. The level of radicals in milk powder samples was correlated with thiobarbituric acid-reactive substances (TBARS) and sensory scores (Stapelfeldt *et al.* 1997), and the level of radicals in dried chicken meat was found to correlate with hexanal in the headspace and sensory evaluation of rancidity (Nissen *et al.* 2000). A similar correlation was observed between TBARS in processed cheese and the level of radicals in freeze-dried samples (Kristensen and Skibsted 1999). The level of stable radicals in dried potatoes was found to be useful as a marker of early events in oxidation (Nissen *et al.* 2002). In freeze-dried raw milk, a negative relation between the content of α-tocopherol in the raw milk and that of free radicals indicated that lipid-based free radicals were the main contributors to the ESR signal (Stapelfeldt *et al.* 1999).

Irradiation of sliced Havarti cheese with light showed that the sensory changes in the cheese were correlated with the changes in the levels of stable radicals detected in freeze-dried samples (Kristensen *et al.* 2000). ESR and headspace-gas chromatography (GC) detection of stable radicals complemented each other in detecting the effects of light and oxygen on the oxidative changes during storage of dry products. Light was found to have the largest effect on the level of free radicals in peanuts, oatmeal, and muesli during storage, whereas oxygen had the largest influence on the formation of hexanal. Opposite effects were observed for pork rinds in which the level of radicals was dependent mainly on the availability of oxygen (Jensen *et al.* 2005).

Spin Trapping

Even though the direct detection of radicals by ESR is limited to lipid systems of very low mobility, detection of lipid-derived radicals can be approached indirectly by the ESR spin-trapping technique. This is based on the reaction of radicals with diamagnetic compounds (spin traps) added to the system to form more stable radicals (spin adducts), which accumulate at detectable concentrations ($>10^{-7}$–10^{-6} M). Detection of these new radical species allows the indirect detection of radicals involved in lipid oxidation. The reader is referred to different texts for a more detailed description of the spin-trapping technique and its applications (Janzen and Haire 1990, Perkins 1980, Rosen *et al.* 1999).

Nitroso compounds and nitrones are the most widely used spin traps; both lead to the formation of nitroxides in which the unpaired electron is located primarily on the nitroxide function (Fig. 6.2). The ESR spectra of nitroxides have a main

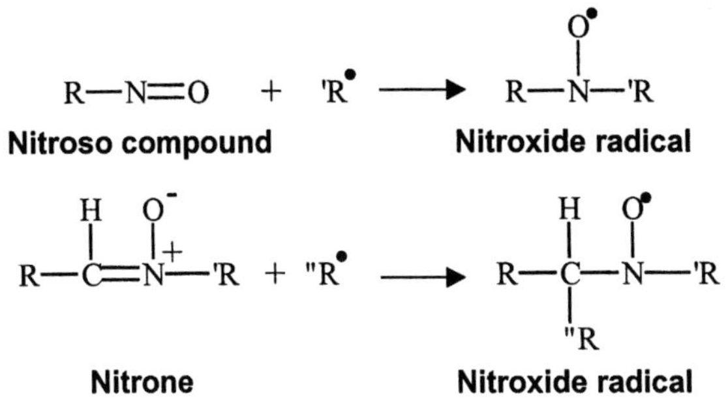

Fig. 6.2. Spin-trapping reactions of nitroso compounds and nitrones to form nitroxide radicals.

triplet splitting due to the interaction of the unpaired electron with the nitrogen nucleus (I = 1) of the nitroxide group. Secondary splittings can arise from other magnetic nuclei in the spin trap and from magnetic nuclei present in the radical trapped. In nitroso spin traps, radicals add directly to the nitrogen, whereas in nitrones, addition takes place to the carbon adjacent to the nitrogen. Thus, radicals trapped by nitroso spin traps can influence the ESR spectrum by the interaction of magnetic nuclei with the unpaired electron, whereas in the case of nitrones, the unpaired electron is more distant and scarcely feels the presence of magnetic nuclei in the radical added; thus, the spectra tend to be rather similar whatever the kind of radical trapped. Even so, the use of nitrones is more frequent because the resulting spin adducts are normally more stable than those of nitroso compounds. 5,5-Dimethylpyrroline-*N*-oxide (DMPO) (**1**) and α phenyl-*N*-*tert*-butylnitrone (PBN) (**2**) are examples of nitrones that have been widely used to detect radicals involved in lipid oxidation in foods and biological material.

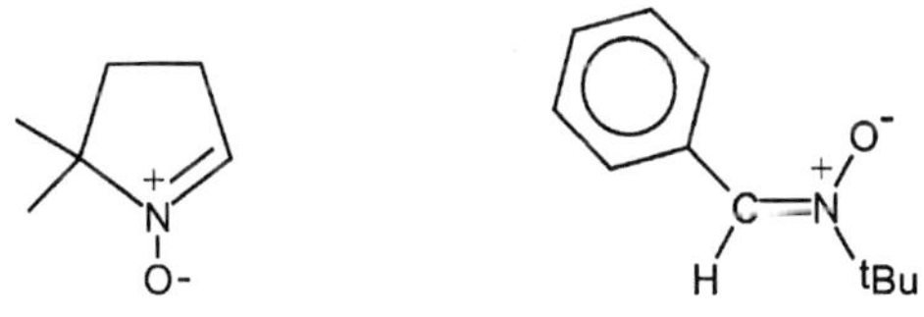

1 **2**

The hyperfine splitting structure of adducts can provide information about the identity of the original radical. Nevertheless, one of the main criticisms of the ESR spin-trapping technique described in the literature is that the identification of radicals in many cases is based on spectral considerations only by the use of coupling constants. In particular, as noted by Dikalov and Mason (1999 and 2001), detection of peroxyl radicals was subjected to such errors. Other methods, such as those based upon the analysis of spin adducts by GC or liquid chromatography (LC) with mass spectrometry (MS) detection (Iwahashi 2000 and 2003, Iwahashi *et al.* 1991, Janzen *et al.* 1990, Qian *et al.* 2002 and 2003) or upon a comparison of the experimental ESR spectral parameters with those of the spin adducts obtained by synthesis (Dikalov and Mason 1999, 2001) are recommended.

It is very important to emphasize that ESR spin trapping is a kinetic method and, as such, detection of radicals depends on kinetic factors. In this respect, many pitfalls may be identified in many studies described in the literature. Few cases exist in which practical limitations and problems with spin traps have been considered in the evaluation of the experimental results. For instance, the success of the detection of a radical species relies on effective trapping and sufficient stability of the resulting spin adduct to be detected within the time scale of the measurement. Effective trapping means that the reaction between a certain radical and the spin trap is fast enough to prevent the radical from participating in subsequent reactions. Therefore, the spectra observed reflect a steady-state situation and depend on the relative rates of the various formation and decay reactions of the spin adducts and competition between the trapping reaction and other reactions of the radicals. Thus, it is clear that not all radicals are trapped equally well by a certain spin trap. Further, trapping also depends on the kind of spin trap used. Failure to detect an expected radical or any ESR signal therefore does not necessarily mean that the radical was not formed. Similarly, detection of a specific radical does not necessarily mean that the radical trapped is a major intermediate of the chemical reaction under study; it could also indicate that it reacts very quickly with the spin trap (Schaich and Borg 1980).

Detection of peroxyl radicals by ESR spin trapping has been a matter of discussion. Even though peroxyl radical adducts are not stable at room temperature (Dikalov and Mason 1999, 2001, Howard and Tait 1978, Janzen *et al.* 1990, Pfab 1978), their detection was reported for model lipid systems and biological samples (Borg and Schaich 1984, Chamulitrat *et al.* 1991, Davies and Slater 1986, Kennedy *et al.* 1989, Niki *et al.* 1983, Ohto *et al.* 1977, Rota *et al.* 1997, Sawa *et al.* 1998, Schaich and Borg 1988, Ueda *et al.* 1996, Yamada *et al.* 1984). By labeling experiments with $^{17}O_2$ ($I = 5/2$), Pfab (1978) suggested that the absence of observable stationary state concentration of peroxyl nitroxides was indicative of a rapid decay of such adducts rather than failure of trapping peroxyl radicals. Janzen *et al.* (1990) reported that peroxyl adducts of PBN are not persistent at temperatures >230K and, on the other hand, that only PBN-alkoxyl radical adducts were detected at temperatures >250K. In experiments with DMPO, detection of peroxyl adducts has been based only on spectral similarity to the DMPO-superoxide radical in conjunction with their insensitivity to superoxide dismutase (Borg and Schaich 1984, Chamulitrat *et al.* 1991, Davies and Slater 1986, Kennedy *et al.* 1989, Rota *et al.* 1997, Sawa *et al.* 1998, Schaich and Borg 1988). Nevertheless, hyperfine coupling constants of the hypothetical peroxyl adducts of DMPO are quite similar to those of synthesized alkoxyl adducts (Dikalov and Mason 1999, Hanna *et al.* 1992). Dikalov and Mason (1999) supported the very early hypothesis expressed by Pfab (1978) and pointed out that peroxyl radicals are trapped by spin traps giving very unstable spin adducts whose decomposition results in the generation of new alkoxyl radicals. The latter react in turn with new molecules of the spin trap to form alkoxyl spin adducts. These authors suggested that this is a general phenomenon that does not depend on the structure of the nitrone spin trap or peroxyl radical. It was concluded that only alkoxyl spin adducts are detected in polyunsaturated lipids at temperatures >250K (Dikalov and Mason 2001). In this context, lipid peroxyl adducts of DMPO were detected by MS as even electron species, i.e., as nonradical compounds, justifying their absence in ESR analyses (Reis *et al.* 2003).

On the other hand, radical species such as alkoxyl aminoxyls were reported to be commonly detected ESR products of reactions of peroxyl radical with nitrones (Janzen and Blackburn 1969, Niki *et al.* 1983, Rosen *et al.* 1980). Oxidized forms of DMPO and PBN, referred to as DMPOx (**3**) and PBNOx (**4**), respectively, are examples of such products and are characterized by easily recognizable ESR signals because of their unusually small coupling constants. Although the mechanism of formation has not yet been established, detection of such species might give evidence of the formation of peroxyl spin adducts during lipid oxidation.

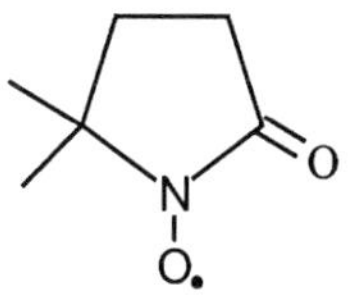

3

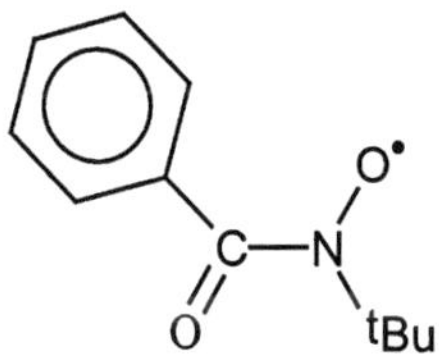

4

Detection of lipid-derived alkyl radicals was also reported for different lipid systems (Iwahashi *et al.* 1991, North *et al.* 1992 and 1994, Novakov *et al.* 2001, Qian *et al.* 2000, 2002, and 2003, Vicente *et al.* 1998). Under conditions of unlimited oxygen supply, however, trapping of alkyl radicals is expected to be less significant because their reaction with oxygen to form peroxyl radicals proceeds at a rate controlled by diffusion, i.e., it is several orders of magnitude faster than their reaction with spin traps (Schaich and Borg 1980).

Another issue of interest is that the spin trap can interfere with the reaction chain, modifying both the pathway and rate of oxidation. There are numerous reports in which added PBN seems to have an important role in protecting biological systems from lipid peroxidation (Ferguson *et al.* 1997, Kalyanaraman *et al.* 1991, Koenig and Meyerhoff 2003, Lee and Park 2003, Li *et al.* 2001, McLellan *et al.* 2003, Milatovic *et al.* 2001, Ondrias *et al.* 1994, Park *et al.* 2002). Barclay and Vinqvist (2000) reported, however, that PBN does not really act as a chain-breaking antioxidant, but rather as a weak retarder of lipid oxidation. In experiments conducted at 0.1 MPa oxygen, different concentrations of PBN did not show any induction period as detected by oxygen consumption. Under the same conditions, the addition of chain-breaking antioxidants inhibits oxygen uptake by the lipid substrate for such a distinct period of time, known as the induction period, until they are completely depleted. Then, the induction period is followed by the return to the uninhibited rate of oxygen consumption. Compared with effective amounts of antioxidants, larger amounts of PBN were necessary to observe only a retardation effect consisting of a decrease in the rate of oxygen uptake. In combination with antioxidants, PBN exhibited only a slight cooperative effect. It was proposed that the effect of PBN on the rate of oxidation results from the trapping of carbon-centered radicals because the reaction between PBN and peroxyl radicals was suggested to be ineffective. Under reduced oxygen partial pressure, larger effects of PBN on the rate of oxidation were observed. Further, experiments combining PBN and antioxidants of the chromanol class showed a definite cooperative effect that was not observed at high oxygen partial pressure. Barclay and Vinqvist (2000) concluded that the effect of PBN consists of trapping initial carbon-centered radicals, especially at low oxygen conditions, and that this spin trap should be considered to be a retarder or preventative antioxidant rather than a chain-breaking antioxidant.

Results obtained in our laboratory, however, revealed that PBN at concentrations of 1 mg/g oil exhibited a strong inhibiting effect on lipid oxidation in oils during storage at 40°C in the presence of air (Velasco *et al.* 2005). The effect of PBN was studied in rapeseed oil (RO), sunflower oil (SO), and fish oil (FO). PBN led to a marked decrease in peroxide value. Further, the remaining concentration of naturally occurring tocopherol was found to be larger during oxidation in samples containing PBN compared with the corresponding control samples. For example, when tocopherol was practically depleted in the control samples, the oils containing PBN had amounts as high as 57, 72, and 90% remaining in RO, SO, and FO, respectively. It was concluded that the effect of PBN on lipid oxidation is dependent on the nature of the oil. Results showed that the effect of PBN on the rate of lipid oxidation increased (RO < SO < FO) as oxidative stability decreased (RO > SO > FO). This effect was suggested to be related to the initial tocopherol content of the oils and thus to be dependent on the ratio between the concentration of PBN and tocopherol because larger effects were observed at larger relative PBN concentrations. Under the conditions assayed, the effect of PBN cannot be attributed to the trapping of only lipid alkyl radicals, as was suggested by Barclay and Vinqvist (2000). As commented above, alkyl radicals react with oxygen at rates controlled by diffusion; therefore, their reaction with PBN is not expected to be significant. Accordingly, the effect of PBN was attributed mainly to its reaction with peroxyl radicals, which in turn depends on the initial tocopherol content. Thus, the effect of PBN increased with the PBN:tocopherol ratio, i.e., the competition of PBN for peroxyl radicals became more significant in the presence of lower amounts of tocopherol. Although this finding may be attributed to differences in reaction rates between PBN and tocopherol, i.e., to the faster scavenging of peroxyl radicals by tocopherol, the different scavenging mechanisms of the two species must also be considered. In this respect, it is known that tocopherol is able to scavenge two radicals per molecule. By contrast, PBN was suggested to react with peroxyl radicals to form spin adducts, which decompose and generate secondary radicals (Dikalov and Mason 1999). In any case, it is evident that PBN at very low concentrations modifies both the pathway and the rate of lipid oxidation.

With respect to the stability of spin adducts, different pathways of radical decay were described in the literature. Disproportionation reactions, reduction to hydroxylamines, oxidation to oxoammonium species, dimerization (Janzen and Haire 1990), and *O*-alkylation (Janzen *et al.* 1990) are representative reactions of nitroxide radicals to form diamagnetic or ESR silent species. In particular, *O*-alkylation represents one of the most important phenomena showing that ESR spin trapping is a technique depending to a great extent on experimental conditions. Janzen *et al.* (1990) reported that the stability of alkoxyl spin adducts of PBN, as well as of DMPO, produced during thermolysis of azo-bis-(isobutyronitrile), depends on the presence of oxygen. Under oxygen-depleted conditions, the ESR signal of alkoxyl spin adducts disappeared, whereas that of alkyl spin adducts remained. These results were attributed to a second trapping of alkyl radicals by alkoxyl spin adducts to produce alkoxylamines (Fig. 6.3). Similar results were obtained in our laboratory during oxidation of nonrefined fish oil at 40°C in the dark (unpublished data). Radical formation was followed by PBN under different oxygen availability conditions. The ESR signal detected under air decreased to a great extent when air became limited as samples were transferred into tubes with lower surface-to-volume ratio and a reduced headspace (Fig. 6.4). In contrast, in refined rapeseed oil and sunflower oil, slight but significant increases in the ESR signal were detected over the oxidation time when oxygen was limited under the same conditions (results not shown). These results emphasize once again that detection of a certain radical species is a very complex phenomenon depending on multiple factors. In this context, Qian *et al.* (2000) observed that separation of DMPO spin adducts from the oxidation reaction using appropriate extraction led to a great increase in their lifetimes, indicating that the short lifetimes normally found are due in part to the reaction with subsequent radicals formed during lipid oxidation.

Despite its limitations, the ESR spin-trapping technique has been applied to evaluate early lipid oxidation events in different kinds of foods, including meat (Carlsen *et al.* 2001 and 2003, Gatellier *et al.* 2000, Lauridsen *et al.* 2000, Monahan *et al.* 1993), cheese (Kristensen and Skibsted 1999), mayonnaise (Thomsen *et al.* 1999 and 2000a), food lipids (Thomsen *et al.* 2000b), raw milk (Kristensen *et al.* 2002), and vegetable oils (Velasco *et al.* 2004).

Monahan *et al.* (1993) applied spin trapping with α-(4-pyridyl-1-oxide) *N-tert*-butyl nitrone (POBN) (**5**) to investigate the effect of dietary oxidized lipid and vitamin

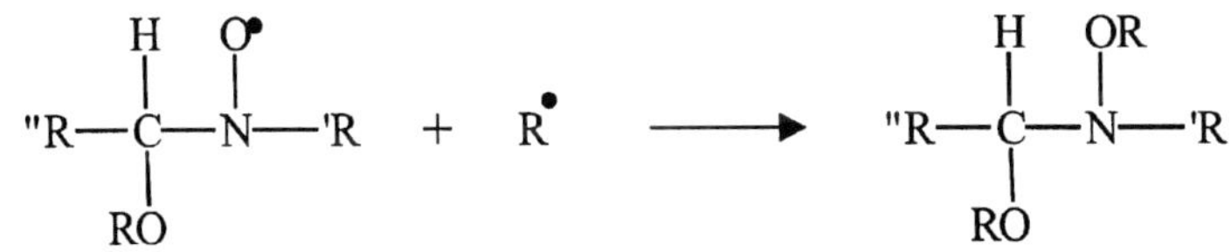

Fig. 6.3. *O*-Alkylation reaction of alkoxynitroxide radicals contributing to spin adduct decay in ESR spin-trapping analysis.

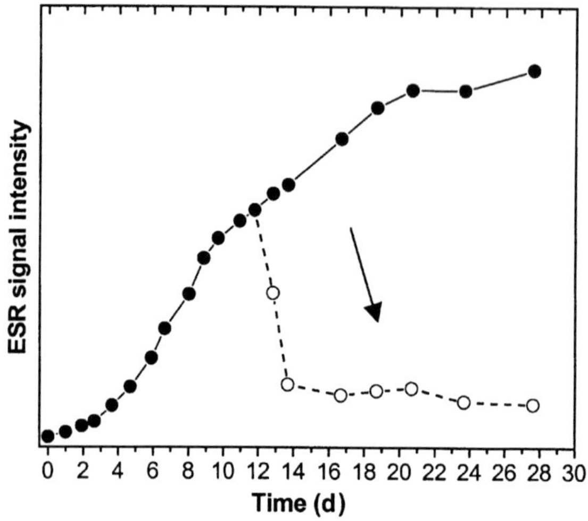

Fig. 6.4. Influence of oxygen availability on radical detection by ESR spin-trapping with PBN in nonrefined fish oil at 40°C in the dark. (●) Samples under high oxygen availability; (○) samples under reduced oxygen availability (unpublished data).

E supplementation on iron-induced radical production in microsomal membrane fractions of porcine muscle. Microsomal suspensions (buffered at pH 7.4) containing POBN and $FeCl_2$ were incubated at 37°C and radicals monitored for a 90-min period. The identity of radicals trapped by POBN was uncertain. On the basis of hyperfine splitting constants, trapping of hydroxyl or superoxide radicals was suggested to be excluded. The authors pointed out that radicals such as alkyl or lipid dienyl radicals were possible radicals trapped, although radicals derived from protein or other molecules in the microsomal preparation could not be ruled out. Despite the uncertainty in the nature of the radicals trapped, results concerning the effect of vitamin E supplementation were in agreement with the TBARS determination. On the other hand, although no effect of dietary oxidized lipid was observed on the development of TBARS, larger radical levels were detected in samples from pigs fed oxidized lipids than those from the corresponding control samples.

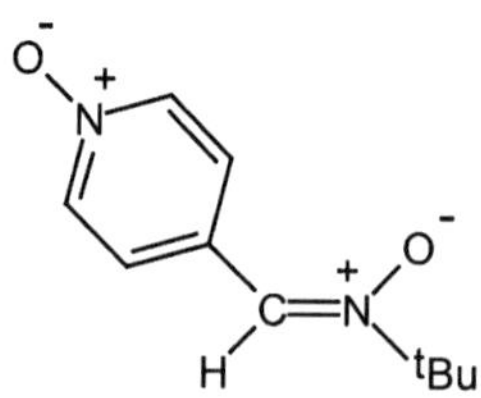

5

Gatellier *et al.* (2000) applied POBN spin trapping in turkey muscle aqueous extracts. ESR measurements were performed after oxidation at 20°C for 1 h by an enzymatic system consisting of NADPH, ADP, and $FeSO_4$/cytochrome P450 reductase. Based only on hyperfine splitting constants, radicals trapped by POBN in such aqueous solutions were identified as ethyl and pentyl radicals arising from the oxidation of n-3 and n-6 membranal fatty acids, respectively. Results obtained by radical trapping were in agreement with TBARS results. With respect to radical identification, the authors referred to previous results obtained by Iwahashi *et al.* (1991), who detected pentyl (or pentenyl) radicals in mixtures of soybean lipoxygenase with linoleic and arachidonic acids (or linolenic acid). The radicals trapped were identified by MS analysis after isolation of spin adducts of POBN by HPLC.

Lauridsen *et al.* (2000) applied POBN spin trapping to study the influence of supranutritional vitamin E and copper on the susceptibility to lipid oxidation of porcine membrane fractions. In microsomal fractions, POBN spin trapping was applied under conditions similar to those applied by Monahan *et al.* (1993), whereas in mitochondrial fractions, incubation was performed in the absence of $FeCl_2$. Under these conditions, the detection of radicals was higher in oxidized mitochondrial fractions than in microsomal fractions oxidized by iron catalysis. It was found that spin trapping of radicals in the mitochondrial membranes provided the same qualitative conclusions as those drawn from the determination of secondary oxidation products in muscle homogenates. In a similar way, spin trapping with POBN was applied to follow radical formation as an indication of lipid oxidation in minced pork meat (Carlsen *et al.* 2001). Meat samples were cooked at different temperatures and analyzed by ESR spin trapping and oxygen consumption. Minced meat was dispersed in aqueous solution buffered at pH 5.7. In the ESR analysis, addition of the spin trap was carried out after homogenization, and the meat dispersions were incubated at 30°C. ESR analysis was performed in the water phase after its separation. Levels of radicals detected after 1 h of incubation were in accordance with results obtained by oxygen consumption. For precooked meat, lipid oxidation during storage was influenced by the temperature applied for cooking. Low-temperature cooking, e.g., 70°C, resulted in slow oxidation as monitored by electrochemical measurement of oxygen consumption, whereas cooking at 95°C resulted in a significantly faster oxidation. Notably, at an even higher temperature (120°C), a slower rate comparable to the rate after cooking at 70°C was observed. The rate of radical formation in the meat system followed the same pattern as that obtained by oxygen consumption, with cooking at 95°C resulting in the highest rate of radical formation. These results confirmed the potential use of the ESR spin-trapping technique in complex systems such as meat. The explanation offered for the slower radical formation and oxidation at the highest cooking temperature was formation of Maillard compounds acting as radical scavengers. Later, the same procedure was applied to study the effect of the addition of protein fractions from pork on oxidative deterioration of meat. ESR analysis was in agreement with the results of peroxide value during storage at 5°C (Carlsen *et al.* 2003).

The effect of storage temperature and the presence of light on radical formation in cream cheese were studied by different ESR techniques, including spin trapping with DMPO. The spin trap was added directly to the cream cheese, and radical formation was monitored during storage. ESR measurements were performed directly on the cheese samples. On the basis of the spectral parameters, signals were assigned to hydroxyl spin adducts. Samples in the dark also showed additional ESR lines that could not be identified. It was suggested that oxidation in processed cream cheese can be monitored by DMPO-spin trapping. Nevertheless, it was recommended that storage experiments should be limited to a few days to avoid the influence of subsequent reactions involved in losses of spin adducts (Kristensen and Skibsted 1999).

In mayonnaise enriched with fish oil, different spin traps, including PBN, DMPO, POBN, 2,4,6,-tri-*tert*-butylnitrosobenzene (BNB), and 2-methyl-2-nitrosopropane (MNP), were tested in fresh samples that were subsequently held at 37°C for periods of 12, 24, and 36 h. The addition of the spin traps was carried out after mayonnaise preparation, and ESR measurements were performed directly on the mayonnaise samples. Detection of radicals was observed only in samples containing either PBN or POBN, although PBN was recommended because of its higher lipophilic nature. The addition of the spin trap before thermal treatment was necessary to be able to obtain detectable ESR signals. Quantification of radicals was approached by external calibration with a stable nitroxyl radical (12-doxylstearic acid), which was added to the mayonnaise under the same conditions used for the spin trap. Suitable conditions concerning concentration of PBN, temperature (37°C), and incubation time (24 h) were selected and applied to mayonnaise samples containing different concentrations of a commercial antioxidant mixture used for mayonnaise. Results were in accordance with those obtained in storage experiments by sensory analyses (Thomsen *et al.* 1999). Later, PBN-spin trapping was applied in storage experiments conducted at 20°C for 4 wk to study the influence of different ingredients in mayonnaise on the detection of radicals initiating lipid oxidation. After sampling, PBN was added and ESR measurements were carried out directly on the mayonnaise samples after an incubation period of 24 h at 37°C. The ESR spin-trapping technique proved valuable in identifying exposure to iron, due to a decrease of pH caused by the addition of either vinegar or lemon juice, as part of the mayonnaise recipe, as the single most important factor determining the initiation of lipid oxidation in mayonnaise enriched with fish oil (Thomsen *et al.* 2000a).

The ESR spin-trapping technique was also applied as a rapid test to determine the oxidative stability of food lipids under mildly accelerated conditions. Rapeseed oil and lipids extracted from mayonnaise, butter, and dairy spread were studied. Oxidative stability was defined as the resistance to formation of radical species as detected by spin trapping with PBN. Radical development exhibited very short induction periods in which radicals were formed very slowly before a sudden linear increase was observed. The induction period and the amount of radicals detected were strongly product and temperature dependent. Different temperatures ranging from 50 to 80°C were necessary for different types of lipids. The parameter consisting of the amount of radicals detected at a fixed time was also found to be suitable, especially in cases in which radical development was poor within the preset experimental oxidation time. In storage experiments of rapeseed oil at 60°C and mayonnaise at 50°C for 4 wk, oxidative stability of rapeseed oil and the lipid fraction of mayonnaise was determined after sampling. As expected, induction periods decreased and the amount of radicals at a fixed time increased with storage time. Although it was not compared directly with other methods, the detection of radicals at very early events of lipid oxidation seemed to be promising as a rapid test to determine oxidative stability (Thomsen *et al.* 2000b). In a recent report, oxidative stability of different vegetable oils, including rapeseed oil, sunflower oil, and different mixtures of the two, was approached by PBN spin trapping at 60°C, and results were compared with those obtained at 100°C by the Rancimat test and by another accelerated method based upon differential scanning calorimetry (DSC). Although results obtained by the ESR method are indicative of the onset of primary oxidation, results obtained by the Rancimat and DSC methods account for the onset of advanced oxidation. In spite of the fact that different aspects of the oxidative process were assessed and that different oxidation conditions were applied, the results obtained by the ESR method showed satisfactory linear correlations with those provided by the Rancimat test ($r = 0.963$) and DSC ($r = 0.979$). These results suggested that oxidative stability can be evaluated as a measure of the resistance to the formation of radicals generated during the early steps of oxidation. Detection of radicals at this stage of the oxidative process allows mild conditions to be applied in a rapid method of oxidative stability. Compared with the Rancimat method and DSC, the ESR method was confirmed to be useful as a method employing milder oxidation conditions and shorter time (Velasco *et al.* 2004).

Radical development was examined in raw milk by PBN spin trapping in a way similar to that used for food lipids. The concentration of the spin trap and the temperature were optimized for suitable detection of radicals within the preset time of the test. Induction periods similar to those obtained in food lipids were found for milk samples during heating at 55°C. The method was applied to samples stored at 5°C in the dark for 3 d. Significant decreases in the induction period, the rate of radical formation after the induction period, and the amount of radicals detected were observed after storage. In experiments aimed at studying the proxidative effect of light, milk samples were stored at 20°C for 2 d under light exposure or protected with aluminum foil. No difference in the induction period was found between the samples exposed to light and those protected from light. However, in samples exposed to light, larger amounts of radicals were detected after the induction period. It was concluded that detection of radicals with PBN at moderate temperature holds the potential of detecting oxidative changes in milk during storage (Kristensen *et al.* 2002).

The ESR spin-trapping technique also finds applications to evaluate antioxidant activity as the ability to scavenge free radicals generated in different systems, including aqueous and organic solutions. Various radicals including hydroxyl (Berkaoui *et al.* 1994, Cynshi *et al.* 1995, Hiramoto *et al.* 1996, Kumari *et al.*

1996, Leonard *et al.* 2002, Madsen *et al.* 1996, Polyakov *et al.* 2001b, Stagko *et al.* 2002, Yoshimura *et al.* 1999) superoxide (Cynshi *et al.* 1995, Kumari *et al.* 1996, Leonard *et al.* 2002, Unno *et al.* 2002, Yoshimura *et al.* 1999, Yun *et al.* 2003), peroxyl (Krainev and Bigelow 1996, Madsen *et al.* 2000, Masaki *et al.* 1995), alkoxyl (Krainev and Bigelow 1996), and methyl (Polyakov *et al.* 2001a, Yoshimura *et al.* 1999) radicals are generated in different ways in the presence of a spin trap. The xanthine/xanthine oxidase system and the Fenton reaction are frequently applied as the source of radical generation. Yoshimura *et al.* (1999) developed a nonenzymatic and non-Fenton type system consisting of H_2O_2/NaOH/DMSO, in which superoxide, hydroxyl, and methyl radicals are generated simultaneously. The evaluation of antioxidant activity is approached in terms of the ability of antioxidants to reduce the ESR signals of the spin adducts formed.

Spin Scavenging

Evaluation of the radical scavenging capacity of antioxidants can also be approached by measuring the loss of a relatively stable radical in aqueous and organic solutions. Direct ESR analysis of radicals such as 1,1-diphenyl-2-picrylhydrazyl (DPPH), the galvinoxyl radical, and potassium nitrosodisulfonate (Fremy's salt) is normally applied. Quiles *et al.* (2002) used the galvinoxyl radical to study the antioxidant capacity of ethanolic extracts from different edible oils subjected to short-term deep frying. It was suggested that the ESR analysis may be used to test the resistance of edible oils to oxidation because of its high sensitivity and because it provides a direct observation of the real capacity of the oils.

Radical formation involved in lipid oxidation processes has been monitored by the depletion of a relatively stable radical reacting at diffusion-controlled rates with the emerging radicals to form ESR silent products. The stable nitroxide radical 2,2,6,6-tetramethyl-1-piperidinyloxyl (TEMPO) was used in several investigations. The effect of temperature and exposure to light on radical formation in cheese cream was approached by different ESR techniques, including spin scavenging with TEMPO (Kristensen and Skibsted 1999). Results were in agreement with those obtained by ESR spin-trapping with DMPO, showing that exposure to light was a more important factor than temperature for the early stages in radical formation. On the other hand, an uneven pattern in radical depletion was shown by TEMPO during storage, which was attributed to reduction to hydroxylamine and subsequent oxidation processes. Grattard *et al.* (2002) applied ESR spin scavenging of TEMPO to follow lipid oxidation of flaxseed oil encapsulated in a solid matrix of maltodextrin. The nitroxide radical was added to the oil before encapsulation; thus, lipid oxidation was monitored directly on the solid sample without the extraction of the lipid phase.

In a very early report, Ueda (1963) used ESR to study the reaction of the radical DPPH with different hydroperoxides, such as *tert*-butyl and cumene hydroperoxide, and peroxy acids in organic solvents. Disappearance of DPPH and the formation of a new radical species were found after the addition of the hydroperoxides. Very different rate constants for both processes were obtained between *tert*-butyl and cumene hydroperoxide, showing that the secondary hydroperoxide had higher reactivity than the tertiary one. The author suggested that the determination of the rate constants could be useful for distinguishing different hydroperoxides in mixtures. However, no stoichiometric relation between DPPH and hydroperoxides was established.

Spin Probing

Oxygen consumption during autoxidation of methyl linoleate in organic solvent was determined by the use of a relatively stable radical (spin probe), added to the system at low concentrations (Pedrielli *et al.* 2001a and 2001b, Pedrielli and Skibsted 2002, Pedulli 1993, Pedulli *et al.* 1996). The ESR line width of the spin probe is broadened in the presence of oxygen in a concentration-dependent way. This effect is caused by nonchemical bimolecular interactions of the spin probe with oxygen in which both molecules exchange the spin orientations of the unpaired electron. ESR oximetry is based on this phenomenon (Swartz and Glockner 1989). Nitroxyl radicals are suitable species for oximetry studies because they do not react chemically with oxygen. Nevertheless, losses of the spin probe can take place in reactions with radical species formed during lipid oxidation, leading to the formation of diamagnetic species (Pedulli 1993, Pedulli *et al.* 1996). Very low concentrations of nitroxyl radicals were used to prevent this kind of loss. In particular, in the case of the spin probe TEMPO, concentrations up to 5×10^{-5} M in organic solvents were reported to be free of this interaction, at least for the experimental time normally used. Similarly, the hydrophobic spin probe 16-doxylstearic acid was useful for determining oxygen permeation in an oil-encapsulating glassy food matrix (Andersen *et al.* 2000). Under oxygen atmosphere, the concentration of oxygen in the encapsulated oil increased during storage, and the rate of oxygen permeation through the glassy matrix increased significantly with temperature below the glass transition temperature of the glassy system, corresponding to an activation-controlled rather than a diffusion-controlled process. It was concluded that the method may allow noninvasive determination of oxygen depletion in dried foods.

Other Methods

Determination of hydroperoxides in edible oils by ESR was approached using 2,2,6,6-tetramethyl-4-piperidone. This secondary amine was found to react with hydroperoxides, giving a stable nitroxide radical easily detected by ESR. Oxidized corn oil and methyl linoleate were incubated in the presence of 2,2,6,6-tetramethyl-4-piperidone for 6 h before measurement. Results showed a good correlation with those obtained by application of the TBA test and by LC with chemiluminescence detection (Yang *et al.* 1991).

Concluding Remarks

ESR spectroscopy was shown to be useful for studying lipid oxidation in a wide range of types of samples, from bulk oils to complex food matrices. Direct detection of radicals by ESR is possible at low temperatures and in dry foods, whereas the low steady-state concentrations of radicals in systems with high molecular mobility often make it necessary to use the spin-trapping technique for indirect detection of radicals. Several potential pitfalls are associated with the spin-trapping technique, and a critical evaluation of the results from spin-trapping experiments is therefore always mandatory. However, the spin-trapping technique was shown by many experiments to be highly useful for the detection of radicals that are associated with lipid oxidation in food-related systems, and the technique seems furthermore to be useful for the prediction of oxidative stability of lipids under relatively mild conditions, which makes it a potentially interesting method for routine quality control of lipid-containing products.

The use of ESR spectroscopy for studying lipid oxidation in foods is still relatively new, and the majority of studies were published only during the last decade. Recent developments in the field of ESR spectroscopy, which include the use of high magnetic fields and microwave frequencies (>200 GHz), pulsed ESR-techniques, and ESR-imaging, will very likely inspire new experiments that lead to new ways of identifying lipid-derived radicals and their reactivity, and will also result in new methods for studying the spatial propagation of lipid oxidation in complex systems. The role of ESR spectroscopy in studies of lipid oxidation will therefore continue to expand in the future.

Acknowledgments

The authors thank the European Community Program "Human Potential (Improving Human Research Potential and the Socio-Economic Knowledge Base)" for supporting Joaquin Velasco with a Marie Curie Fellowship under contract number "HPMF-CT-2002-01652." The financial support by LMC-Centre for Advanced Food Studies is also acknowledged.

References

Andersen, M.L., and Skibsted, L.H. (1998) Electron Spin Resonance Spin Trapping Identification of Radicals Formed During Aerobic Forced Aging of Beer, *J. Agric. Food Chem. 46*, 1272–1275.

Andersen, M.L., and Skibsted, L.H. (2002) Detection of Early Events in Lipid Oxidation by Electron Spin Resonance Spectroscopy, *Eur. J. Lipid Sci. Technol. 104*, 65–68.

Andersen, A.B., Risbo, J., Andersen, M.L., and Skibsted, L.H. (2000) Oxygen Permeation Through an Oil-Encapsulating Glassy Food Matrix Studied by ESR Line Broadening Using a Nitroxyl Spin Probe, *Food Chem. 70*, 499–508.

Barclay, L.R.C., and Vinqvist, M.R. (2000) Do Spin Traps Also Act as Classical Chain-Breaking Antioxidants? A Quantitative Kinetic Study of Phenyl *tert*-Butylnitrone (PBN) in Solution and in Liposomes, *Free Radic. Biol. Med. 28*, 1079–1090.

Becker, D., Yanez, J., Sevilla, M.D., Alonso-Amigo, M.G., and Schlick, S. (1987) An Electron Spin Resonance Investigation of the Motion of Lipid Peroxyl Radicals, *J. Phys. Chem. 91*, 492–496.

Berkaoui, M., Souchard, J.P., Massol, M., and Nepveu, F. (1994) Hydroxyl Radical Scavenging Activity of Compounds with Pharmaceutical Interest—A Quantitative Analysis by ESR Spectroscopy, *J. Chim. Phys. PCB 91*, 1799–1808.

Borg, D.C., and Schaich, K.M. (1984) Cytotoxicity from Coupled Redox Cycling of Autoxidizing Xenobiotics and Metals—A Selective Critical Review and Commentary on Work-in-Progress, *Isr. J. Chem. 24*, 38–53.

Carlsen, C.U., Andersen, M.L., and Skibsted, L.H. (2001) Oxidative Stability of Processed Pork, Assay Based on ESR Detection of Radicals, *Eur. Food Res. Technol. 213*, 170–173.

Carlsen, C.U., Rasmussen, K.T., Kjeldsen, K.K., Westergaard, P., and Skibsted, L.H. (2003) Pro- and Antioxidative Activity of Protein Fractions from Pork (*Longissimus dorsi*), *Eur. Food Res. Technol. 217*, 195–200.

Chamulitrat, W., Hughes, M.F., Eling, T.E., and Mason, R.P. (1991) Superoxide and Peroxyl Radical Generation from the Reduction of Polyunsaturated Fatty Acid Hydroperoxides by Soybean Lipoxygenase, *Arch. Biochem. Biophys. 290*, 153–159.

Chiba, T., and Kaneda, T. (1984) ESR Spectra of Peroxy Radicals Derived from Unsaturated Esters, *Agric. Biol. Chem. 48*, 2593–2594.

Cynshi, O., Takashima, Y., Katoh, Y., Tamura, K., Sato, M., and Fujita, Y. (1995) Action of Phenolic Antioxidants on Various Active Oxygen Species, *J. Biolumin. Chemilumin. 10*, 261–269.

Davies, M.J., and Slater, T.F. (1986) Studies on the Photolytic Breakdown of Hydroperoxides and Peroxidized Fatty Acids by Using Electron Spin Resonance Spectroscopy, *Biochem. J. 240*, 789–795.

Dikalov, S.I., and Mason, R.P. (1999) Reassignment of Organic Peroxyl Radical Adducts, *Free Radic. Biol. Med. 27*, 864–872.

Dikalov, S.I., and Mason, R.P. (2001) Spin Trapping of Polyunsaturated Fatty Acid-Derived Peroxyl Radicals: Reassignment to Alkoxyl Radical Adducts, *Free Radic. Biol. Med. 30*, 187–197.

Eaton, S.S., and Eaton, G.R. (1997) Electron Paramagnetic Resonance, in *Analytical Instrumentation Handbook* (Ewing, G.W., ed.), p. 767, Marcel Dekker, New York.

Ferguson, E., Singh, R.J., Hogg, N., Kalyanaraman, B. (1997) The Mechanism of Apolipoprotein B-100 Thiol Depletion During Oxidative Modification of Low-Density Lipoprotein, *Arch. Biochem. Biophys. 341*, 287–294.

Fischer, H. (1973) Structure of Free Radicals by ESR Spectroscopy, in *Free Radicals* (Kochi, J.K., ed.), Vol II, pp. 435–491, Wiley-Interscience Publication, John Wiley and Sons, New York.

Gallez, B., Baudelet, C., and Debuyst, R. (2000) Free Radicals in Licorice-Flavored Sweets Can Be Detected Noninvasively Using Low Frequency Electron Paramagnetic Resonance After Oral Administration to Mice, *J. Nutr. 130*, 1831–1833.

Gatellier, P., Mercier, Y., Rock, E., and Renerre, M. (2000) Influence of Dietary Fat and Vitamin E Supplementation on Free Radical Production and on Lipid and Protein Oxidation in Turkey Muscle Extracts, *J. Agric. Food Chem. 48*, 1427–1433.

Geoffroy, M., Lambelet, P., and Richert, P. (2000) Role of Hydroxyl Radicals and Singlet Oxygen in the Formation of Primary Radicals in Unsaturated Lipids: A Solid State Electron Paramagnetic Resonance Study, *J. Agric. Food Chem. 48*, 974–978.

Gonis, J., Hewitt, D.G., Troup, G., Hutton, D.R., and Hunter, C.R. (1995) The Chemical Origin of Free Radicals in Coffee and Other Beverages, *Free Radic. Res. 23*, 393–399.

Grattard, N., Salaun, F., Champion, D., Roudaut, G., Le Meste, M. (2002) Influence of Physical State and Molecular Mobility of Freeze-Dried Maltodextrin Matrices on the Oxidation Rate of Encapsulated Lipids, *J. Food Sci. 67*, 3002–3010.

Hanna, P.M., Chamulitrat, W., and Mason, R.P. (1992) When are Metal Ion-Dependent Hydroxyl and Alkoxyl Radical Adducts of 5,5-Dimethyl-1-pyrroline *N*-Oxide Artifacts? *Arch. Biochem. Biophys. 296*, 640–644.

Hiramoto, K., Ojima, N., Sako, K., and Kikugawa, K. (1996) Effect of Plant Phenolics on the Formation of the Spin-Adduct of Hydroxyl Radical and the DNA Strand Breaking by Hydroxyl Radical, *Biol. Pharm. Bull. 19*, 558–563.

Hofmann, T., Bors, W., and Stettmaier, K. (2002) CROSSPY: A Radical Intermediate of Melanoidin Formation in Roasted Coffee, in *Free Radicals in Food, Chemistry, Nutrition, and Health Effects* (Morello, M.J., Shahidi, F., and Ho, C.-T., eds.), pp. 49–68, ACS Symposium Series 807, American Chemical Society, Washington.

Howard, J.A., and Tait, J.C. (1978) Electron Paramagnetic Resonance Spectra of the *tert*-Butylperoxy and *tert*-Butoxy Adducts to Phenyl *tert*-Butyl Nitrone and 2-Methyl-2-nitrosopropane: Oxygen-17 Hyperfine Coupling Constants, *Can. J. Chem. 56*, 176–178.

Iwahashi, H. (2000) Some Polyphenols Inhibit the Formation of Pentyl Radical and Octanoic Acid Radical in the Reaction Mixture of Linoleic Acid Hydroperoxide with Ferrous Ions, *Biochem. J. 346*, 265–273.

Iwahashi, H. (2003) Identification of the Several New Radicals Formed in the Reaction Mixture of Oxidized Linoleic Acid with Ferrous Ions Using HPLC-ESR and HPLC-ESR-MS, *Free Radic. Res. 37*, 939–945.

Iwahashi, H., Albro, P.W., McGown, S.R., Tomer, K.B., and Mason, R.P. (1991) Isolation and Identification of α-(4-Pyridyl-1-Oxide)-*N-tert*-Butylnitrone Radical Adducts Formed by the Decomposition of the Hydroperoxides of Linoleic Acid, Linolenic Acid, and Arachidonic Acid by Soybean Lipoxygenase, *Arch. Biochem. Biophys. 285*, 172–180.

Janzen, E.G., and Blackburn, B.J. (1969) Detection and Identification of Short-Lived Free Radicals by Electron Spin Resonance Trapping Techniques (Spin Trapping): Photolysis of Organolead, -Tin, and -Mercury Compounds, *J. Am. Chem. Soc. 91*, 4481–4490.

Janzen, E.G., and Haire, L. (1990) Two Decades of Spin Trapping, in *Advances in Free Radical Chemistry* (Tanner, D.D., ed.), Vol. 1, pp. 253–295, JAI Press, Greenwich, CT.

Janzen, E.G., Krygsman, P.H., Lindsay, D.A., and Haire, L. (1990) Detection of Alkyl, Alkoxyl, and Alkylperoxyl Radicals from the Thermolysis of Azobis(isobutyronitrile) by ESR/Spin Trapping: Evidence for Double Spin Adducts from Liquid-Phase Chromatography and Mass Spectroscopy, *J. Am. Chem. Soc. 112*, 8279–8284.

Jensen, P.N., Danielsen, B., Bertelsen, G., Skibsted, L.H., and Andersen, M.L. (2005) Storage Stability of Pork Scratchings, Peanuts, Oatmeal and Muesli: Comparison of ESR Spectroscopy, Headspace-GC and Sensory Evaluation for Detection of Oxidation in Dry Foods, *Food Chem. 91*, 25–38.

Kalyanaraman, B., Joseph, J., and Parthasarathy, S. (1991) The Spin Trap, α-Phenyl *n-tert*-Butylnitrone, Inhibits the Oxidative Modification of Low-Density-Lipoprotein, *FEBS Lett. 280*, 17–20.

Kennedy, C.H., Dyer, J.M., Church, D.F., Winston, G.W., and Pryor, W.A. (1989) Radical Production in Liver Mitochondria by Peroxidic Tumor Promoters, *Biochem. Biophys. Res. Commun. 160*, 1067–1072.

Koenig, M.L., and Meyerhoff, J.L. (2003) *In Vitro* Neuroprotection Against Oxidative Stress by Pre-treatment with a Combination of Dihydrolipoic Acid and Phenyl-butyl Nitrones, *Neurotox Res. 5*, 265–272.

Krainev, A.G., and Bigelow, D.J. (1996) Comparison of 2,2′-Azobis(2-amidinopropane) Hydrochloride (AAPH) and 2,2′-Azobis(2,4-dimethylvaleronitrile) (AMVN) as Free Radical Initiators: A Spin-Trapping Study, *J. Chem. Soc. Perkin Trans. 2*, 747–754.

Kristensen, D., and Skibsted, L.H. (1999) Comparison of Three Methods Based on Electron Spin Resonance Spectrometry for Evaluation of Oxidative Stability of Processed Cheese, *J. Agric. Food Chem. 47*, 3099–3104.

Kristensen, D., Orlien, V., Mortensen, G., Brockhoff, P., and Skibsted, L.H. (2000) Light-Induced Oxidation in Sliced Havarti Cheese Packaged in Modified Atmosphere, *Int. Dairy J. 10*, 95–103.

Kristensen, D., Andersen, M.L., and Skibsted, L.H. (2002) Prediction of Oxidative Stability of Raw Milk Using Spin Trapping Electron Spin Resonance Spectroscopy, *Milchwissenschaft 57*, 255–258.

Kumari, M.V.R., Yoneda, T., and Hiramatsu, M. (1996) Scavenging Activity of "β Catechin" on Reactive Oxygen Species Generated by Photosensitization of Riboflavin, *Biochem. Mol. Biol. Int. 38*, 1163–1170.

Lauridsen, C., Jensen, S.K., Skibsted, L.H., and Bertelsen, G. (2000) Influence of Supranutritional Vitamin E and Copper on α-Tocopherol Deposition and Susceptibility to Lipid Oxidation of Porcine Membranal Fractions of *M. Psoas major* and *M. Longissimus dorsi*, *Meat Sci. 54*, 377–384.

Lee, J.H., and Park, J.W. (2003) Protective Role of α-Phenyl-*N*-*t*-butylnitrone Against Ionizing Radiation in U937 Cells and Mice, *Cancer Res. 63*, 6885–6893.

Leonard, S.S., Cutler, D., Ding, M., Vallyathan, V., Castranova, V., and Shi, X.L. (2002) Antioxidant Properties of Fruit and Vegetable Juices: More to the Story than Ascorbic Acid, *Ann. Clin. Lab. Sci. 32*, 193–200.

Li, L., Shou, Y., Borowitz, J.L., and Isom, G.E. (2001) Reactive Oxygen Species Mediate Pyridostigmine-Induced Neuronal Apoptosis: Involvement of Muscarinic and NMDA Receptors, *Toxicol. Appl. Pharmacol. 177*, 17–25.

Madsen, H.L., Nielsen, B.R., Bertelsen, G., and Skibsted, L.H. (1996) Screening of Antioxidative Activity of Spices: A Comparison Between Assays Based on ESR Spin Trapping and Electrochemical Measurement of Oxygen Consumption, *Food Chem. 57*, 331–337.

Madsen, H.L., Andersen, C.M., Jorgensen, L.V., and Skibsted, L.H. (2000) Radical Scavenging by Dietary Flavonoids: A Kinetic Study of Antioxidant Efficiencies, *Eur. Food Res. Technol. 211*, 240–246.

Masaki, H., Atsumi, T., and Sakurai, H. (1995) Peroxyl Radical Scavenging Activities of Hamamelitannin in Chemical and Biological-Systems, *Free Radic. Res. 22*, 419–430.

McLellan, M.E., Kajdasz, S.T., Hyman, B.T., and Bacskai, B.J. (2003) *In Vivo* Imaging of Reactive Oxygen Species Specifically Associated with Thioflavine S-Positive Amyloid Plaques by Multiphoton Microscopy, *J. Neurosci. 23*, 2212–2217.

McNamee, M.G. (1984) Electron Paramagnetic Resonance Spectroscopy, in *Food Analysis. Principles and Techniques II. Physicochemical Techniques* (Gruenwedel, D.W., and Whitaker, J.R., eds.), pp. 343–386, Marcel Dekker, New York.

Milatovic, D., Zivin, M., Gupta, R.C., and Dettbarn, W.D. (2001) Alterations in Cytochrome c Oxidase Activity and Energy Metabolites in Response to Kainic Acid-Induced Status Epilepticus, *Brain Res. 912*, 67–78.

Monahan, F.J., Gray, J.I., Asghar, A., Haug, A., Shi, B., and Buckley, D.J. (1993) Effect of Dietary Lipid and Vitamin E Supplementation on Free Radical Production and Lipid Oxidation in Porcine Muscle Microsomal Fractions, *Food Chem. 46*, 1–6.

Niki, E., Yokoi, S., Tsuchiya, J., and Kamiya, Y. (1983) Spin Trapping of Peroxy Radicals by Phenyl-*N*-(*tert*-butyl)nitrone and Methyl-*N*-burylnitrone, *J. Am. Chem. Soc. 105*, 1498–1503.

Nissen, L.R., Månsson, L., Bertelsen, G., Huynh-Ba, T., and Skibsted, L.H. (2000) Protection of Dehydrated Chicken Meat by Natural Antioxidants as Evaluated by Electron Spin Resonance Spectrometry, *J. Agric. Food Chem. 48*, 5548–5556.

Nissen, L.R., Huynh-Ba, T., Petersen, M.A., Bertelsen, G., and Skibsted, L.H. (2002) Potential Use of Electron Spin Resonance Spectroscopy for Evaluating the Oxidative Status of Potato Flakes, *Food Chem. 79*, 387–394.

North, J.A., Spector, A.A., and Buettner, G.R. (1992) Detection of Lipid Radicals by Electron-Paramagnetic Resonance Spin Trapping Using Intact-Cells Enriched with Polyunsaturated Fatty-Acid, *J. Biol. Chem. 267*, 5743–5746.

North, J.A., Spector, A.A., and Buettner, G.R. (1994) Cell Fatty-Acid Composition Affects Free-Radical Formation During Lipid-Peroxidation, *Am. J. Phys. 267*, 177–188.

Novakov, C.P., Feierman, D., Cederbaum, A.I., and Stoyanovsky, D.A. (2001) An ESR and HPLC-EC Assay for the Detection of Alkyl Radicals, *Chem. Res. Toxicol. 14*, 1239–1246.

Ohto, N., Niki, E., and Kamiya, Y. (1977) Study of Autoxidation by Spin Trapping: Spin Trapping of Peroxyl Radicals by Phenyl *N*-*t*-Butyl Nitrone, *J. Chem. Soc. Perkin Trans. II 13*, 1770–1774.

Ondrias, K., Misik, V., Stasko, A., Gergel, D., and Hromadova, M. (1994) Comparison of Antioxidant Properties of Nifedipine and Illuminated Nifedipine with Nitroso Spin Traps in Low-Density Lipoproteins and Phosphatidylcholine Liposomes, *Biochim. Biophys. Acta 1211*, 114–119.

Park, J.E., Yang, J.H., Yoon, S.J., Lee, J.H., Yang, E.S., and Park, J.W. (2002) Lipid Peroxidation-Mediated Cytotoxicity and DNA Damage in U937 Cells, *Biochemistry 84*, 1199–1205.

Pedrielli, P., and Skibsted, L.H. (2002) Antioxidant Synergy and Regeneration Effect of Quercetin, (–)-Epicatechin, and (+)-Catechin on Alpha-Tocopherol in Homogeneous Solutions of Peroxidating Methyl Linoleate, *J. Agric. Food Chem. 50*, 7138–7144.

Pedrielli, P., Holkeri, L.M., and Skibsted, L.H. (2001a) Antioxidant Activity of (+)-Catechin: Rate Constant for Hydrogen-Atom Transfer to Peroxyl Radicals, *Eur. Food Res. Technol. 213*, 405–408.

Pedrielli, P., Pedulli, G.F., and Skibsted, L.H. (2001b) Antioxidant Mechanism of Flavonoids: Solvent Effect on Rate Constant for Chain-Breaking Reaction of Quercetin and Epicatechin in Autoxidation of Methyl Linoleate, *J. Agric. Food Chem. 49*, 3034–3040.

Pedulli, G.F. (1993) Stable Radicals as Probes of the Oxygen Concentration in Autoxidation Studies, in *Free Radicals and Antioxidants in Nutrition* (Corongiu, F., Banni, S., Dessi, M.A., and Rice-Evans, C., eds.) pp. 169–185, Richelieu Press, London.

Pedulli, G.F., Lucarini, M., Pedrielli, P., Sagrini, M., and Cipollone, M. (1996) The Determination of the Oxygen Consumption in Autoxidation Studies by Means of EPR Spectroscopy, *Res. Chem. Intermediates 22*, 1–14.

Perkins, M.J. (1980) Spin Trapping, *Adv. Phys. Org. Chem. 17*, 1–64.

Pfab, J. (1978) Alkylperoxynitroxides in the Photo-Oxidation of C-Nitrosoalkanes and the 'Spin Trapping' of Peroxy Radicals by C-Nitroso Compounds, *Tetrahedron Lett. 19*, 843–846.

Polyakov, N.E., Kruppa, A.I., Leshina, T.V., Konovalova, T.A., and Kispert, L.D. (2001a) Carotenoids as Antioxidants: Spin Trapping EPR and Optical Study, *Free Radic. Biol. Med. 31*, 43–52.

Polyakov, N.E., Leshina, T.V., Konovalova, T.A., and Kispert, L.D. (2001b) Carotenoids as Scavengers of Free Radicals in a Fenton Reaction: Antioxidants or Pro-oxidants? *Free Radic. Biol. Med. 31*, 398–404.

Qian, S.Y., Wang, H.P., Schafer, F.Q., and Buettner, G.R. (2000) EPR Detection of Lipid-Derived Free Radicals from PUFA, LDL, and Cell Oxidations, *Free Radic. Biol. Med. 29*, 568–579.

Qian, S.Y., Tomer, K.B., Yue, G.H., Guo, Q., Kadiiska, M.B., and Mason, R.P. (2002) Characterization of the Initial Carbon-Centered Pentadienyl Radical and Subsequent Radicals in Lipid Peroxidation: Identification *via* On-Line High Performance Liquid Chromatography/Electron Spin Resonance and Mass Spectrometry, *Free Radic. Biol. Med. 33*, 998–1009.

Qian, S.Y., Yue, G.H., Tomer, K.B., and Mason, R.P. (2003) Identification of All Classes of Spin-Trapped Carbon-Centered Radicals in Soybean Lipoxygenase-Dependent Lipid Peroxidations of ω-6 Polyunsaturated Fatty Acids *via* LC/ESR, LC/MS, and Tandem MS, *Free Radic. Biol. Med. 34*, 1017–1028.

Quiles, J.L., Ramírez-Tortosa, M.C., Gómez, J.A., Huertas, J.R., and Mataix, J. (2002) Role of Vitamin E and Phenolic Compounds in the Antioxidant Capacity, Measured by ESR, of Virgin Olive, Olive and Sunflower Oils After Frying, *Food Chem. 76*, 461–468.

Reis, A., Domingues, M.R.M., Amado, F.M.L., Ferrer-Correia, A.J.V., and Domingues, P. (2003) Detection and Characterization by Mass Spectrometry of Radical Adducts Produced by Linoleic Acid Oxidation, *J. Am. Soc. Mass Spectrom. 14*, 1250–1261.

Rosen, G.M., Rauckman, E.J., and Finkelstein, E. (1980) Spin Trapping of Radical Species Involved in the Propagation of Lipid Peroxidation, in *Autoxidation in Food and Biological Systems* (Simic, M.G., and Karel, M., eds.), pp. 45–87, Plenum Press, New York.

Rosen, G.M., Britigan, B.E., Halpern, H.J., and Pou, S. (1999) *Free Radicals: Biology and Detection by Spin Trapping*, Oxford University Press, New York.

Rota, C., Barr, D.P., Martin, M.V., Guengerich, F.P., Tomasi, A., and Mason, R.P. (1997) Detection of Free Radical Produced from the Reaction of Cytochrome P-450 with Linoleic Acid Hydroperoxide, *Biochem. J. 328*, 565–571.

Sawa, T., Akaike, T., Kida, K., Fukushima, Y., Takagi, K., and Maeda, H. (1998) Lipid Peroxyl Radicals from Oxidized Oils and Heme Iron: Implication of a High-Fat Diet in Colon Carcinogenesis, *Cancer Epidemiol. Biomark. Prev. 7*, 1007–1012.

Schaich, K.M. (2002) Free Radical Generation During Extrusion: A Critical Contributor to Texturization, in *Free Radicals in Food, Chemistry, Nutrition, and Health Effects* (Morello, M.J., Shahidi, F., and Ho, C.-T., eds.), pp. 35–48, ACS Symposium Series 807, American Chemical Society, Washington.

Schaich, K.M., and Borg, D.C. (1980) EPR Studies in Autoxidation, in *Autoxidation in Food and Biological Systems* (Simic, M.G., and Karel, M., eds.), pp. 45–87, Plenum Press, New York.

Schaich, K.M., and Borg, D.C. (1988) Fenton Reactions in Lipid Phases, *Lipids 23*, 570–579.

Schaich, K.M., and Rebello, C.A. (1999) Extrusion Chemistry of Wheat Flour Proteins: I. Free Radical Formation, *Cereal Chem. 76*, 748–755.

Stagko, A., Liptakova, M., Malik, F., and Misik, V. (2002) Free Radical Scavenging Activities of White and Red Wines: An EPR Spin Trapping Study, *Appl. Magn. Reson. 22*, 101–113.

Stapelfeldt, H., Nielsen, B.R., and Skibsted, L.H. (1997) Towards Use of Electron Spin Resonance Spectrometry in Quality Control of Milk Powder: Correlation Between Sensory Score of Instant Whole Milk Powder and Concentration of Free Radicals and 2-Thiobarbituric Acid Reactive Substances, *Milchwissenschaft 52*, 682–685.

Stapelfeldt, H., Nyholm Nielsen, K., Krogh Jensen, S., and Skibsted, L.H. (1999) Free Radical Formation in Freeze-Dried Raw Milk in Relation to Its α-Tocopherol Level, *J. Dairy Res. 66*, 461–466.

Swartz, H.M., and Glockner, J.F. (1989) Measurements of the Concentration of Oxygen in Biological Systems Using EPR Techniques, in *Advanced EPR: Application in Biology and Biochemistry* (Hoff, A.J., ed.), pp. 753–784, Elsevier, Amsterdam.

Thomsen, M.K., Vedstesen, H., and Skibsted, L.-H. (1999) Quantification of Radical Formation in Oil-in-Water Food Emulsions by Electron Spin Resonance Spectroscopy, *J. Food Lipids 6*, 149–158.

Thomsen, M.-K., Jacobsen, C., and Skibsted, L.H. (2000a) Mechanism of Initiation of Oxidation in Mayonnaise Enriched with Fish Oil as Studied by Electron Spin Resonance Spectroscopy, *Eur. Food Res. Technol. 211*, 381–386.

Thomsen, M.K., Kristensen, D., and Skibsted, L.-H. (2000b) Electron Spin Resonance Spectroscopy for Determination of Oxidative Stability of Food Lipids, *J. Am. Oil Chem. Soc. 77*, 725–730.

Uchida, M., and Ono, M. (1996) Improvement for Oxidative Flavor Stability of Beer—Role of OH-Radical in Beer Oxidation, *J. Am. Soc. Brew. Chem. 54*, 198–204.

Ueda, J. (1963) Application of Electron Spin Resonance to the Determination of Hydroperoxides, *Anal. Chem. 35*, 2213–2214.

Ueda, J., Saito, N., and Ozawa, T. (1996) ESR Spin Trapping Studies on the Reaction of Hydroperoxides with Cu(II) Complex, *J. Inorg. Biochem. 64*, 197–206.

Unno, T., Sugimoto, A., and Kakuda, T. (2000) Scavenging Effect of Tea Catechins and Their Epimers on Superoxide Anion Radicals Generated by a Hypoxanthine and Xanthine Oxidase System, *J. Sci. Food Agric. 80*, 601–606.

Utsumi, H., and Yamada, K. (2003) *In Vivo* Electron Spin Resonance-Computed Tomography/Nitroxyl Probe Technique for Non-Invasive Analysis of Oxidative Injuries, *Arch. Biochem. Biophys. 416*, 1–8.

Velasco, J., Andersen, M.L., and Skibsted, L.H. (2004) Evaluation of Oxidative Stability of Vegetable Oils by Monitoring the Tendency to Radical Formation. A Comparison of Electron Spin Resonance Spectroscopy with the Rancimat Method and Differential Scanning Calorimetry, *Food Chem. 85*, 623–632.

Velasco, J., Andersen, M.L., and Skibsted, L.H. (2005) ESR Spin-Trapping for Analysis of Lipid Oxidation in Oils: Inhibiting Effect of the Spin Trap α-Phenyl-*N-tert*-butylnitrone (PBN) on Lipid Oxidation, *J. Agric. Food Chem. 53*, 1328–1336.

Vicente, M.L., Empis, J.A., Deighton, N., Glidewell, S.M., Goodman, B.A., and Rowlands, C.C. (1998) Use of EPR and ENDOR Spectroscopy in Conjunction with the Spin Trapping Technique to Study the High-Temperature Oxidative Degradation of Fatty Acid Methyl Esters, *J. Chem. Soc. Perkin Trans. 2*, 449–454.

Weil, J.A., Bolton, J.R., and Wertz, J.E. (1994) *Electron Paramagnetic Resonance: Elementary Theory and Practical Applications*, John Wiley & Sons, Inc., New York.

Yamada, T., Niki, E., Yokoi, S., Tsuchiya, J., Yamamoto, Y., and Kamiya, Y. (1984) Oxidation of Lipids. XI. Spin Trapping and Identification of Peroxy and Alkoxy Radicals of Methyl Linoleate, *Chem. Phys. Lipids 36*, 189–196.

Yanez, J., Sevilla, C.L., Becker, D., and Sevilla, M.D. (1987) Low-Temperature Autoxidation in Unsaturated Lipids: An Electron Spin Resonance Study, *J. Phys. Chem. 91*, 487–491.

Yang, G.C., Qiang, W., Morehouse, K.M., Rosenthal, I., Ku, Y., and Yurawecz P. (1991) Determination of Hydroperoxides in Edible Oils by Electron Spin Resonance, Thiobarbituric Acid Assay, and Liquid Chromatography-Chemiluminescence Techniques, *J. Agric. Food Chem. 39*, 896–898.

Yoshimura,Y., Inomata, T., Nakazawa, H., Kubo, H., Yamaguchi, F., and Ariga, T. (1999) Evaluation of Free Radical Scavenging Activities of Antioxidants with an H_2O_2/NaOH/DMSO System by Electron Spin Resonance, *J. Agric. Food Chem. 47*, 4653–4656.

Yun, Y.S., Nakajima, Y., Iseda, E., and Kunugi, A. (2003) Determination of Antioxidant Activity of Herbs by ESR, *J. Food Hyg. Soc. Jpn. 44*, 59–62.

Zhu, J., and Sevilla, M.D. (1990) Kinetic Analysis of Free-Radical Reactions in the Low Temperature Autoxidation of Triglycerides, *J. Phys. Chem. 94*, 1447–1452.

Chapter 7

Analysis of Lipid Oxidation by Differential Scanning Calorimetry

Grzegorz Litwinienko

Department of Chemistry, Warsaw University, Pasteura 1, 02-093 Warsaw, Poland

Introduction

During every chemical process, heat is either absorbed or evolved; therefore, the course of the chemical processes can be followed by monitoring the sample temperature or heat exchange when the reacting system is subjected to a controlled temperature program. Thermal analysis (TA) compromises a group of methods based on measurements of physical and chemical properties of a substance, or a mixture of reacting substances, as a function of temperature or time. Two main thermoanalytical techniques, differential thermal analysis (DTA) and differential scanning calorimetry (DSC), might be applied to investigate the behavior of organic and inorganic materials with respect to temperature change and the heat flow, respectively. For some processes, the mass of the reacting system is a parameter that can be monitored to follow the reaction course. Thermogravimetric analysis (TG) is the method based on measurements of a change in the mass. TGA, DTA, and DSC are the most widely applied thermoanalytical techniques, but several other methods that were developed from modifying these techniques sometimes included modification of the equipment design. TGA, DTA, and DSC together with methods for monitoring change in other physical parameters (Table 7.1) serve as parent techniques for the family of TA methods. Some of the TA methods can be applied in combination, for example, TG in combination with evolved gas analysis (EGA). Moreover, growing attention has been paid recently to possibilities of combinations of TA techniques with other analytical methods [e.g., DSC + Fourier transform infrared (FTIR)].

Several important advantages of TA make this branch of chemistry a still growing area of analysis. These advantages include the small amount of sample needed for successful measurement (0.1–20 mg), the short time of analysis (from several minutes for a single measurement), the lack of requirements regarding the state of the sample, i.e., any nonvolatile liquid or solid material can be analyzed (homogenous as well as nonhomogeneous), and the possibility of the application of wide temperature ranges and various heating programs. Simplicity of operation is also a considerable benefit to an analyst; however, much attention must be paid to the interpretation of the results because TA experiments yield indirect data.

TABLE 7.1
Main Groups of Thermoanalytical Methods[a]

Parameter measured	Name	Abbreviation
Mass	Thermogravimetry	TG
Temperature difference	Differential thermal analysis	DTA
Heat flow (enthalpy)	Differential scanning calorimetry	DSC
Volume of gaseous products	Evolved gas analysis	EGA
Dimension (length, volume) changes	Dilatometry	DIL
Mechanistic properties, deformations	Thermomechanical analysis	TMA
	Dynamic mechanical analysis	DMA

[a]*Source:* Hatakeyama and Quinn (1999).

Differential Thermal Analysis

A general scheme of the DTA apparatus is shown in Figure 7.1. In classical DTA, both sample (S) and reference (R) materials are heated by a single heat source. Temperatures are measured separately for S and R, and a function of temperature difference, $\Delta T = T_S - T_R$, is recorded against time, τ, or programmed temperature ($T = T_o + \beta\tau$, where T_o is an initial temperature and β is a heating rate in K/min). An idealized DTA curve is presented in Figure 7.2. Sectors AB and EF are straight lines (baseline), indicating that there is no difference between temperatures T_S and T_R, hence no exo- or endothermic process occurs. When a chemical reaction or phase transition begins, the baseline changes into a peak (here BC and FG); then the DTA curve returns to the baseline. Depending on the thermal character of the

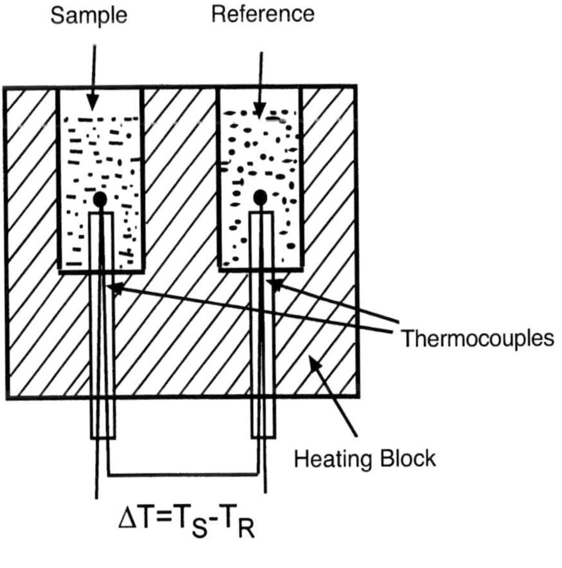

Fig. 7.1. Schematic diagram of differential thermal analysis cell.

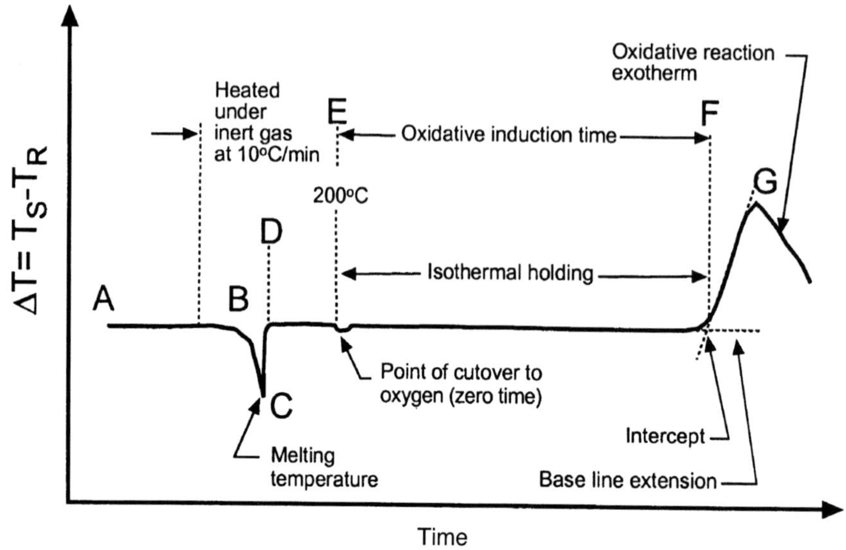

Fig. 7.2. Example of typical differential thermal analysis curve. *Source:* Howard (1973). Reprinted with permission of John Wiley & Sons.

process occurring, the recorded peak is endothermic (i.e., $T_S < T_R$, such as peak BCD in Fig. 7.2), or exothermic ($T_S > T_R$, e.g., peak FG). If a programmed heating rate is employed in the experiment, the temperature difference between S and R is usually plotted as a function of increasing temperature, whereas in isothermal mode (T = constant), the DTA curve is plotted vs. time.

The magnitude of ΔT is proportional to changes in enthalpy and heat capacity, and to the total thermal resistance to heat flow. The last factor depends on the physical nature of the sample (size, geometry, and the way it is packed in the sample vessel); thus, in modern TA equipment, the size of the vessels is minimized to reduce and standardize the influence of these physical factors. Despite these improvements, DTA systems are not very suitable for precise calorimetric measurements. Their disadvantages were overcome in another TA method, i.e., DSC.

Differential Scanning Calorimetry

General Scheme of DSC Equipment

Figure 7.3 presents a schematic diagram of power-compensation DSC. The system is heated (or kept at T = constant) similarly to the DTA method, but in this case, the sample and reference vessels are additionally heated with individual heaters to keep $\Delta T = 0$ at any moment of the experiment. If an exo- or endothermal reaction starts in S, the power input compensates the ΔT, and the amount of heat provided

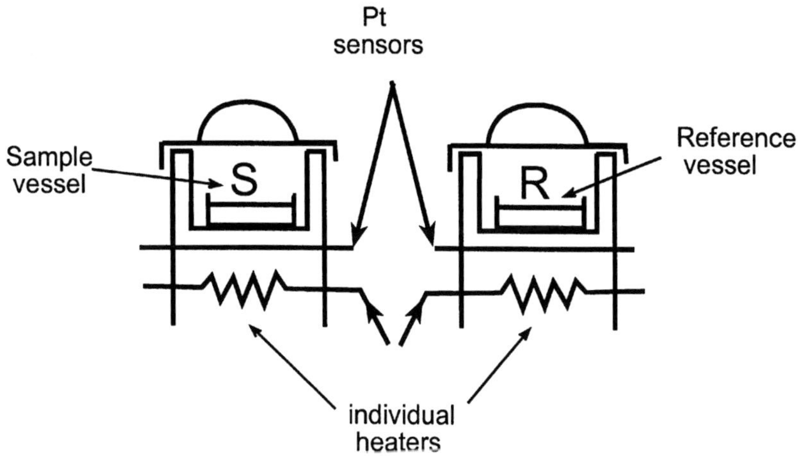

Fig. 7.3. Power compensation differential scanning calorimetry.

to S (or R) is monitored and recorded as heat flow, d*H*/dτ (Watson 1964). The temperature range of power-compensation DSC depends on the model of the equipment. For example, for organic/polymer materials, typical equipment is able to run from –180°C (if a cooling accessory is applied) up to ~800°C.

Another technical type of DSC system is a Heat-Flux DSC. Its scheme is shown in Figure 7.4. In this system, which could be considered as an intermediate step between DTA and power-compensation DSC, the temperature difference between the sample and reference is measured, and the difference is proportional to the change in the heat flux. Unlike in DTA systems, the thermocouples are attached to the base of the sample and reference holders. Absorption or emission of heat by the sample causes a variation in heat flux through the heat-sensitive plate. The temperature difference between the heat-sensitive plate and the furnace is automatically recalculated into the enthalpy of the ongoing process. The theoretical basis and descriptions of various DSC systems (heat-flux DSC, power compensation DSC, temperature-modulated DSC and high-sensitivity DSC) were presented in recent monographs (Gallagher 1997, Hatakeyama and Quinn 1999, Wunderlich 1997).

The optimal weight of samples analyzed by the DSC method should be <10 mg to form a flat, thin layer on the bottom of the sample pan (vessel). As a reference material, the same size sample pan should be used. Open vessels made of aluminium are recommended for DSC measurements if the lipid and polymer oxidation is analyzed.

Figure 7.5 presents an idealized example of a DSC curve for a polymer heated at a constant heating rate. When the system reaches the desired start temperature, the equilibration of the baseline takes place; just after the equilibration, the differential heat flow is close to zero (section AB). Small deviations of the heat flow can be caused by a difference between the heat capacities of the sample analyzed and the reference material. A rapid decrease in the baseline level occurs at point B. In

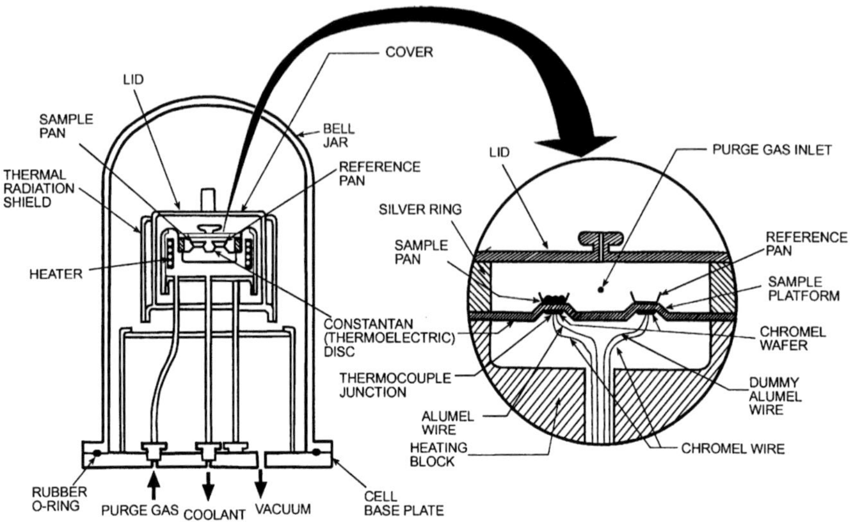

Fig. 7.4. Scheme of heat-flux differential scanning calorimetry cell (reprinted with permission of TA Instruments).

the case of polymers, this change in the baseline would be interpreted as the glass transition. The peak observed between points C and E is due to an exothermal process (for example, polymer or lipid crystallization), and the peak between points F and H is an endothermic phase transition. The height of the DSC peak is expressed as heat flow and the units are mW or cal/s (sometimes given in mW/g, i.e., power unit per gram of the sample) and an area under the DSC curve equals

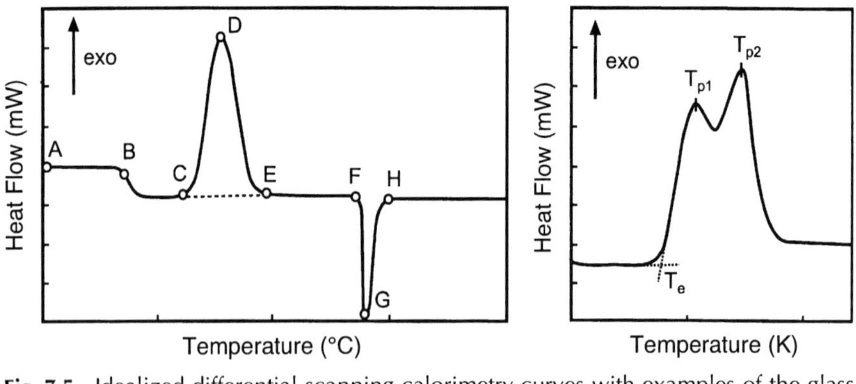

Fig. 7.5. Idealized differential scanning calorimetry curves with examples of the glass transition, and exo- and endothermal effects (left panel), definition of temperatures of extrapolated onset, T_e, and peaks, T_{p1} and T_{p2} (right panel). The arrows on the plots indicate the sign of the thermal effect (exo- or endo-).

enthalpy of the process in J/g. Usually heat flow is plotted as a function of the temperature. The distinctive points of a peak that allow it to be distinguished from other peaks are as follows: the extrapolated temperature of the start (T_e, see Fig. 7.5) and the temperature of the maximum heat flow (peak maximum, T_p, or if there is more than one peak maximum: T_{p1}, T_{p2}, and so on).

The thermal curves presented in Figures 7.2 and 7.5 are idealized; under applied experimental conditions, baseline levels observed before and after reaction can be different. A shift in baseline (Fig. 7.6, plot A) is observed when the process occurs with a change of heat capacity. A change of thermal conductivity (thermal resistance) during this process is manifested by an additional difference in the baseline slope, as presented in Figure 7.6, plot B. The effect of the change in thermal resistance is greatly reduced by minimization of the weight of the sample analyzed (usually 3–10 mg); thus, the response of the system is faster.

Calibration of the Calorimeter

Every DSC instrument must be calibrated before measurements are made. The heat of fusion, ΔH_f, or heat of transition $solid_1 \rightarrow solid_2$, ΔH_t, of the standard materials are parameters used for the calorimeter calibration that are easily measured. A good standard should be chemically stable during physical change, and its vapor pressure should be low to minimize the thermal effect of vaporization during the change. The choice of the standard substance for the temperature calibration and for the calorimeter calibration depends on the temperature range of the TA experiments. Although organic standards are recommended for experiments in which organic materials are studied, high-purity indium is the standard most frequently used for calibration. Some standard reference materials are listed in Table 7.2. Usually one calibration experiment is sufficient to obtain both temperature calibration and calorimetric constant. However, for precise correction of the temperature scale, a "two different standards" procedure is required.

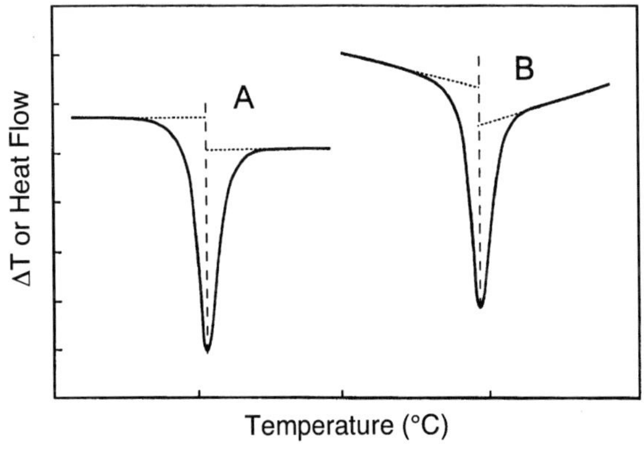

Fig. 7.6. Effect of the change of heat capacity (C_p) before and after the reaction (Curve A) and the effect of the change of thermal conductivity (Curve B).

TABLE 7.2
Standards Used for DTA and DSC Calibration[a]

Standard	Temperature/°C	ΔH_f (J/g)
Azoxybenzene	34.6	21.6
Benzophenone	48.2	23.5
Diphenyl ether[b]	26.9	
Palmitic acid	62.5	51.2
Biphenyl	69.8	28.7
Benzoic acid	121.8	33.9
KNO_3	128	12.86 (ΔH_t)
In[b]	156.4	6.79
$AgNO_3$	212	17.7
Sn[b]	231.9	14.40
Pb[b]	327.50	
Zn[b]	419.58	
Ag_2SO_4[b]	430	

[a]See Table 7.1 for abbreviations.
[b]Proposed by National Bureau of Standards (NBS) and The International Confederation of Thermal Analysis (ICTA).

Figure 7.7 presents a DSC curve recorded for the process of melting high-purity indium. From the DSC peak, the transition temperatures (T_e and T_p) and transition enthalpy (area of the peak, ΔH_{exper}) are determined. Commercial software for calibration is normally provided with DSC equipment; therefore the recorded transition temperature is automatically compared with the known literature data and the difference is corrected. The calorimetric cell constant is the ratio of the measured experimental heat of fusion (ΔH_{exper}, in J/g) to the literature value, ΔH_{ref}, that is, $K_{cal} = (\Delta H_{exper})/(\Delta H_{ref})$. Usually, the calculation procedure is performed automatically by DSC data analysis software.

Assessment of Kinetic Parameters from DSC Data

The interpretation of the kinetic parameters obtained must be conducted carefully because certain kinetic parameters themselves (without knowledge of the mechanism of reaction) may lead to incorrect conclusions, especially during analyses for which the observed process is an effect of several reactions (Arnold *et al.* 1981 and 1982). Therefore, results of thermal analysis, i.e., analytical signals and measured kinetic parameters, are usually compared with and fitted to theoretical models to verify the validity of the kinetic description of the observed process (such as reaction order, activation energy, rate constants).

The rate of the reaction is defined as the rate of disappearance of a reactant (or appearance of a product). In thermal analyses, the parameters measured are the mass of the sample (in TG), or difference between temperatures (in DTA), or heat flow (in DSC). The determination of the kinetic parameters is based on the calculation of the degree of conversion, α, as a function of time (in isothermal mode) or as

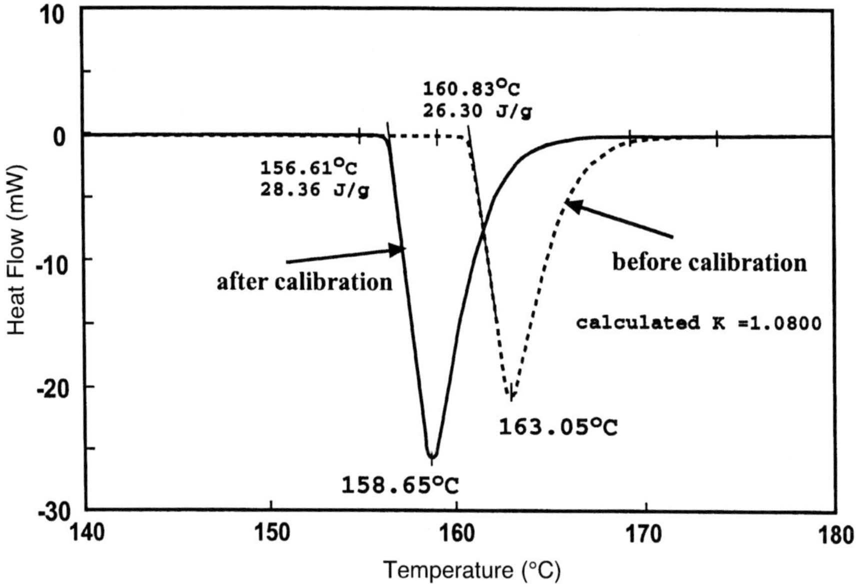

Fig. 7.7. Differential scanning calorimetry (DSC) curves of pure indium melting. The first plot obtained (dashed line) was used for temperature and calorimeter calibration. The solid line represents DSC curve obtained after calibration.

a function of linearly increasing temperature (when nonisothermal mode is employed); α is a ratio of the amount of the reactant consumed from the beginning of the process to time τ (m_τ or ΔH_τ) to the total amount of the reactant (m_∞ or ΔH_∞, see Fig. 7 8):

$$\alpha = \frac{\Delta H_\tau}{\Delta H_\infty} = \frac{\Delta m_\tau}{\Delta m_\infty} \qquad [1]$$

Because, according to the above definition, the starting amount of reagent is normalized the amount of nonreacted substance at time τ is $(1 - \alpha)$. In DSC, the rate of the process is measured with respect to the amount of heat released or absorbed during the course of the reaction:

$$\frac{d\alpha}{d\tau} = \frac{1}{\Delta H_\infty} \times \frac{dH}{d\tau} \qquad [2]$$

where $dH/d\tau$ is the heat flow at the time τ.

A general expression describing the reaction rate is as follows:

$$\frac{d\alpha}{d\tau} = k(\mathrm{T}) \times f(\alpha) \qquad [3]$$

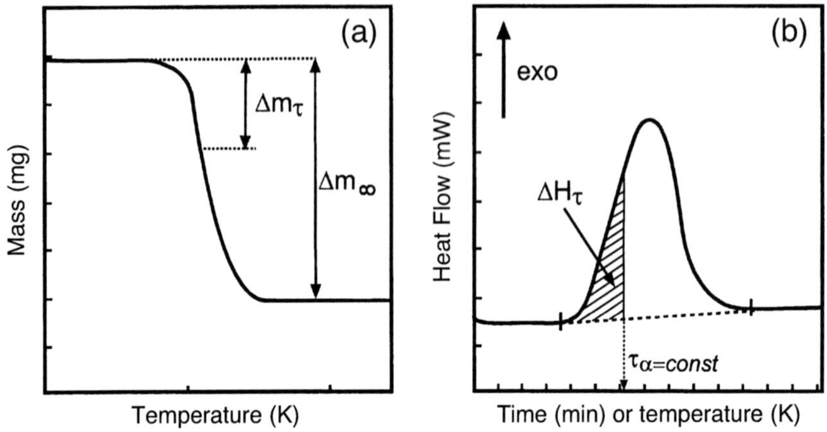

Fig. 7.8. Determination of the degree of conversion (α) of the reacting system calculated on the basis of the mass change in a thermogravimetric experiment (a), and calculated on the basis of the heat released within the period from the start of the process to the time τ, in a differential scanning calorimetry experiment (b).

where $f(\alpha)$ is a function describing a kinetic model of the reaction and the rate constant $k(\mathrm{T})$ is given by the Arrhenius equation:

$$k(\mathrm{T}) = A \times e^{\frac{-\mathrm{E_a}}{\mathrm{RT}}} \qquad [4]$$

Part $f(\alpha)$ in [3] is a function of the degree of conversion and is the mathematical representation of kinetic model of the observed process. Determination of the proper form of $f(\alpha)$ is not a trivial task. One of the most frequently used models is: $f(\alpha) = (1 - \alpha)^n$ where n is order of the reaction. Some examples of $f(\alpha)$ are presented in Table 7.3. Although experimental data can be described by more than one form of $f(\alpha)$, the aim of kinetic analysis is to find the simplest model that would describe the process in good agreement with experimental data. Using the selected model, the kinetic parameters obtained for particular peaks on the DSC curve can be assigned to the individual processes. Because the DSC curve can be an effect of complex processes, e.g., consecutive, competitive, or parallel, the kinetic analysis of experimental thermoanalytical data must always be accompanied by detailed interpretation of the DSC curve. Otherwise, the kinetic data calculated will not allow us to predict the behavior of the system at desired thermal conditions. The relatively simple forms of $f(\alpha)$ for a model of an nth-order reaction (Table 7.3) have $n = 0, 1, 2$ or simply 1/2, 2/3 and so on.

By combining Equations 3 and 4, Equation 5 is obtained in logarithmic form:

$$\ln \frac{\frac{d\alpha}{d\tau}}{f(\alpha)} = \ln A + \left(-\frac{\mathrm{E_a}}{\mathrm{R}}\right)\left(\frac{1}{T}\right) \qquad [5]$$

TABLE 7.3
Examples of Kinetic Models of Reactions Used in Thermal Analysis

Name of model	Factor limiting rate of the process	$f(\alpha)$
*n*th-order reaction	Chemical reaction	$(1-\alpha)^n$
Šestak-Berggren	Chemical reaction	$\alpha(1-\alpha)^n$
Johnson-Mehl-Avrami	Chemical reaction, diffusion	$n(1-\alpha)[-\ln(1-\alpha)]^{1-1/n}$
Jander	Diffusion	$3/2(1-\alpha)^{2/3}/[1-(1-\alpha)^{2/3}]$
Ginstling-Brounshtein	Diffusion	$3/2[(1-\alpha)^{-1/3}-1]$

The calculation of the kinetic parameters E_a and A is therefore based on the plotting of a linear dependence $\ln[(d\alpha/d\tau)/f(\alpha)]$ as a function of $1/T$. For measurements under isothermal conditions, the following procedure might be applied: from several DSC curves recorded at different temperatures, the points of constant conversion $\alpha = const$ (for example for 50% conversion) are calculated by comparison of partial and total heat evolved (τ_α as it is shown in Fig. 7.8B), and E_a can be calculated from the following equation:

$$\frac{\Delta \ln \alpha}{\Delta(1/T)} = -\frac{E_a}{R} + \frac{\Delta \ln Af(\alpha)}{\Delta(1/T)} \qquad [6]$$

Induction time τ_{ind} (defined as shown in Fig. 7.2) and time of the maximal rate of the process (i.e., time of the maximal heat flow, τ_{max}) can be easily determined without calculation of ΔH_τ and ΔH_∞. For these two points, the conversion is constant at any isothermal temperature; therefore, τ_{ind} and τ_{max} can be used for the calculation of the kinetic parameters by a simple version of the isothermal method. For example, from several DSC curves of oxidation conducted at various temperatures (Fig. 7.9A), the times of maximal heat flow are determined and logarithms of τ_{max} are plotted as a function of reciprocal of absolute temperature of oxidation, as presented in Figure 7.9B. In this method the value of activation energy is calculated as follows:

$$\frac{d \ln \tau_{max}}{dT^{-1}} = \frac{E_a}{R} + const \qquad [7]$$

This method does not require information concerning the form of $f(\alpha)$.

When measurements are carried out under nonisothermal conditions, [4] is modified to describe the changes of α as a function of a linear increase of temperature β:

$$\frac{d\alpha}{dT} = \frac{1}{\beta}\frac{d\alpha}{d\tau} = \frac{A}{\beta}\exp\left(\frac{-E_a}{RT}\right) f(\alpha) \qquad [8]$$

Equation 8 is widely used in computer simulations to compare the calculated degree of conversion with values obtained experimentally and to verify the kinetic parameters calculated. Separation of the variables in Equation 8 leads to Equation 9:

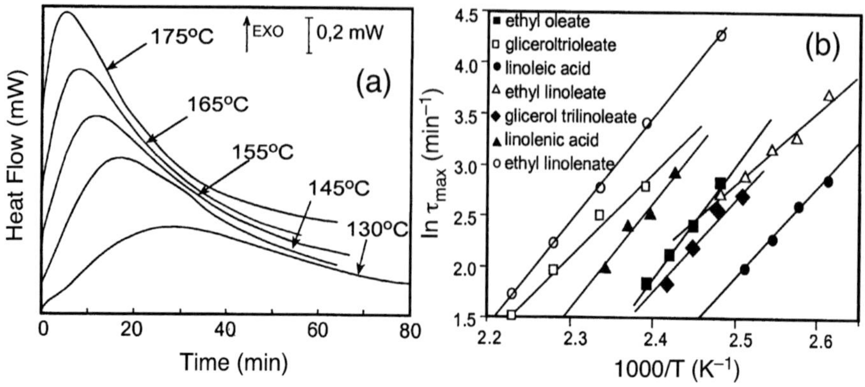

Fig. 7.9. Differential scanning calorimetry curves of isothermal oxidation of ethyl linoleate in the temperature range from 130 to 175°C (panel a). Time of the maximal heat flow, τ_{max}, was determined from each curve. These time points were used to construct the straight line function: $\ln(\tau_{max})$ vs. 1000/T. Panel b shows examples of that function calculated for several unsaturated fatty acids and their esters. *Source:* Litwinienko (2001). Reprinted with permission of Kluwer Academic Publishers.

$$\frac{d\alpha}{f(\alpha)} = \frac{A}{\beta}\exp\left(\frac{-E_a}{RT}\right)dT \qquad [9]$$

From this equation, two kinds of methods, differential and integral, can be evaluated for determination of kinetic parameters. In differential methods the experimental data are put directly into Equation (9), whereas integration of that equation is the basis of the integral methods.

Differential Methods. By taking the logarithm of [9] and assuming one of the forms of $f(\alpha)$, the parameters A and E_a can be calculated from:

$$\ln\frac{\frac{d\alpha}{dT}}{f(\alpha)} = \ln\frac{A}{\beta} - \left(\frac{E_a}{R}\right)\left(\frac{1}{T}\right) \qquad [10]$$

According to the method of Carroll and Manche (1970), a quantity $d\alpha/dT$ is determined for a constant α in a series of experiments carried out for various heating rates (β). The activation energy (E_a) is calculated from the slope of $\ln (d\alpha/dT)_{\alpha=const}$ vs. $(T^{-1})_{\alpha=const}$. In the method of Freeman and Carroll (1958 and 1969), based on Equations 3, 4, and 5, the parameters are $(d\alpha/dT)$, $(1-\alpha)$, and T^{-1}. E_a is calculated from the logarithmic function of the reaction rate for a given $\alpha = const$:

$$\Delta\ln\frac{d\alpha}{dT} = \ln\ f(\alpha) - \ln\ A - \frac{E_a}{R}\Delta\left(\frac{1}{T}\right) \qquad [11]$$

For a given α, the value $f(\alpha)$ is constant and for a kinetic model of nth-order reaction, $f(\alpha) = (1 - \alpha)^n$, Equation [11] will be:

$$\Delta \ln \frac{d\alpha}{dT} = n \ln (1-\alpha) - \ln A - \frac{E_a}{R} \Delta\left(\frac{1}{T}\right) \quad [12]$$

The influence of the mass and the size of the sample, the heating rates, and the range of α on n, E_a, and A calculated in the polymerization and degradation processes was examined carefully by Van Dooren and Müller (1983). The best results were obtained for α between 0.2 and 0.8. The above equations allow the calculation of n, E_a, and A by using $d\alpha/dT$ as a parameter to follow the reaction course. However, the presence of α and $d\alpha/dT$ in the same equation is controversial because of the possibility of autocorrelation (Agrawal 1992). Flynn and Wall (1966) showed that the results obtained by means of the differential methods are scattered and can sometimes lead to completely wrong values.

The integration of Equation 9 gives:

$$g(\alpha) = \int_{T_0}^{T} \frac{d\alpha}{f(\alpha)} = \int_{T_0}^{T} \left[\frac{A}{\beta} \exp\left(-\frac{E_a}{R}\right)\right] dT \quad [13]$$

where T_0 is the initial temperature of the process. Calculation of the integral form of $g(\alpha)$ with substitution $x = E_a/RT$ yields the expression:

$$\frac{1}{\beta} \int_{T_0}^{T} \exp\left(\frac{-E}{RT}\right) dT = \frac{E}{R} \int_{x_0}^{x} \frac{\exp(-x)}{x^2} dx = \left[\frac{\exp(-x)}{x} - \int_{x}^{\infty} \frac{\exp(-x)}{x^2} dx\right]_{x_0}^{x} = p(x) \quad [14]$$

Vand (1943) proposed the notation of the integral $p(x) = [\exp(-x)/x]\ \pi(x)$ where $\pi(x)$ is a dimensionless correction factor usually used in the form of Taylor Polynomials. There are several expressions and semi-empirical approximations of $p(x)$. Agraval (1987) gave the general equation of the integral in the form: $p(x) = \exp(-x)/x^2\ [(1 - 2/x)/(1 - mx^{-2})]$ where $m = 0$ for that method (Coats and Redfern 1965), $m = 4$, 5, and 6 for the methods of Gorbatchev, Agraval, and Lee, respectively. The error of the approximation depends on x; $[(1- 2/x)/(1 - mx^{-2})]$ is a correction factor (CF), and Arrhenius kinetic parameters can be determined from the linearized equation:

$$\ln\left[\frac{g(\alpha)}{T^2}\right] = \ln \frac{AR}{\beta E_a} CF - \frac{E_a}{RT} \quad [15]$$

A plot of $g(\alpha)/T^2$ vs. T^{-1} gives a straight line with a slope $-E_a/R$. The use of CF does not affect the calculated activation energy; however, CF is important for the magnitude of the preexponential factor. For $50 \geq x \geq 20$, the approximation with CF = 1 yields the Kissinger-Akahiro-Sunose method:

$$\ln\frac{\beta}{T^2} = -\frac{E_a}{RT} + const \quad [16]$$

The method was proposed by Kissinger for temperatures at the maximum rate of reaction (T_p) because DTA curves always reach the same constant α at the peak maximum (Kissinger 1957). However, this method was extended by Akahira and Sunose (1971) on other temperatures of constant conversion, $T_{\alpha = const}$. For $x > 20$, the approximation $\log p(x) = -2.315 - 0.4567x$ was proposed by Doyle (1961 and 1965); it was introduced in [13] by Ozawa (1965 and 1970), and, independently, by Flynn and Wall (1966). According to the Ozawa-Flynn-Wall method, for a given degree of conversion, a plot of $\log\beta$ vs. T^{-1} should be a straight line $\log\beta = (-0.4567\ E_a/RT) - 2.315 + \log(AE_a/R)$. Activation energy can be calculated from the slope ($-0.4567\ E_a/R$), and the preexponential factor A can be calculated from the intercept.

Interpretations of the Results. For complex processes, the kinetic analysis obeys the calculation of the kinetic parameters, but a verification of the parameters should also be made by comparing the experimental data (e.g., characteristic temperatures T_e, T_p of DSC curves or changes of α as function of T) with the same data obtained from the theoretical curves predicted from E_a, A, and k with the assumption of a kinetic model of the complex process. Disagreement between the results obtained from experimental and predicted thermoanalytical curves proves that there was a false interpretation of the kinetics.

Currently, limited literature exists about interpreting the shape of a thermoanalytical signal. Studies of the influence of reaction order and kinetic parameters on symmetry of a peak showed that peak width increases with increasing order of the reaction and decreases with increasing E_a (Flynn and Wall 1966, Kissinger 1957). Therefore, drawing a direct conclusion about activation energy and reaction order on the basis of single DSC curve can be misleading.

Only a few works concern theoretical considerations of nonisothermal kinetics of complex (parallel and competitive) processes (Agrawal 1986 and 1988, Criado *et al.* 1988, Flynn 1980, Ozawa 1975 and 1976). It was shown previously that variation in the heating rate can separate some overlapping DSC peaks but only for considerable differences in activation energies of the reactions. Another observation suggests that processes of a lower energy barrier are predominating when the heating rate is low, whereas for higher β reactions of higher E_a are prevailing.

If a complex process is the sum of parallel reactions occurring at comparable rates, an overall activation energy is a mean value of individual reactions. However,

TABLE 7.4
Values of Overall Activation Energy Calculated for Various Degrees of Conversion[a]

α	0.02	0.05	0.1	0.3	0.4	0.6	0.8	0.9	0.95
E_a kJ/mol	204.37	202.59	202.16	204.99	209.03	229.22	249.93	254.59	256.44

[a]*Source:* Opfermann and Kaisersberger (1992). The two-step consecutive process: a → b → c described by parameters: $E_{a \to b}$ = 200.0 kJ/mol, $\log A_{a \to b}$ = 11.6 and $E_{b \to c}$ = 260.0 kJ/mol, $\log A_{b \to c}$ = 15.0 was modeled.

for significant differences in the rates, the overall E_a represents the faster process (Agrawal 1988). Opfermann and Kaisersberger (1992) studied complex reactions using isoconversional methods for calculation of the overall E_a for different degrees of conversion. For a single-step process, the calculated E_a is the same in all ranges of α. A change of E_a with increasing α indicates a more complex mechanism for the observed process. For example, in Table 7.4, values of overall E_a calculated for different α for a two-step process: a → b → c are listed. The results clearly demonstrate a change in kinetic parameters with an increase in the degree of conversion (Opfermann and Kaisersberger 1992).

Analysis of kinetic data to obtain unique kinetic parameters must consider the compensation effect. Because the parameters E_a and A are linked, an increase in the slope (E/R) causes an increase in the intercept and *vice versa* (Exner 1964). As a consequence, one process can be described by more than one pair of E_a and A. Mistakes caused by the compensation effect can be excluded by verifying the kinetic parameters calculated, for example, by comparison of experimental and modeled rates of the process ($d\alpha/dT$) for various heating rates. If the determined kinetic parameters are false, the experimental and calculated curves ($d\alpha/dT$) vs. T will show different sensitivity to the change in β.

The isokinetic temperature is another issue that should always be considered during the analyses of complex processes. Figure 7.10 presents a comparison of

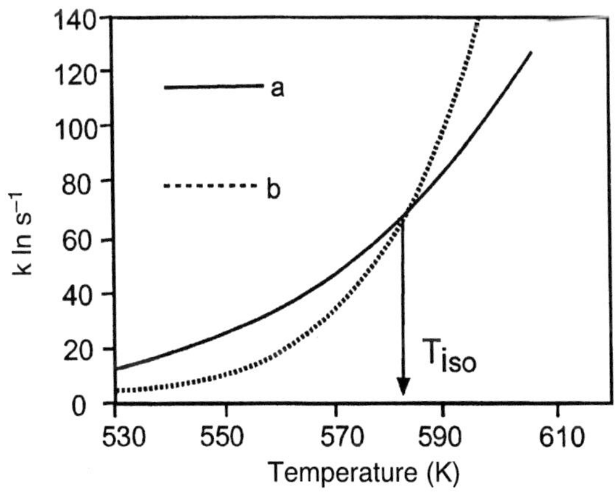

Fig. 7.10. Plots of the first-order rate constants vs. absolute temperature calculated for two reactions: a) E = 80.0 kJ/mol, A = 1.0 × 10^9 s^{-1}, b) E = 150.0 kJ/mol, A = 1.85 × 10^{15} s^{-1}. The arrow indicates the isokinetic temperature, for which the rate constants of both reactions are equal, despite the different activation energies and preexponential factors.

Arrhenius dependencies of $k(T)$ vs. T for two reactions. The isokinetic temperature is the temperature at which rate constants for both processes are the same, i.e., $A_1\exp(-E_1/RT_{iso}) = A_2\exp(-E_2/RT_{iso})$. For temperatures $<T_{iso}$, the low E_a reaction is faster. At a temperature above the isokinetic temperature, the high E_a reaction is faster. Thus, the heating rate can be the parameter determining the shape and range of the peaks during the analysis of complex processes by a nonisothermal DSC method. For competitive processes, one of the reactions may predominate for lower β and another process may be faster for higher β. For multistep processes, various reactions can be a rate-limiting step, depending on T, β, and, of course, on value of T_{iso}.

A number of kinetic and analytical calculations can be made using the commercially available software that is usually supplied with thermoanalytical equipment. The commercial software is useful for the determination of the heat capacity, calculations of purity and enthalpy, and for determining the reaction rate and rate constants. However, it should be stressed that any application of commercial software without understanding the mechanism of the process studied and without knowledge of the kinetic model used for these calculations can lead to mistakes and artifacts. One of the most common errors is an incorrect interpretation of a DSC curve obtained for complex processes, mainly because the overlapping peaks are difficult to distinguish from each other. Results of DSC analyses are often presented mechanistically without deeper reflection concerning the compensating effects and without calculation of the isokinetic temperature. Similarly, kinetic parameters may be incorrectly assigned due to misinterpretation of the kinetic mechanism of the complex process studied.

Thermoanalytical Investigations of Lipid Autoxidation

Kinetic Description of Autoxidation Process

Thermogravimetry, differential thermogravimetry, DTA, and DSC can be applied in the studies of autoxidation and oxidative decomposition processes. Some specific problems can be solved by the use of TA methods only (Adonyi 1972). Application of TA in studies of oxidation is possible if some kinetic assumptions are made. Therefore, in the first part of this section, some remarks on the mechanism of autoxidation will be presented to explain the methodological basis of TA applicability as a valuable tool for oxidation research.

Autoxidation is a chain process that can be described by the following sequence of reactions:

Initiation	Initiator	$\rightarrow$	$R^\bullet$	R_i; k_i	[17]
Propagation	$R^\bullet + O_2$	$\rightarrow$	$ROO^\bullet$	k_{p1}	[18]

	$ROO^{\bullet} + RH$	$\rightarrow$	$R^{\bullet} + ROOH$	k_{p2}	[19]
Termination	$ROO^{\bullet} + ROO^{\bullet}$	$\rightarrow$	$(ROO)_2$	$v_{t1}; k_{t1}$	[20]
	$R^{\bullet} + ROO^{\bullet}$	$\rightarrow$	R-OO-R	$v_{t2}; k_{t2}$	[21]
	$R^{\bullet} + R^{\bullet}$	$\rightarrow$	R-R	$v_{t3}; k_{t3}$	[22]

where R_i is a rate of initiation and v_i, k_i denote rate and rate constant of i-th reaction, respectively. If a source of free radicals is the initiation process only and a termination is a chain-breaking process only (i.e., for unbranched chain reactions), the stationary state is reached. Thus, the rate of initiation (R_i) and a sum of rates of terminations are equal

$$R_i = \Sigma(v_t)_i \qquad [23]$$

and if the kinetic chains are long enough, the ratio of alkyl to peroxyalkyl radicals concentration is:

$$\frac{[R^{\bullet}]}{[R_2^{\bullet}]} = \frac{k_{p2}[RH]}{k_{p1}[O_2]} \qquad [24]$$

For low oxygen pressure, the ratio $[R^{\bullet}]/[ROO^{\bullet}]$ increases and [23] could be presented as follows:

$$R_i = 2\,k_{t1}[ROO^{\bullet}]^2 + 2\,k_{t2}\,[R^{\bullet}]\,[ROO^{\bullet}] + 2\,k_{t3}\,[R^{\bullet}]^2 \qquad [25]$$

The rate of oxidation can be described as:

$$v = \frac{k_{p1}k_{p2}[RH][O_2]\sqrt{R_i}}{\sqrt{2k_{t1}k_{p1}^2[O_2]^2 + 2k_{t2}k_{p1}k_{p2}[O_2][RH] + 2k_{t3}k_{p2}^2[RH]^2}} \qquad [26]$$

For relatively high oxygen pressure (>13 kPa), the first term in the square root dominates considerably over the others:

$$k_{t1}k_{p1}^2[O_2]^2 >> k_{t2}k_{p1}k_{p2}[O_2][RH] + k_{t3}k_{p2}^2[RH] \qquad [27]$$

The rate of oxidation [26] can be simplified to the form:

$$v = k_{p2}\sqrt{\frac{R_i}{2k_{t1}}}[RH] \qquad [28]$$

The abstraction of a hydrogen from a lipid is a process limiting the oxidation rate, and the autoxidation is a first-order process with respect to lipid. According to this equation, a determination of any parameter related to ν can be used for monitoring the oxidative stability of lipids. Therefore, the heat evolved during oxidation and the changes in the mass of the oxidized system are valuable tools with which to follow the course of oxidation. During the initial stage of oxidation, [RH] is assumed to be constant and the rate of initiation R_i is effectively constant. Therefore,

$$k_{p2}\sqrt{\frac{R_i}{2k_{t1}}} = k = const \qquad [29]$$

and k is the global (overall) first-order reaction rate constant (Garcia-Ochoa *et al.* 1989, Jensen *et al.* 1981).

During the progress of oxidation, the hydroperoxides decompose to ketones, alcohols, and fatty acids:

$$RCH(OOH)R_1 \rightarrow R(C{=}O)R_1 \qquad k_{d1} \qquad [30]$$

$$RCH(OOH)CH_2R_1 \rightarrow RCOOH + R_1COOH \qquad k_{d2} \qquad [31]$$

$$ROOH \rightarrow RO^{\bullet} + {}^{\bullet}OH \qquad k_{d3} \qquad [32]$$

$$RO^{\bullet} + RH \rightarrow ROH + R^{\bullet} \qquad k_{d4} \qquad [33]$$

Because $k_{d3} << k_{d4}$, the rate of formation of alcohols (reaction [33]) is determined by the rate of the process [32]; thus, the overall decomposition of hydroperoxides (reactions [30]–[32]) is assumed to be a first-order process (Blaine and Savage 1992):

$$\nu_{decomp} = (k_{d1} + k_{d2} + k_{d3})\,[ROOH] \qquad [34]$$

In conclusion, both the formation of peroxides and their decomposition are first-order processes, according to Equations 28 and 34, respectively. Several methods are based on this relatively simple description of overall autoxidation kinetics. Studies done as early as in the 1960s and 70s showed a real possibility of TA as a group of methods that potentially could replace many expensive or time-consuming classical methods of analysis.

Studies of Oils and Fats Autoxidation by Thermogravimetry

Nieschlag and co-workers (1977) made use of a commercial thermogravimetric system for continuous monitoring of mass changes in samples oxidized in nonisothermal mode. The optimal mass of the samples was 1.5–2 mg and volatilization during isothermal stabilization at baseline (before the start of temperature program, 5 K/min) was <0.2% of the starting weight. Although the results were very limited, the authors concluded that TG parameters, especially T_i (temperature at the start of weight gain, i.e., permanent positive change from baseline, which indicates the end of the induction period), are more sensitive indicators of sample conditions than measurements of conjugate diene absorbance at 233 nm.

The use of derivatography (thermogravimetric equipment coupled with DTA) for assessing the oxidative stability of sunflower and rapeseed oils confirmed the applicability of these methods for studies carried out under isothermal and non-isothermal conditions (Buzas *et al.* 1977, 1978, 1979). Figure 7.11 presents exam-

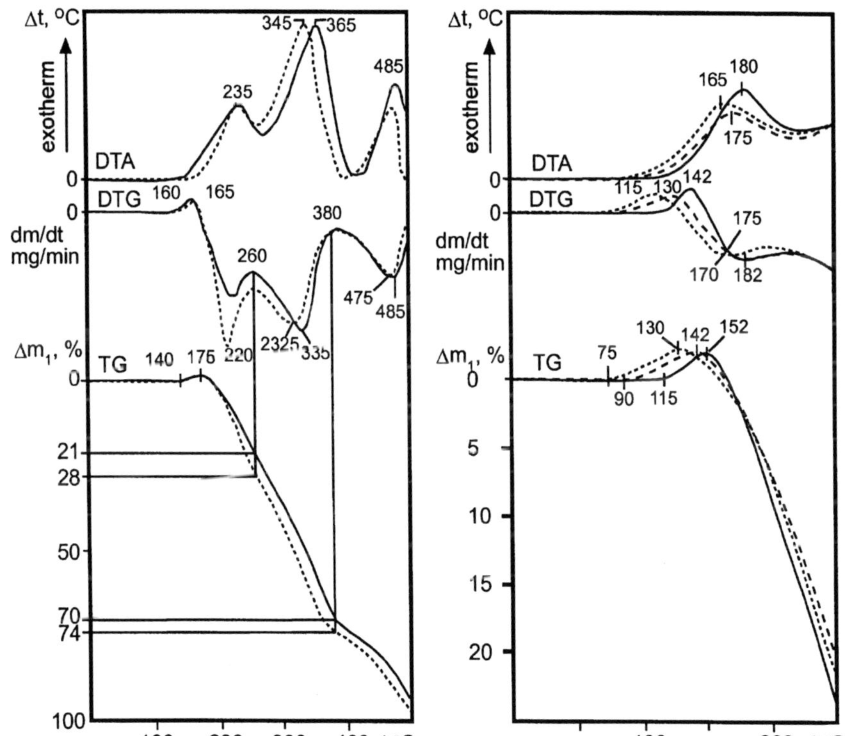

Fig. 7.11. Thermogravimetric (TG), differential thermogravimetric (DTG), and differential thermal analysis (DTA) curves of nonisothermal oxidation and decomposition of rapeseed oil. *Source:* Buzas *et al.* (1979). Reprinted with permission of American Oil Chemists' Society. Left panel: decomposition of fresh sunflower (solid line) and rapeseed oil (dashed line). Heating rate 5 K/min. Right panel: decomposition of fresh (solid line) and thermally treated (fried) sunflower oils. Heating rate 2.5 K/min. Dashed and dotted lines denote fried oil used 10 and 20 times, respectively.

ples of TG, DTG, and DTA curves of nonisothermal oxidation reprinted from their work. Buzas and co-workers (1977, 1978, and 1979) distinguished three steps in the process. In the first step, the oxidation of the oil manifests as a weight increase (>140°C with a maximum DTA peak at 160°C) followed by the degradation process recorded as a weight decrease in the temperature range from 175 to 260°C. The next step occurs within the temperature range 260 to 380°C. Above this temperature, the complete decomposition of the sample takes place. In conclusion, the isothermal method for rapid indication of the oxidative stability was proposed with monitoring of the following parameters: length of induction period, time of maximum weight gain, time of maximum rate of weight increase, and time of maximum heat flow (Buzas *et al.* 1979).

Similar methods were applied to investigate the oxidative stability of wax esters (Hagemann and Rotfus 1979). In that study, apart from T_i and weight gain (in %), the rate of oxygen uptake was calculated in μg O_2/°C. The oils studied were sperm whale oil, jojoba oil, several wax esters, and behenyl arachidate. More recently, good correlations of T_i with peroxide values (PV) and with results of the Rancimat method were reported for TG studies of oxidative stability of inhibited and noninhibited linseed oil (Rudnik *et al.* 2001). Gennaro and co-workers (1998) applied thermogravimetry as a tool to monitor the autoxidation of olive oil inhibited by chain-breaking antioxidants: 2,6-di-*tert*-butyl-4-methylphenol (BHT) and 2,6-di-*tert*-butyl-4-methoxyphenol (BHA), and some natural phenols. The initial temperature and the temperature of maximal weight gain were measured during the heating of the samples with $\beta = 2$ K/min. Another example of TG application is continuous monitoring of thermal oxidation of a thin film (17–310 μm) of unsaturated triacylglycerols (Takaoka *et al.* 1994).

Wesolowski and his group presented another approach to TG of industrial (Wesolowski 1985, 1986c, and 1986d, Wesolowski *et al.* 1998), edible (Wesolowski 1986b, 1987a, and 1987b), and pharmaceutical oils (Wesolowski 1986a) under nonisothermal oxidative conditions. The temperatures of 1, 5, 15,....100% oxidative decomposition of oils during the heating in derivatograph were correlated with physicochemical properties of the oils studied such as viscosity, water contents, ignition temperature, refractometric index, and acid number, saponification number, and iodine number. In his studies, principal component analysis was applied to find the relation between the chemical and thermoanalytical variables as was shown for rapeseed oil (Wesolowski and Erecinska 1998).

Depending on the phenomenon analyzed (increase or decrease of the mass), results using the TG method can be compared with two classes of conventional methods. Monitoring of the weight increase makes the TG technique similar to methods based on peroxide number (PN) measurements and oxygen consumption methods. Other TG methods, based on the measurements of oxidative degradation, can be compared with conventional methods based on the determination of volatile degradation products. The advantages of the TG method over these techniques [active oxygen method (AOM), oxygen bomb, oven test] include the much shorter time of analysis, the small sample amount (3–12 mg) required, good precision, and continuous monitoring of the oxidation process. All of these advantages make the TG method more convenient than studying oxidation by periodic measurements of oil sample weight during the oxidation course.

Thermogravimetry is a useful tool in determining not only qualitative parameters but also quantitative kinetic parameters. Measurements of oxidation rate, kinetic chain length, and oxidizability parameter:

$$O_x = k_{p2}/\sqrt{2k_{t1}} \qquad [35]$$

where k_{p2} and k_{t1} are the rate constants of propagation and termination (reactions [18], [20], and Equation 28), respectively, gave results comparable to literature data (Litwinienko and Dabrowska 2001). Calculated kinetic chain lengths (ν) during the induction period for ethyl linoleate oxidation ranged from >2700 (at 35°C) to >3900 (at 50°C) and the rate of the oxidation was 1.76×10^{-5} mol dm^{-3} s^{-1} (35°C). The rate of the oxidation was three orders of magnitude lower and ν was two orders of magnitude lower for the oxidation inhibited by hindered phenols. However, as shown in Figure 7.12, the isothermal TG experiments at temperatures between 35 and 70°C for the single measurement require >1 h.

Studies Using Isothermal DSC Methods

The first DSC study of fats and oils oxidation was conducted by Cross (1970). The parameter directly determined from the thermoanalytical curve was induction time (τ_{ind}), defined as the extrapolated time of the start of the exothermal oxidation effect. The induction times were correlated with results of the AOM, but they were too scattered to be proposed as a method replacing AOM using a single DSC experiment. This difficulty disappeared when pressure differential scanning calorimeter (PDSC) was applied to oxidation studies (Hassel 1976). Correlation of PDSC with AOM was 0.95, and τ_{ind} was shorter than for a typical AOM analysis, varying from 20 to 140 min depending on the temperature of isothermal oxidation.

Nonisothermal pressure DTA was used to assess the oxidative stability of palmitic (18:0), oleic (18:1), linoleic (18:2), and linolenic (18:3) acid methyl esters and their triglycerides (Yamazaki *et al.* 1980). The induction times determined correlated with the number of double bonds in the carbon chains of the acyl groups. For the same methyl esters of 18:0, 18:1, 18:2, and 18:3 acids, and several plant oils, Raemy *et al.* (1987) used isothermal DSC in the temperature range 80–160°C. Because τ_{ind} was not detected for 18:2 and 18:3 at temperatures >100°C, the time of maximal heat flow (τ_{max}) was used to assess the oxidative stability of the esters. Additionally, experiments carried out under an atmosphere of oxygen-free argon gave straight calorimetric lines without any thermal effects of evaporation, decomposition, polymerization, or isomerization, indicating that the effects that could alter the shape of the DSC oxidation curve were not detected.

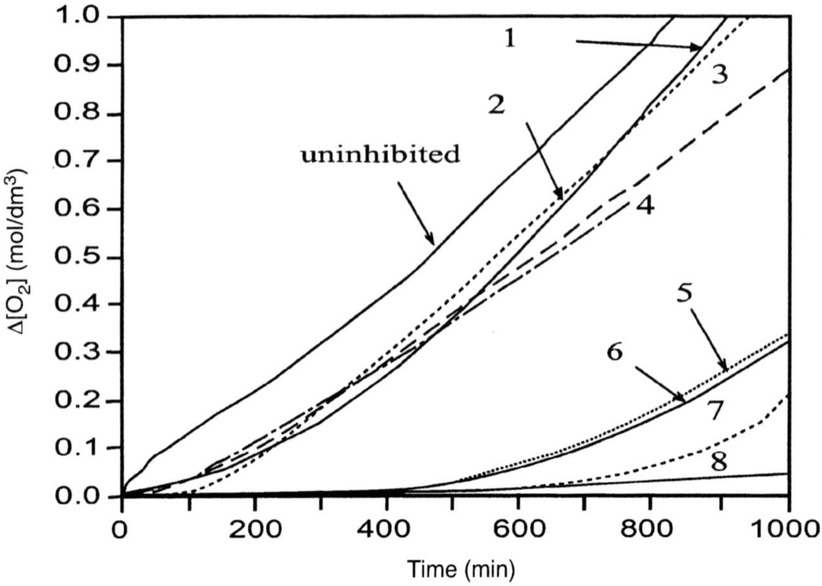

Fig. 7.12. Plots of oxygen adsorption $\Delta[O_2]$ vs. time during noninhibited and inhibited oxidation of ethyl linoleate at 40°C. Concentration of inhibitors 0.001 M, concentration of initiator [α,α′-azoisobutyronitrile (AIBN)] = 0.04 M. Numbers denote: (1) 2-hydroxyphenylacetic acid, (2) α-tocopherol, (3) β-carotene, (4) 2-hydroxyacetophenone, (5) 2,2′-methylene-bis-(4-methyl-6-*tert*-butylphenol), (6) 2-*tert*-butyl-6-methylphenol (BMP), (7) caffeic acid and (8) 2,6-di-*tert*-butyl-4-methylphenol. *Source:* Litwinienko and Dabrowska (2001). Reprinted with permission of Kluwer Academic Publishers.

Several plant oils were analyzed with respect to τ_{ind} measured by a combination of dynamic and isothermal DSC (Pereira and Das 1990). The samples analyzed were initially heated with β = 20 K/min up to 170°C, and then kept under isothermal conditions. That procedure prevented the occurrence of the uncontrolled jump in temperature from ambient to 170°C. Recently, a comparison of results of the studies of oxidative stability of 12 edible oils by isothermal DSC with the results using oxidative stability index method (OSI) showed good correlation between tind and OSI values (Tan *et al.* 2002).

An advantage of TA over accelerated tests is not only the possibility to determine τ_{ind} as a parameter describing oxidative stability of studied lipid system, but also to calculate Arrhenius kinetic parameters of oxidation. Historically, DSC was used for determination of the kinetic parameters of oxidation of hydrocarbons, because hydrocarbon oxidation was more intensively studied in the past. A summary of the analysis of petrochemical product oxidation carried out with the use of TA techniques before 1980 was reviewed by Wesolowski (1981). More recent reviews summarizing the general applications of TA include only a few references to the autoxidation studies (Dollimore 1994, 1996, Dollimore and Lerdkanchanaporn 1998).

Analysis of these early studies of hydrocarbon autoxidation might be important to understand the development of TA methods for determining kinetic parameters for lipid oxidation. These early DSC studies of isothermal autoxidation of long-chain saturated hydrocarbons showed that induction time is proportional to increasing concentration of the inhibitor (Noel and Cranton 1974). These authors, taking into account earlier works of Barret (1967) and Duswalt (1968), proposed simpler notation of the effective (overall) rate constant of autoxidation (k) containing rate constants of the initiation, propagation, and termination steps (see [29]). Assuming that an effective process follows first-order kinetics, the thermal effect observed (a sum of all processes occurring at the moment) can be described by the equation:

$$-\ln\left(\frac{H_T - H_\tau}{H_T}\right) = -\ln\left(\frac{P_T - P_\tau}{P_T}\right) = k\tau \qquad [36]$$

where H_T is the total heat of the process, P_T is the total area under DSC curve, and $H\tau$ is the heat of the process released to time τ proportional to the area P_T under the part of the DSC peak from the start of the process to time τ. Values of k calculated from isothermal experiments were applied to construct a plot of lnk vs. $1/T$. The overall activation energy, calculated in that way, was 26.5 kcal/mol for noninhibited autoxidation and agreed with E_a in the literature for simple saturated hydrocarbons. Similar work conducted by Cranton (1976) confirmed the applicability of the TA method for measurement of the rate constants of inhibited systems. Studies on the role of oxygen pressure in thermal effects and kinetics of nonisothermal oxidation of liquid hydrocarbons showed that some peaks overlapping under low p_{O2} can be separated for higher partial pressure of O_2 (Vossoughi and El-Shoubary 1990). The methodological aspects of hydrocarbon oxidation presented are valid for studies of lipid autoxidation.

DSC methods based on the monitoring of heat released during isothermal oxidation combine the classical accelerated tests and the calorimetric method of control of a reaction course. This compilation allows us to determine induction time but also the time in which an oxidized system reaches a known, desired degree of conversion. The time of extrapolated onset, τ_{on}, and the time at which oxidation occurs with the maximal rate (time of the maximal heat flow, τ_{max}) correspond to the constant α (at various temperatures). Therefore, both of these factors can be used to calculate kinetic parameters (Kowalski 1989 and 1992). PDSC studies on the oxidation of soybean, rapeseed, corn, and sunflower oils showed that experimental temperatures and oxygen pressure determine the shape of the isothermal calorimetric curve (Kowalski 1989). Parameters τ_{on} and τ_{max} depend on oxidative stability, but of these two, only τ_{on} is practically independent from the mass of the oxidized sample. Therefore, if τ_{max} is to be used for calculation, the approximately constant mass of the samples should be used in one set of measurements.

In isothermal methods, the kinetic parameters are calculated on the basis of the assumptions and equations discussed earlier in this chapter, with kinetic model $f(\alpha)$ for an nth-order reaction (Noel and Cranton 1974):

$$\frac{d\alpha}{d\tau} = \frac{1}{H_T}\frac{dH}{d\tau} = \mathrm{k}\left(\frac{H_T - H_\tau}{H_T}\right)^n = Ae^{-\mathrm{E/RT}}(1-\alpha)^n \qquad [37]$$

After separation of the variables and integration, the linear Arrhenius-type functions are obtained:

$$\ln(\tau_{max}) = aT^{-1} + b \qquad [38]$$

$$\ln(\tau_{on}) = a_1T^{-1} + b_1 \qquad [39]$$

where T is temperature in K and a, b, a_1, and b_1 are parameters of linear equations. Activation energy of autoxidation is calculated from the slopes (= E_a/R) of the straight lines. Examples of activation energy values for isothermal oxidation of some edible oils are listed in Table 7.5. A comparison of oxidative behavior of several edible and pharmaceutical oils based on the parameters τ_{on} and τ_{max} and on calculated values of E_a led to an interesting observation, i.e., the range of relative oxidative stabilities at temperatures <110°C was different from the sequence for higher temperatures (Kowalski 1989). However, the kinetic parameters calculated from the times τ_{on} and τ_{max} gave (in the temperature range 70–140°C) rate con-

TABLE 7.5
Literature Values of Activation Energy, E_a, for Isothermal Oxidation of Several Edible Oils

Oil	E_a (kJ/mol)	Reference
Soybean	67.2	Kowalski 1989
	80.8	Tan *et al.* 2001c
Sunflower	84.6	Kowalski 1989
Corn	86.2–98.6	Kowalski 1989, Kasprzycka-Guttman *et al.* 1994, Tan *et al.* 2001a
Linseed	56.8–57.2	Kasprzycka-Guttman *et al.* 1994; Kasprzycka-Guttman and Odzeniak 1993
Canola	86.0	Tan *et al.* 2001c
Castor	85.7-89.7	Kasprzycka-Guttman *et al.* 1994, Kasprzycka-Guttman and Odzeniak 1993
Olive	59.8	Kasprzycka-Guttman *et al.*, 1994, Kasprzycka-Guttman and Odzeniak 1993
Cod-liver	75.2	Kasprzycka-Guttman and Odzeniak 1993
Peanut	68.2	Kasprzycka-Guttman and Odzeniak 1994,
	99.1	Tan *et al.* 2001c

stants that allowed the prediction of such inversions of oxidative stability. This peculiarity can be explained if one realizes that for the oils studied, the isokinetic temperature (see above in this chapter) was ~100°C. Thus, above and below T_{iso}, the systems studied have different relative oxidative stabilities. This observation demonstrates the drawbacks of classical accelerated tests in which induction times are determined at high temperatures, whereas their results are usually extrapolated to lower temperatures. In contrast, kinetic parameters calculated by means of the PDSC method allow us to predict the rate of oxidation over a wide temperature range.

Values of τ_{max} and rate constants of oxidation of several plant oils were compared with the fatty acid composition of these fats (Kowalski *et al.* 1993). This comparison showed that above 100°C, the range of oxidative stability agreed with the iodine number. Similar correlation studies for parameters τ_{on} and τ_{max} with PN gave linear dependence for PN ≤30 mmol O_2^{2-}/kg (Kowalski *et al.* 1997). Times τ_{on} and τ_{max} are more reliable than measurements of PN because for advanced stages of autoxidation, a decrease in PN is observed due to decomposition of peroxides. Such an observation is also applicable for reprocessed oils, i.e., oils heated under N_2 to decrease PN.

The pressure DSC method is relatively fast, the experimental conditions are easy to repeat, results are acquired with good precision, and the kinetic parameters can be extrapolated to lower temperatures. However, parameter τ_{on} is not recommended for PDSC measurements of noninhibited oxidation of oils because oxidation may occur during equilibration, before a constant temperature is reached. Thus, τ_{on} can be misleading and usually is too short to be determined. For such easily oxidizable systems, more reliable results can be obtained with nonisothermal (dynamic) DSC.

Modeling of the kinetics of isothermal oxidation was the subject of a few studies in which the degree of conversion was calculated from Equation 1 (see Fig. 7.8). Differentiation of α with respect to the time and introduction of obtained $d\alpha/d\tau$ to Arrhenius Equation 3 gives simple dependence of the rate of oxidation as a function of α. Calculations performed for sunflower oil showed that for low degrees of conversion ($\alpha \leq 0.16$), the best agreement of experimental data was observed for the model rate equation $d\alpha/d\tau = (k_a\alpha + k_c)(1 - \alpha)^2$, where k_a and k_c are the rate constants of catalytic and autocatalytic process, respectively, and k_c was ~4–10 times lower than k_a (Kowalski 1992). Values of activation energy calculated for this model were consistent with values for E_a calculated from parameters τ_{on} and τ_{max} measured directly from PDSC exotherms. For $0.2 \leq \alpha \leq 0.5$ the autocatalytic model gave the best fit to experimental data. This difference between kinetic models for lower and higher degrees of conversion was interpreted as a consequence of the change of the kinetics and reaction order, perhaps due to increasing concentration of hydroperoxides and decreasing concentration of nonoxidized lipid (Kowalski 1992). In other papers (Kasprzycka-Guttman *et al.* 1994, Kasprzycka-Guttman and Odzeniak 1994), experimental data for the autocatalytic model were used for direct calculation of E_a from the equation: $\log(d\alpha/d\tau) = \log k + n \log[(1 - \alpha)\ \alpha^{m/n}]$. Similar modeling of the kinetics for noninhibited oxidation of linseed, olive, castor, and cod-liver oil confirmed the autocatalytic model of isothermal oxidation (Kasprzycka-Guttman and Odzeniak 1993).

Nonisothermal (Dynamic) DSC Measurements

In a typical, nonisothermal experiment, the sample is heated at a programmed linear heating rate β. Because various physical and chemical changes (e.g., melting, evaporation, polymerization) may occur when the temperature increases, it is crucial to be sure that thermal effect does not interfere with the processes of oxidation studied. Nonisothermal measurements of heat capacities (C_p) performed for a number of liquid fats (under oxygen-free atmosphere) proved that C_p does not vary significantly with increasing temperature (from 70 to 140°C), meaning that no effect of baseline disturbance occurs (Kowalski 1988a, Kasprzycka-Guttman and Odzeniak 1991). In similar experiments under N_2, but over a wider temperature range (80–350°C), the endothermic effects that might be caused by evaporation and decomposition were not found (Kowalski 1988b, Raemy *et al.* 1987). Also, the possibility of self-ignition under normal conditions, i.e., nonisothermal mode with β <25 K/min and oxygen atmosphere, was excluded. For the majority of fats, more aggressive, nonstationary conditions (40 K/min < β <90 K/min, oxygen pressure 800–2800 kPa) are required for spontaneous ignition to occur (Kowalski 1990, Raemy *et al.* 1983, 1985, Raemy and Loeliger 1985). Therefore, all of the studies described above showed that the heat flow monitored for lipid systems in an oxygen atmosphere, for heating rates <40°K/min, can be interpreted as demonstrating the thermal effect of oxidation. Similarly, in isothermal experiments, thermoanalytical procedures for the determination of kinetic parameters of oxidation can be applied after simplification of the formal description of the process and introduction of the overall rate constant (see Equation 29).

In the first nonisothermal DSC studies, calculations of the kinetic parameters of lipid oxidation were based on temperatures of the maximal heat flow (T_p) determined from the oxidation curves. This method was based directly on the hypothesis (valid for simple one-step reactions) that peak maximum serves as an easy to measure point at which the reacting system always reaches the same degree of conversion (Duswalt 1968, Kissinger 1957, Ozawa 1965 and 1970). The Kissinger method and Ozawa-Flynn-Wall method (see above) were used for calculation of the parameters E_a and A for various oils and fats. Some examples of the parameters are presented in Table 7.6. The activation energies of oxidation determined were in good agreement with those given for autoxidation of linoleic acid and its esters and ranged between 15.2 and 17.2 kcal/mol (63.7–72.1 kJ/mol) determined by the manometric method (Korycka-Dahl and Richardson 1980). However, disagreement was observed between the values of E_a calculated from temperatures

TABLE 7.6
Values of E_a and logA for Oxidation of Selected Fats and Oils Determined from Nonisothermal Differential Scanning Calorimetry Measurements on the Basis of Parameters T_p and T_e

Oil	E_a (kJ/mol)	logA (min^{-1})	Reference
Rapeseed	62.3–73.4	5.69–6.66	Kowalski 1988, 1991a
Soybean	62.8–63.6	5.65–6.07	Kowalski 1988, 1991a, Kowalski and Kot, 1989
	79.66	8.6	Adhvaryu *et al.* 2000[a]
Sunflower	59.8–64.0	5.56–6.14	Kowalski 1988, Kowalski and Kot, 1989
	63.8	7.06	Adhvaryu *et al.* 2000[a]
High-oleic sunflower	86.5	9,26	Adhvaryu *et al.* 2000[a]
Corn	47.4	4.27	Kasprzycka-Guttman and Odzeniak, 1992
	77.78		Adhvaryu *et al.* 2000[a]
Linseed	56.3 (44.7[b])	5.80	Kasprzycka-Guttman and Odzeniak, 1992
Cottonseed	63.30	6.7	Adhvaryu *et al.* 2000[a]
Safflower	75.27	8.28	Adhvaryu *et al.* 2000[a]
High-oleic safflower	88.70	9.50	Adhvaryu *et al.* 2000[a]
High-linoleic safflower	73.51	8.03	Adhvaryu *et al.* 2000[a]
Castor	77.0 (87.2[b])	6.60	Kasprzycka-Guttman and Odzeniak 1992
Olive	74.9 (84.8[b])	4.30	Kasprzycka-Guttman and Odzeniak 1992
Cod-liver	59.7 (64.4[b])	4.00	Kasprzycka-Guttman and Odzeniak 1992
Animal fat	90.3–93.4	8.33–8.57	Kowalski 1991[a]

[a]Calculated from temperatures T_e.
[b]Calculated by the Kissinger method; other values calculated by the Ozawa-Flynn-Wall method.

T_{p2} (temperature of the greatest peak on the DSC curve) and the values E_a obtained from temperatures T_e (Litwinienko *et al.* 1995). Thus, the problem arose of interpreting these inconsistent results. One of the first interpretations of a nonisothermal DSC curve of lipid oxidation was made by Kaisersberger in his study on the oxidation of several edible oils and fats (Kaisersberger 1989). A comparison of the DSC oxidation profile of highly unsaturated oil (sunflower oil) with saturated fat (hardened coconut containing only 1% of linolenic acid) led to the hypothesis that the exothermic peak in the range of 150–220°C "has to be related to the oxidation of unsaturated fatty acids" (Kaisersberger 1989). That interpretation was accepted by other researchers despite the fact that Kaisersberger's hypothesis was not confirmed either by an experiment or by modeling of the nonisothermal oxidation. Moreover, in some works, the further oxidation peaks on DSC curve (i.e., occurring at higher temperatures) were mechanistically assigned to the oxidation of the saturated components of fats.

Interpretation of the DSC Curve of Nonisothermal Oxidation

The problem of interpretation of DSC was solved by the modeling of the nonisothermal DSC curve and a comparison of experimental and theoretical data (Litwinienko and Kasprzycka-Guttman 1998a). Among the complex reaction schemes, e.g., competitive, multiple (parallel), sequential, and sequential with autocatalytic start of the process, the best fit of experimental and predicted degrees of conversion was obtained for the last-mentioned, i.e., for a process of general scheme:

$$a \xrightarrow{b} b \longrightarrow c \qquad [40]$$

where the first step, $a \rightarrow b$, is the process catalyzed by b. In addition, an excellent agreement was achieved between the calculated and experimentally determined characteristic temperatures, T_e, T_{p1}, and T_{p2} only for this mechanism. According to this scheme, the first observed exothermal effect is caused by the autoxidation of the lipid. In other words, the first exothermal peak on the DSC curve is assigned to the formation process of peroxides. The second peak results from the decomposition of the peroxides to further products. The above interpretation was confirmed experimentally by several other studies including oxidation of partially oxidized oils (Litwinienko 2001). Figure 7.13 shows DSC oxidation curves of linseed oil containing various initial concentrations of peroxides formed during the isothermal preheating in air.

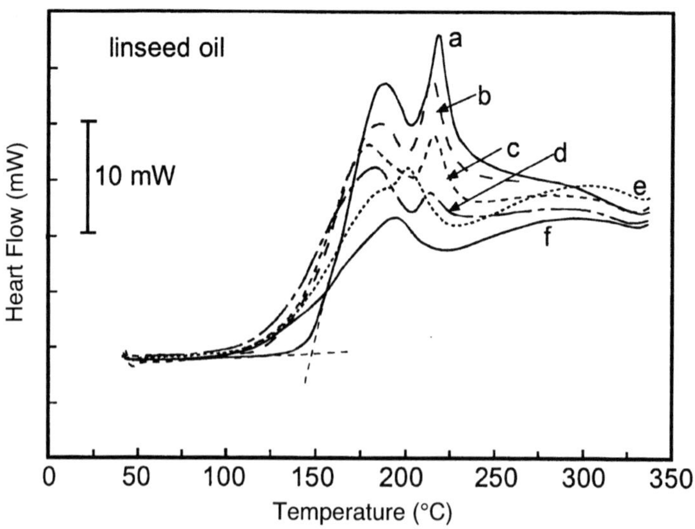

Fig. 7.13. Differential scanning calorimetry curves of nonisothermal (with heating rate 10 K/min) oxidation of linseed oil supplemented with various concentrations of peroxides. Peroxide numbers are: (a) 31.3, (b) 119.8, (c) 180.3, (d) 252.7, (e) 349.9 and (f) 383.3 mmol O_2/kg.

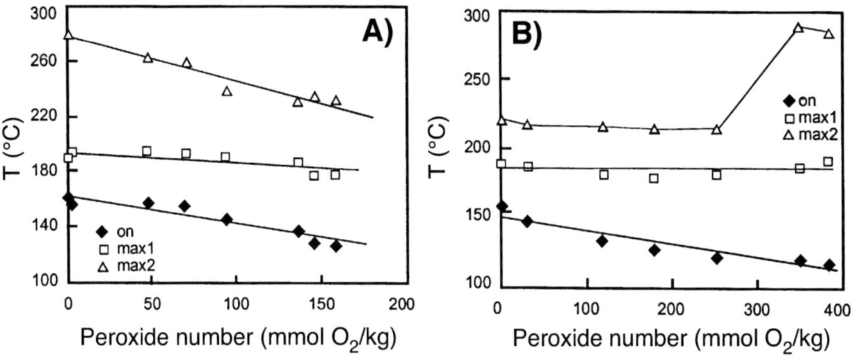

Fig. 7.14. The influence of the initial peroxide content on temperatures of extrapolated onset point (on), and temperatures of first and second peak (max1 and max2, respectively) for: (A) corn oil, (B) linseed oil. *Source:* Litwinienko (2001). Reprinted with permission of Kluwer Academic Publishers.

Indeed, the start of the oxidation process is accelerated by increasing amount of peroxides and, as can be seen in Figure 7.14, the extrapolated onset temperatures are correlated with the initial PN of the oxidized lipid. The next argument for the validity of the proposed kinetic model is a comparison of the activation energies calculated for the initial stages of the nonisothermal oxidation with values of E_a determined for isothermal oxidation (the isothermal process proved to be autocatalytic and only one exothermal peak of autoxidation is observed). The values for some unsaturated fatty acids and their esters are listed in Table 7.7; for each lipid analog listed, both E_a values are coherent. It strongly suggests that only for

TABLE 7.7
Comparison of Activation Energies of Isothermal and Nonisothermal Oxidation of Unsaturated Fatty Acids and Their Esters

	E_a (kJ/mol)	
Lipid	Isothermal[a]	Nonisothermal[b]
Oleic acid	90.6 ± 5.2	89.6 ± 4.4
Ethyl oleate	85.5 ± 1.1	88.4 ± 4.7
Glycerol trioleate	85.1 ± 13.0	95.0 ± 4.7
Erucic acid	79.6 ± 11.5	91.8 ± 13.3
Linoleic acid	72.9 ± 8.5	72.0 ± 2.9
Ethyl linoleate	67.6 ± 6.5	76.4 ± 5.0
Glycerol trilinoleate	52.1 ± 7.1	74.3 ± 3.0
Linolenic acid	59.9 ± 6.5	62.4 ± 3.7
Ethyl linolenate	73.5 ± 2.5	74.5 ± 8.2

[a]Litwinienko (2001).
[b]Litwinienko and Kasprzycka-Guttman (2000).

the initial stage of nonisothermal oxidation (first peak) are the kinetics the same as for isothermal oxidation. Analogously, the same conclusion must be stated if a comparison of the isothermal and "initial nonisothermal" E_a for saturated fatty acids and their esters is considered on the basis of Table 7.8.

The overall activation energies and overall rate constants for oxidation of saturated lipid analogs calculated on the basis of temperatures T_e were compared with data in the literature for hydrocarbon oxidation in the liquid phase obtained by other methods. This comparison revealed good agreement between these two methods of analysis (Litwinienko *et al.* 2000). In the same work, the kinetic data for nonisothermal oxidation were confirmed by gas chromatography (GC) measurements of the reaction rates for isothermal oxidation of saturated fatty acid. Thus, from several characteristic temperatures of a nonisothermal DSC curve, only these related to the first exothermal peak (the start and maximum heat flow) corresponded to the autoxidation process. Therefore, these temperatures are recommended for studies on autoxidation kinetics. DSC investigations of the inhibited autoxidation described in the next section confirmed the proposed interpretation.

Thermoanalytical Investigations of Inhibited Autoxidation

The first comparison of inhibited and noninhibited oxidation was carried out under isothermal conditions (Raemy *et al.* 1987). The addition of propyl gallate (PG) to animal fat increased induction time from 15 min for noninhibited fat to 489 min for the sample containing 500 ppm of PG. The measurements were carried out at 120°C under oxygen atmosphere. Results of these DSC experiments were in good agreement with induction times obtained by the AOM method (Rancimat) for the whole range of PG concentration (Raemy *et al.* 1987). The same method was used to study antioxidant activity of flavonoids (Pereira and Das 1990). The lipid sam-

TABLE 7.8
Comparison of Activation Energies of Isothermal and Nonisothermal Oxidation of Saturated Fatty Acids and Their Esters[a]

	E_a (kJ/mol^{-1})	
Lipid	Isothermal	Nonisothermal
Palmitic acid	125.1 ± 11.2	125.3 ± 3.6
Ethyl palmitate	126.6 ± 5.0	124.5 ± 4.5
Glycerol tripalmitate	105.3 ± 7.7	108.0 ± 4.3
Stearic acid	134.3 ± 12.0	115.4 ± 4.8
Ethyl stearate	128.5 ± 2.6	106.0 ± 7.5
Glycerol tristearate	102.5 ± 10.9	117.8 ± 17.7
Lauric acid	97.3 ± 7.3	116.7 ± 1.7
Ethyl laurate	127.3 ± 6.4	118.7 ± 19.6
Ethyl myristate	117.1 ± 5.4	119.0 ± 12.0

[a]*Source:* Litwinienko *et al.* (1999a).

ples were heated from ambient temperature to 170°C with $\beta = 20$ K/min and then kept under isothermal temperature. The antioxidant efficiencies of the phenolic compounds analyzed (~30 μM), which were dissolved in refined, bleached, and deodorized palm oil, ranged with respect to the length of τ_{ind}: BHA < α-tocopherol < retinol < kaempferol < PG, quercetin < myricetin < morin. Isothermal DSC was employed not only to follow the oxidation course of inhibited autoxidation, but also to characterize some physical and chemical properties of the antioxidants themselves, i.e., to measure their volatilization and thermal decomposition (Kowalski 1991).

PDSC studies of isothermal oxidation of rapeseed oil containing BHT, BHA, and PG demonstrated the following range of increasing induction time: BHA, BHT < PG (Kowalski 1993). An interesting observation was presented in that study. Although the induction time was longer due to the presence of the antioxidants, the values of the activation energies were not substantially different from E_a for noninhibited oxidation of the oil. This peculiarity was explained by the presence of a small amount of natural antioxidants dissolved in the oil. The same observation was explained differently by Kasprzycka-Guttman and Odzeniak (1994) who analyzed peanut oil containing lignin and its fractions. The similar values of E_a calculated for inhibited and noninhibited oxidation were noticed because in both experiments, the same parts of the DSC curves were used for the calculation of E_a. Indeed, the shapes of the DSC peaks of inhibited and noninhibited autoxidation were similar and the only differences were induction times. Therefore, due to the same kinetic profile of DSC peaks, the method based on measurements of $d\alpha/d\tau$ can be misleading. Instead of the $d\alpha/d\tau$ measurement, methods using the dependency of $\log\tau_{ind}$ vs. $1/T$ should be used (Equations 38 and 39). Isothermal DSC was also used to study the effect of BHT and PG on oxidative stability of corn, linseed, and castor oils (Kasprzycka-Guttman *et al.* 1994). Tan *et al.* (2001a) successfully used refined, bleached, deodorized palm olein as a lipid matrix in isothermal DSC investigations of antioxidant activity of α-tocopherol and sage and rosemary extracts.

The antioxidant activity of several phenolic compounds as chain-breaking antioxidants was investigated using the nonisothermal DSC method. The first study on the applicability of this method for evaluation of antioxidant activity was limited to rapeseed oil and lard containing BHT, BHA, and PG (Kowalski 1991). The high temperatures (>180°C) at the start of noninhibited oxidation of the fats studied was the main reason that only PG showed considerable antioxidant effect. BHT and BHA showed weaker antioxidant effect and the author concluded that these phenols are too volatile to be monitored by nonisothermal DSC. However, when a less resistant lipid matrix was applied, the results were more promising (Litwinienko *et al.* 1995 and 1997, Litwinienko and Kasprzycka-Guttman 1998b). As a lipid analog, linolenic acid (18:3) was chosen due to its low oxidative stability (start of oxidation at temperatures below 100°C) and its clear calorimetric effect of oxidation.

The study of the nonisothermal oxidation inhibited by chain-breaking antioxidants confirmed the interpretation of the DSC curve described earlier in this chapter, i.e., that the first peak of the DSC curve is caused by autoxidation (formation of peroxides) and that decomposition of the peroxides occurs in the second and further peaks. On the basis of this interpretation, one can predict that the addition of an antioxidant should be manifested as an increase of temperatures corresponding to the first peak (T_e and T_{p1}), whereas the parameter T_{p2} should not be correlated directly with the inhibitory effect. Typical DSC curves of the nonisothermal oxidation of linolenic acid containing various concentrations of 1,2-dihydroxybenzene are presented in Figure 7.15; as can be seen, the inhibiting effect is expressed as a delay in the start of oxidation (first peak), and the shifts of T_e and T_{p1} are proportional to the increasing concentration of antioxidant (Litwinienko and Kasprzycka-Guttman 1998b). A more detailed analysis of the values of T_e, T_{p1}, and T_{p2} for the DSC curves in the oxidation of linolenic acid inhibited by BHT, at a concentration range of 0–10 mmol/mol of lipid (see Fig. 7.16), shows that temperature T_e is more sensitive parameter than T_{p1}. Moreover, the values of T_{p2} are not correlated with the concentration of the inhibitor (Litwinienko and Kasprzycka-Guttman 1998b).

The influence of the concentration of simple phenols (within the range of 0–15 mmol phenol/mol lipid) showed that E_a increased from 70.4 kJ/mol for pure linolenic acid oxidation to 96, 101, and 91 kJ/mol for linolenic acid oxidation inhibited by

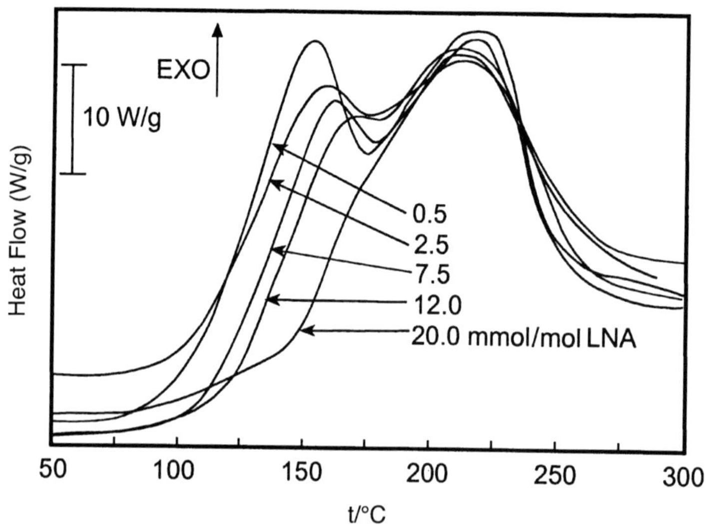

Fig. 7.15. Differential scanning calorimetry curves of linolenic acid (LNA) oxidation initiated by 0.04 M α,α′-azoisobutyronitrile. All scans were obtained for the same heating rate (β = 10 K/min). Various concentrations of 1,2-dihydroxybenzene were used, from 0.5 to 20.0 mmol of catechol/mol linolenic acid, as indicated by arrows. *Source:* Litwinienko *et al.* (1999b). Reprinted with permission of Elsevier.

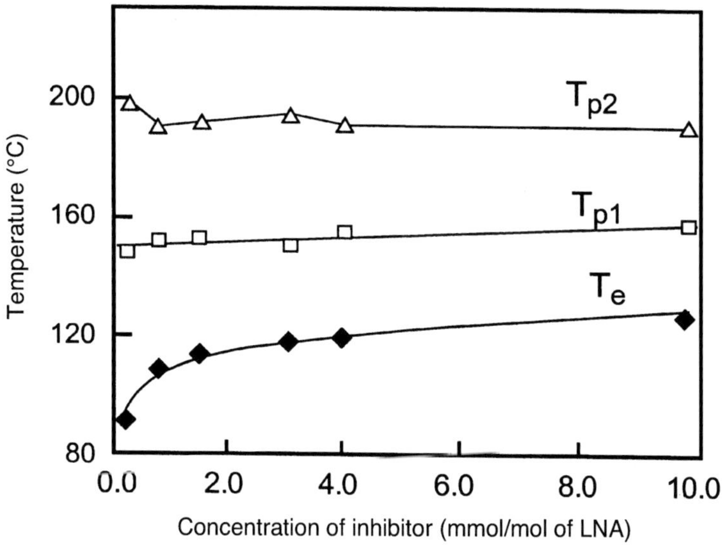

Fig. 7.16. Influence of the concentration of 2,6-di-*tert*-butyl-4-methylphenol (BHT) [in mmol of phenol/mol linolenic acid (LNA)] on the temperatures determined from differential scanning calorimetry curves of LNA oxidation. All scans were obtained for the same heating rate, 5 K/min. *Source:* Litwinienko and Kasprzycka-Guttman (1998b). Reprinted with permission of Kluwer Academic Publishers

BHT, *ortho*-cresol, and 2,4,6-trimethylphenol, respectively. The values of E_a obtained during the analysis of linolenic acid containing dimethylphenols did not exceed 86 kJ/mol. Analysis of the phenols examined showed that the decrease in inhibitory effect occurred at concentrations >10 mmol/mol (Litwinienko *et al.* 1997). Diminution of antioxidant activity, when higher concentrations of phenols were added, was also noted for dihydroxybenzenes (see Fig. 7.17) (Litwinienko *et al.* 1999b).

The nonisothermal DSC method can be used not only to determine the temperatures of the start of inhibited oxidation but also to measure the kinetic parameters of the process. Because the temperatures T_e and T_{p1} correlate with the inhibitory effect, the isoconversional methods are particularly valuable in studies of antioxidant activity. In addition, the kinetic parameters of inhibited autoxidation are considered to have more universal character than induction times (or induction temperatures). A good example of the universal meaning of the kinetic parameters is the comparison of T_e values determined for linolenic acid (LNA) oxidation initiated with azo-initiator (α,α'-azoisobutyronitrile, AIBN) plotted in Figure 7.18 (data from Litwinienko *et al.* 1999b). The T_e temperatures are higher for the system LNA + AIBN than for the system LNA + AIBN + dihydroxybenzene (all three isomers). According to this observation, pure LNA seems to be more resistant to oxidation than LNA containing dihydroxybenzenes. Conventional accelerated tests

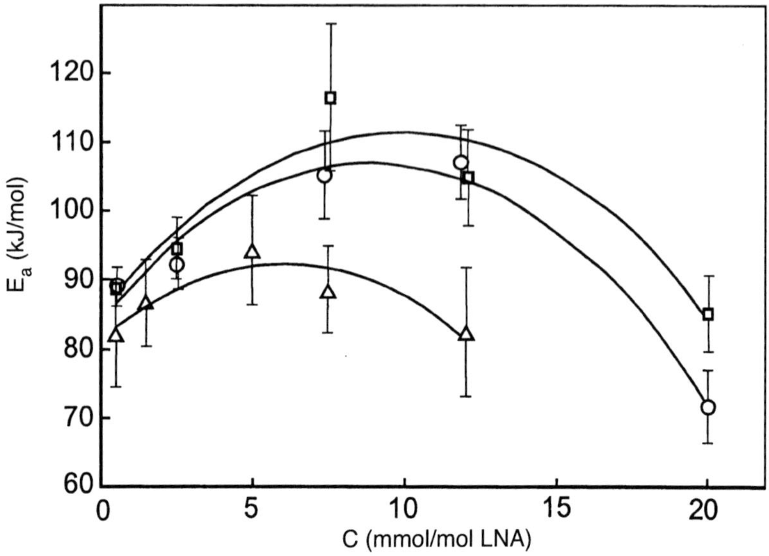

Fig. 7.17. Plots of activation energies of linolenic acid (LNA) autoxidation inhibited by 1,2-dihydroxybenzene (□), 1,3-dihydroxybenzene (○), and 1,4-dihydroxybenzene (△) at various concentrations. Source: Litwinienko *et al.* (1999b). Reprinted with permission of Elsevier.

(usually carried out at temperatures >100°C) would, therefore, show the prooxidative effect for hydroquinone, resorcinol, and (partially, for low concentrations) catechol. That result seems to be unusual because these three dihydroxybenzenes are moderate or strong antioxidants. Rate constants calculated from E_a and A, parameters for oxidation, confirm the prooxidative effects at 100°C (Fig. 7.19); however, it is possible to extrapolate rate constants to lower temperatures (Litwinienko *et al.* 1999b). Figure 7.20 presents log k values calculated for oxidation occurring at 25°C; at concentrations <10 mmol/mol of lipid, all dihydroxyphenols demonstrated an antioxidative effect. Such inversion of antioxidative properties with the change in temperature can be explained by the isokinetic temperature, i.e., the temperature at which the rate constants of two processes are equal, despite their different E_a and A parameters. In this situation, E_a and A for oxidation of pure LNA are 74.6 KJ/mol and 1.97×10^8 s^{-1}, respectively. For oxidation of LNA containing 7.5 mmol of 1,3-dihydroxybenzene, the parameters are $E_a' = 104.9$ KJ/mol and $A' = 9.35 \times 10^{12}$ s^{-1} (Litwinienko *et al.* 1999). At the isokinetic temperature, T_{iso}, the rate constants of both processes are equal. From the equation $A \exp[-E_a'/RT] = A' \exp[-E_a'/RT]$, the temperature T_{iso} can be calculated as: $T = (E - E')/[R \log (A/A')] = 338$ K, i.e., 65°C. If the temperature is above T_{iso}, the higher rate will be observed for a process of higher activation energy, whereas at $T < T_{iso}$, the process of higher E_a will be slower. The inversion of the reaction rates presented is a clear

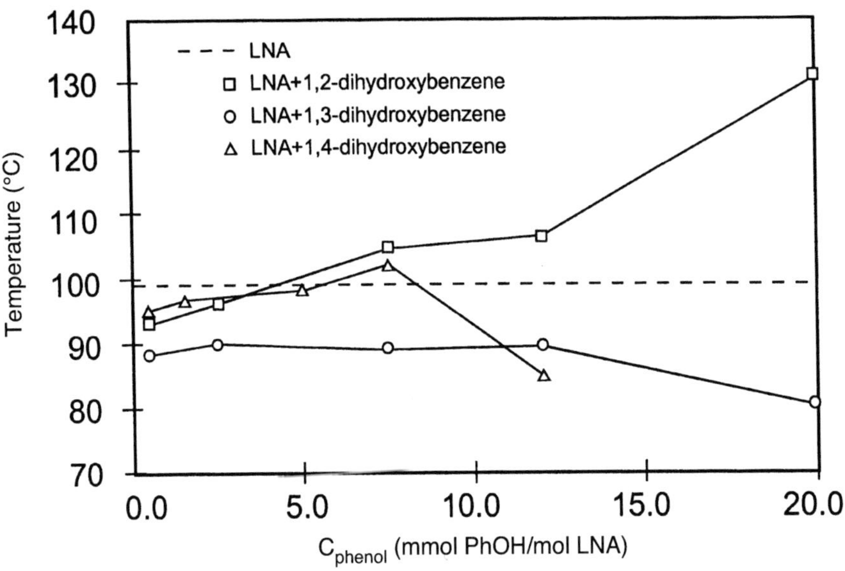

Fig. 7.18. Values of temperatures T_e measured for nonisothermal oxidation of linolenic acid (LNA) containing isomeric dihydroxybenzenes. Dashed line indicates the value of T_e for noninhibited autoxidation of LNA. All presented values were measured for $\beta = 5$ K/min.

argument that the DSC method is a valuable tool in assessing the oxidative stability of lipid systems as well as in assessing the antioxidant activity of added compounds. Assuming the process will be governed by the same mechanism at both room temperature and 120°C, the prediction/extrapolation of the rate constants obtained from DSC measurements is a more reliable method than the methods based on measurements of τ_{ind}. Unfortunately, it is common practice that qualitative results of accelerated tests (Rancimat, OSI, or Shaal test) obtained at temperatures >100°C are extrapolated directly to lower temperatures. The results presented above for dihydroxybenzenes as inhibitors showed that the conclusions made with induction parameters can be misleading.

Concluding Remarks

An attempt was made in this chapter to give readers a brief introduction to the methods of TA used in studies on the kinetics of lipid autoxidation. However, beyond the aim of this review are the problems of the application of DSC as an analytical method to monitor heat-related phenomena other than the oxidation process *per se*. Readers interested in the use of DSC as a method of identification of vegetable oils and fats and who are interested in the melting and crystallization properties of lipid systems will find more on this subject in the review by Toro-

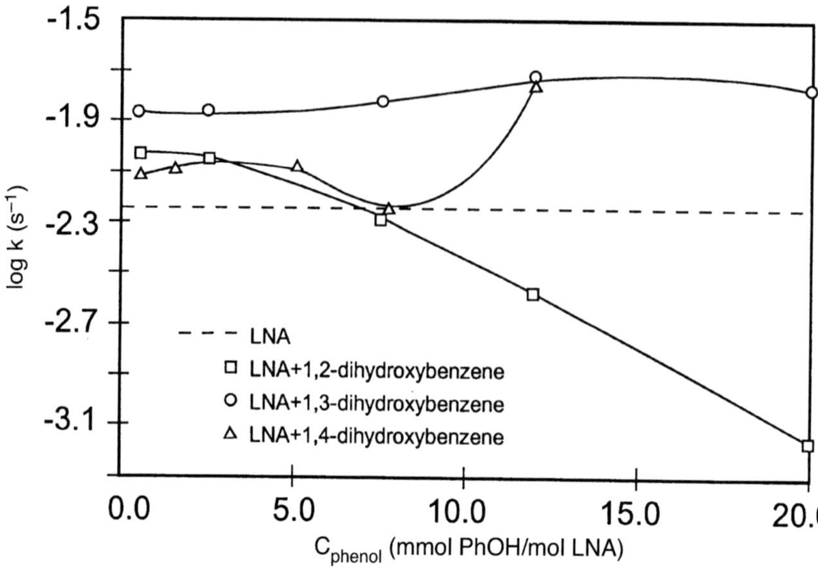

Fig. 7.19. Values of the logarithm of the rate constants (log k) of autoxidation of linolenic acid (LNA) containing isomeric dihydroxybenzenes calculated for the temperature 100°C. The dashed line indicates the value of log k for noninhibited autoxidation of LNA.

Vazquez *et al.* (2002). Another problem relevant to lipid oxidation, i.e., an application of DSC for measurements of melting and crystallization properties of oxidized oils, was reviewed by Tan and Che Man (2002). The authors found good correlation between melting temperature and PV for some thermally oxidized plant oils (Tan *et al.* 2001b, Tan and Che Man 1999). However, applied methods based on phase equilibrium measurements cannot be used directly for determination of the kinetic parameters of lipid autoxidation.

The main goal of this review was to present the application of DSC as an accelerated test to determine the oxidative stability of oils and fats. The methods described in this chapter allow us to obtain kinetic parameters of autoxidation within a relatively short time. A single isothermal DSC experiment takes from several minutes up to 1 h, and the kinetic parameters can be calculated on the basis of at least five measurements, each at a different temperature, to obtain the plot of $\ln\tau_{ind}$ as a function of $1/T$. For nonisothermal DSC methods, the time for a single analysis is even shorter, i.e., from 10 to 30 min depending on the heating rate. In addition, at least five measurements should be made to obtain a reliable straight line dependency of $\log\beta$ vs. $1/T_e$ or ($1/T_{p1}$). Therefore, the complete DSC analysis is significantly faster than other time-consuming methods for assessing oxidative stability (several hours to measure the single induction time of oxidation). Moreover, when large numbers of samples are measured, it is possible to

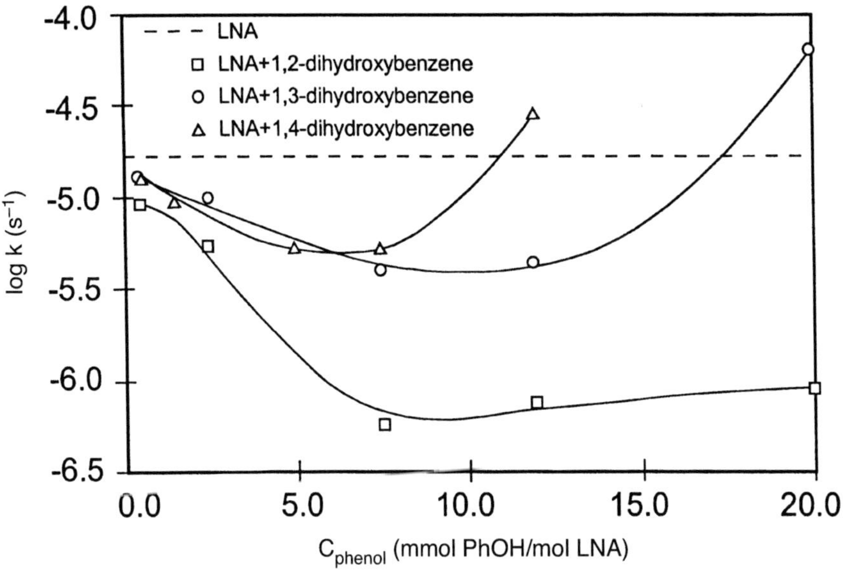

Fig. 7.20. The same parameters as in Figure 7.19, but calculated for a temperature of 25°C. The dashed line indicates the value of log *k* for noninhibited autoxidation of linolenic acid (LNA) at 25°C.

improve the calorimetry equipment to function in a fully automated manner. The samples can be handled and placed in the DSC chamber and removed after the measurement with the help of commercially available robot arms.

If only qualitative measurements are required, a single measurement is sufficient to measure induction parameters, i.e., parameter τ_{ind} (isothermal method) and T_c (non-isothermal method). However, as was described in the last part of this chapter, the induction times and temperatures can be used as qualitative parameters and should not be extrapolated to lower temperatures. In other words, if the oxidative stability of two or more lipid systems is compared with respect to their induction times at 120°C, the range of their stabilities at lower temperatures might be partly or completely different. DSC methods are strongly recommended to determine the Arrhenius kinetic parameters of oxidation to avoid this "extrapolation" problem.

There are several aspects of the perspectives of DSC as an analytical tool for analysis of lipid autoxidation. Technical progress has led to the development of a constantly increasing number of combinations of DSC or TG with other analytical techniques. To the author's knowledge, no scientific paper on lipid autoxidation describing the usage of simultaneous TA systems has been published to date. A combination of DSC (or TG) with equipment that can instantaneously detect the reaction products (systems with IR, mass spectrometry, or GC) would be helpful in lipid research. Other groups of simultaneous TA methods combine DTA with polarizing light microscopy and X-ray diffractometry. However, these methods are usually applied in studies on phase behavior and their relevance to accelerated tests of oxidative stability measurements is rather poor.

As a result of the continuous development of DSC techniques, the sensitivity and precision of the instruments continue to increase, Therefore DSC is becoming very competitive in other areas of the analytical chemistry of fats and oils. The small amount sample needed, the short time of analysis, the fairly straightforward procedures of evaluation of kinetic parameters, and the high repeatability of the results are clear advantages of DSC techniques. However, the cost of DSC instruments is too high to make this method competitive with conventional accelerated tests. On the other hand, due to the small costs of utilization and the "clean chemistry" procedures (no chemicals and solvents are used for analysis), these methods are promising as alternatives to other conventional methods currently used to determine the oxidative stability of oils.

Acknowledgment

The author thanks The Foundation for Polish Science for financial support (grant No. 3/04).

References

Adhvaryu, A., Erhan, S.Z., Liu, Z.S., and Perez, J.M. (2000) Oxidation Kinetic Studies of Oils Derived from Unmodified and Genetically Modified Vegetables Using Pressurized Differential Scanning Calorimetry and Nuclear Magnetic Resonance Spectroscopy, *Thermochim. Acta 364*, 87–97.

Adonyi, Z. (1972) Thermoanalysis of Petroleum-Products. Criticism to Conradson-Number and a Possibility to Determine the Flash Point, *Period. Polytech. Chem. Eng. 16*, 285–298.

Agrawal, R.K. (1986) Kinetic Analysis of Complex Reactions, *J. Therm. Anal. 31*, 1253–1262.

Agrawal, R.K. (1987) A New Equation for Modeling Nonisothermal Reactions, *J. Therm. Anal. 32*, 149–156.

Agrawal, R.K. (1988) Analysis of Irreversible Complex Chemical Reactions and Some Observations on Their Overall Activation Energy, *Thermochim. Acta 128*, 185–208.

Agrawal, R.K. (1989) The Compensation Effect—a Fact or a Fiction, *J. Therm. Anal. 35*, 909–917.

Agrawal, R.K. (1992) Analysis of Nonisothermal Reaction-Kinetics. 1. Simple Reactions, *Thermochim. Acta 203*, 93–110.

Akahira, T., and Sunose, T. (1971) Research Report, Chiba Institute of Technology, Vol. 16, p. 22.

Arnold, M., Veress, G.E., Paulik, J., and Paulik F. (1981) The Applicability of the Arrhenius Model in Thermal-Analysis, *Anal. Chim. Acta 124*, 341–351.

Arnold, M., Veress, G.E., Paulik, J., Paulik, F. (1982) Critical Reflection upon the Application of the Arrhenius Model to Non-Isothermal Thermogravimetric Curves, *Thermochim. Acta 52*, 67–81.

Barret, K.E.J. (1967) Determination of Rates of Thermal Decompositions of Polymerization Initiators with a Differential Scanning Calorimeter, *J. Appl. Polym. Sci. 11*, 1617–1626.

Blaine, S., and Savage, P.E. (1992) Reaction Pathways in Lubricant Degradation. 3. Reaction Model for *n*-Hexane Autoxidation, *Ind. Eng. Chem. Res. 31*, 69.

Buzas, I., Simon, J., and Hollo, J. (1977) Effect of the Experimental Conditions on the Thermooxidative Behavior of Vegetable Oils, *J. Therm. Anal. 12*, 397–405.

Buzas, I., Kurucz-Lusztig, E., and Hollo, J. (1978) Study of the Oxidative Stability of Vegetable Oils by Means of the Derivatograph, *Acta Aliment. 7*, 335–342.

Buzas, I., Kurucz, E., and Hollo, J. (1979) Study of the Thermooxidative Behavior of Edible Oils by Thermal Analysis, *J. Am. Oil Chem. Soc. 56*, 685–688.

Carroll, B., and Manche, E.P., (1970) Kinetic Parameters from Thermogravimetric Data. Comments, *Anal. Chem. 42*, 1296–1297.

Coats, A.W., and Redfern, J.P. (1965) Kinetic Parameters from Thermogravimetric Data, *J. Polym. Sci. B Polym. Lett. 3*, 917–920.

Cranton, G.E. (1976) Composition and Oxidation of Petroleum Fractions, *Thermochim. Acta 14*, 201–208.

Criado, J.M., Gonzales, M., Ortega, A., and Real, C. (1988) Discrimination of the Kinetic Model of Overlapping Solid State Reactions from Non-Isothermal Data, *J. Therm. Anal. 34*, 1387–1396.

Cross, C.K. (1970) Oil Stability—A DSC Alternative for Active Oxygen Method, *J. Am. Oil Chem. Soc. 47*, 229–230.

Dollimore, D. (1994) Thermal Analysis, *Anal. Chem. 66*, 17R–25R.

Dollimore, D. (1996) Thermal Analysis, *Anal. Chem. 68*, 63R–71R.

Dollimore, D., and Lerdkanchanaporn, S. (1998) Thermal Analysis, *Anal. Chem. 70*, 27R–35R.

Doyle, C.D. (1961) Kinetic Analysis of Thermogravimetric Data, *J. Appl. Polym. Sci. 5*, 285–292.

Doyle, C.D. (1965) Series Approximations to Equation of Thermogravimetric Data, *Nature 207*, 290–291.

Duswalt, A.A. (1968) Analysis of Highly Exothermic Reactions by DSC, in *Analytical Calorimetry* (Porter, R.S., Johnson, J.F., eds.), pp. 313–317, Plenum Press, New York.

Exner, O. (1964) Concerning Isokinetic Relationship, *Nature 20*, 488–490.

Flynn, J.H. (1980) The Effect of Heating Rate Upon the Coupling of Complex-Reactions. 1. Independent and Competitive Reactions, *Thermochim. Acta 37*, 225–238.

Flynn, J.H., and Wall, L.A. (1966) A Quick Direct Method for Determination of Activation Energy from Thermogravimetric Data, *J. Polym. Sci. B Polym. Lett. 4*, 323–328.

Freeman, E.S., and Carroll, B. (1958) The Application of Thermoanalytical Techniques to Reaction Kinetics—the Thermogravimetric Evaluation of the Kinetics of the Decomposition of Calcium Oxalate Monohydrate, *J. Phys. Chem. 62*, 394–397.

Freeman, E.S., and Carroll, B. (1969) Interpretation of the Kinetics of Thermogravimetric Analysis, *J. Phys. Chem. 73*, 751–752.

Gallagher, P.K. (1997) Thermoanalytical Instrumentation, Techniques, and Methodology, in *Thermal Characterization of Polymeric Materials* (Turi, E., ed.), Vol. 1, pp. 2–205, Academic Press, New York.

Garcia-Ochoa, F., Romero, A., and Querol, J. (1989) Modeling of the Thermal *n*-Octane Oxidation in the Liquid Phase, *Ind. Eng. Chem. Res. 28*, 43.

Gennaro, L., Piccoli-Bocca, A., Modesti, D., Masella, R., and Coni, E. (1998) Effect of Biophenols on Olive Oil Stability Evaluated by Thermogravimetric Analysis, *J. Agric. Food Chem. 46*, 4465–4469.

Hagemann, J.W., and Rotfus, J.A. (1979) Oxidative Stability of Wax Esters by Thermogravimetric Analysis, *J. Am. Oil Chem. Soc. 56*, 629–631.

Hassel, R.L. (1976) Thermal-Analysis—Alternative Method of Measuring Oil Stability, *J. Am. Oil Chem. Soc. 53*, 179–181.

Hatakeyama, T., and Quinn, F.X. (1999) *Thermal Analysis. Fundamentals and Applications to Polymer Science*, 2nd ed., pp. 5–24, John Wiley & Sons, Chichester.

Howard, J.B. (1973) DTA for Control of Stability in Polyolefin Wire and Cable Compounds, *Polym. Eng. Sci. 13*, 429–434.

Jensen, R.K., Korcek, S., Mahoney, L.R., and Zinbo, M. (1981) Liquid Phase Autoxidation of Organic Compounds at Elevated Temperatures. 2. Kinetics and Mechanisms of the Formation of the Cleavage Products in *n*-Hexadecane Autoxidation at 120–180°C, *J. Am. Chem. Soc. 103*, 1742–1749.

Kaisersberger, E. (1989) DSC Investigations of the Thermal Characterization of Edible Fats and Oils, *Thermochim. Acta 151*, 83–90.

Kasprzycka-Guttman, T., and Odzeniak, D. (1991) Specific-Heats of Some Oils and Fats, *Thermochim. Acta 191*, 41–45.

Kasprzycka-Guttman, T., and Odzeniak, D. (1992) Thermoanalytical Investigation of Edible Oil, *Thermochim. Acta 204*, 303–310.

Kasprzycka-Guttman, T., and Odzeniak, D. (1993) Iso-Thermal DSC Investigation of the Kinetics of Thermoxidative Decomposition of Some Edible Oils, *J. Therm. Anal. 39*, 217–220.

Kasprzycka-Guttman, T., and Odzeniak, D. (1994) Antioxidant Properties of Lignin and Its Fractions, *Thermochim. Acta 231*, 161–168.

Kasprzycka-Guttman, T., Odzeniak, D., and Supera, M. (1994) Thermokinetic Properties of Inhibited Vegetable-Oils, *Thermochim. Acta* 237, 207–211.

Kissinger, H.E. (1957) Reaction Kinetics in Differential Thermal Analysis, *Anal. Chem. 29*, 1702–1706.

Korycka-Dahl, M., and Richardson T. (1980) Initiation of Oxidative Changes in Foods, *J. Dairy Sci. 63*, 1181–1198.

Kowalski, B. (1988a) Determination of Specific-Heats of Some Edible Oils and Fats by Differential Scanning Calorimetry, *J. Therm. Anal. 34*, 1321–1326.

Kowalski, B. (1988b) Thermoanalytical Investigations of Edible Oils and Fats. I. Kinetics of Thermal-Oxidative Decomposition of Rapeseed Oil, *Acta Aliment. Pol. 14*, 195.

Kowalski, B. (1989) Determination of Oxidative Stability of Edible Vegetable Oils by Pressure Differential Scanning Calorimetry, *Thermochim. Acta 156*, 347–358.

Kowalski, B. (1990) Determination of Spontaneous Ignition Temperatures of Edible Oils and Fats by Pressure Differential Scanning Calorimetry, *Thermochim. Acta 173*, 117–127.

Kowalski, B. (1991) Evaluation of the Stability of Some Antioxidants for Fat-Based Foods, *Thermochim. Acta 177*, 9–14.

Kowalski, B. (1991a) Thermal-Oxidative Decomposition of Edible Oils and Fats—DSC Studies, *Thermochim. Acta 184*, 49–57.

Kowalski, B. (1992) Thermokinetic Analysis of Vegetable Oil Oxidation as an Autocatalytic Reaction, *Pol. J. Food Nutr. Sci. 1/42*, 51–59.

Kowalski, B. (1993) Evaluation of Activities of Antioxidant in Rapeseed Oil Matrix by Pressure Differential Scanning Calorimetry, *Thermochim. Acta 213*, 135–146.

Kowalski, B. and Kot, B. (1989) Thermoanalytical Investigations of Edible Oils and Fats. II. Kinetics of Thermooxidative Decomposition of Soybean and Sunflower Oil, *Acta Aliment. Pol. 15*, 55–62.

Kowalski, B., Pieńkowska, H., and Zadernowski, R. (1993) Characterization of Lipids from Vegetable Bio-Oils. Part II. Thermokinetic Evaluation of Oxidative Stabilities of Some Vegetable Oils, *Pol. J. Food Nutr. Sci. 2/43*, 61–68.

Kowalski, B., Ratusz, K., Miciula, A., Krygier, K. (1997) Monitoring of Rapeseed Oil Autoxidation with a Pressure Differential Scanning Calorimeter, *Thermochim. Acta 307*, 117–121.

Litwinienko, G. (2001) Autoxidation of Unsaturated Fatty Acids and Their Esters, *J. Therm. Anal. Calorim. 65*, 639–646.

Litwinienko, G., and Kasprzycka-Guttman, T. (1998a) A DSC Study on Thermoxidation Kinetics of Mustard Oil, *Thermochim. Acta 319*, 185–191.

Litwinienko, G., and Kasprzycka-Guttman, T. (1998b) The Influence of Some Chain-Breaking Antioxidants on Thermal-Oxidative Decomposition of Linolenic Acid, *J. Therm. Anal. 54*, 203.

Litwinienko, G., and Kasprzycka-Guttman, T. (2000) Study on Autoxidation Kinetics of Fats by Differential Scanning Calorimetry. 2. Unsaturated Fatty Acids and Their Esters, *Ind. Eng. Chem. Res. 39*, 13–17.

Litwinienko, G., and Dabrowska, M. (2001) Thermogravimetric Investigation of Antioxi-dant Activity of Selected Compounds in Lipid Oxidation, *J. Therm. Anal. Calorim. 65*, 411–417.

Litwinienko, G., Kasprzycka-Guttman, T., and Jarosz-Jarszewska, M. (1995) Dynamic and Isothermal DSC Investigation of the Kinetics of Thermoxidative Decomposition of Some Edible Oils, *J. Therm. Anal. 45*, 741–750.

Litwinienko, G., Kasprzycka-Guttman, T., and Studzinski, M. (1997) Effects of Selected Phenol Derivatives on the Autoxidation of Linolenic Acid Investigated by DSC Non-Isothermal Methods, *Thermochim. Acta 307*, 97–106.

Litwinienko, G., Daniluk, A., and Kasprzycka Guttman, T. (1999a) A Differential Scanning Calorimetry Study on the Oxidation of Saturated C_{12}-C_{18} Fatty Acids and Their Esters, *J. Am. Oil Chem. Soc. 76*, 655–657.

Litwinienko, G., Kasprzycka-Guttman, T., and Jamanek, M. (1999b) DSC Study of Antioxidant Properties of Dihydroxyphenols, *Thermochim. Acta 331*, 79–86.

Litwinienko, G., Daniluk, A., and Kasprzycka-Guttman, T. (2000) Study on Autoxidation Kinetics of Fats by Differential Scanning Calorimetry. 1. Saturated C_{12}–C_{18} Fatty Acids and Their Esters, *Ind. Eng. Chem. Res. 39*, 7–12.

Nieschlag, H.J., Hagen, J.W., Rothfus, J.W., and Smidt, D.I. (1977) Rapid Thermogravimetric Estimation of Oil Stability, *Anal. Chem. 46*, 2215–2217.

Noel, F., and Cranton, G.E. (1974) Application of Scanning Calorimetry to Petroleum Oil Oxidation Studies, in *Analytical Calorimetry* (Porter, R.S., and Johnson, J.F., eds.), Vol. 3, pp. 305–320, Plenum Press, New York.

Opfermann, J., and Kaisersberger, E. (1992) An Advantageous Variant of the Ozawa-Flynn-Wall Analysis, *Thermochim. Acta 203*, 167–175.

Ozawa, T. (1965) A New Method of Analyzing Thermogravimetric Data, *Bull. Chem. Soc. Jpn. 38*, 1881–1885.

Ozawa, T. (1970) Kinetic Analysis of Derivative Curves in Thermal Analysis, *J. Therm. Anal. 2*, 301–324.

Ozawa, T. (1975) Critical Investigations of Methods for Kinetic Analysis of Thermoanalytical Data, *J. Therm. Anal. 7*, 607–617.

Ozawa, T. (1976) Some Demonstrations of the Effect of Heating Rate on Thermoanalytical Curves, *J. Therm. Anal. 9*, 217–227.

Pereira, T.A., and Das, N.P. (1990) The Effects of Flavonoids on the Thermal Autoxidation of Palm Oil and Other Vegetable Oils Determined by Differential Scanning Calorimetry, *Thermochim. Acta 165*, 129–137.

Raemy, A., and Loeliger, J. (1985) Self-Ignition of Powders Studied by High-Pressure Differential Thermal Analysis, *Thermochim. Acta 95*, 343–346.

Raemy, A., Hurrell, R.F., and Loeliger J. (1983) Thermal-Behavior of Milk Powders Studied by Differential Thermal-Analysis and Heat-Flow Calorimetry, *Thermochim. Acta 65*, 81–92.

Raemy, A., Lambelet, P., and Loeliger, J. (1985) Thermal-Analysis and Safety in Relation to Food-Processing, *Thermochim. Acta 95*, 441–446.

Raemy, A., Froelicher, I., and Loeliger, J. (1987) Oxidation of Lipids Studied by Isothermal Heat Flux Calorimetry, *Thermochim. Acta 114*, 159–164.

Rudnik, E., Szczucinska, A., Gwardiak, H., Szulc, A., and Winiarska, A. (2001) Comparative Studies of Oxidative Stability of Linseed Oil, *Thermochim. Acta 370*, 135–140.

Takaoka, K., Takasago, M., Kobayashi, K., and Taru, Y. (1994) Study of Thermal Oxidation of Thin Film of Unsaturated Triacylglycerols. VII. Vacuum Thermogravimetric Analysis of Trilinoleoylglycerol, *Yukagaku 43*, 484–489.

Tan, C.P., and Che Man, Y.B. (1999) Qualitative Differential Scanning Calorimetric Analysis for Determining Total Polar Compounds in Heated Oils, *J. Am. Oil Chem. Soc. 76*, 1047–1057.

Tan, C.P., and Che Man, Y.B. (2002) Recent Developments in Differential Scanning Calorimetry for Assessing Oxidative Deterioration of Vegetable Oils, *Trends Food Sci. Technol. 13*, 312–318.

Tan, C.P., Che Man, Y.B., Selemat, J., and Yusoff, M.S.A. (2001a) Efficacy of Natural and Synthetic Antioxidants in RBD Palm Olein by Differential Scanning Calorimetry, in *Micronutrients and Health: Molecular Biological Mechanism* (Nesaretnam, K., and Packer, L., Eds.), pp. 108–118, AOCS Press, Champaign, IL.

Tan, C.P., Che Man, Y.B., Jinap, S., and Yusoff, M.S.A. (2001b) Effects of Microwave Heating on Changes in Chemical Thermal Properties of Vegetable Oils, *J. Am. Oil. Chem. Soc. 76*, 1227–1232.

Tan, C.P., Che Man, Y.B., Selemat, J., and Yusoff, M.S.A. (2001c) Application of Arrhenius Kinetics to Evaluate Oxidative Stability in Vegetable Oils by Isothermal Differential Scanning Calorimetry, *J. Am. Oil Chem. Soc. 78*, 1133–1138.

Tan, C.P., Che Man, Y.B., Selemat, J.., and Yusoff, M.S.A. (2002) Comparative Studies of Oxidative Stability of Edible Oils by Differential Scanning Calorimetry and Oxidative Stability Index Method, *Food Chem. 76*, 385–389.

Toro-Vazquez, J.F., Dobildox-Alvarado, E., Herrera-Coronado, V., and Charo-Alonso, M. (2002) Triglyceride Crystallization, in Vegetable Oils, in *Engineering and Food for the 21st Century* (Welti-Chanes, J., Barbosa-Canovas, G.V., and Aguilera, J.M., eds.), pp. 105–123, CRC Press, Boca Raton, FL.

Vand,V. (1943) A Theory of Irreversible Electrical Resistance Changes of Metallic Films Evaporated in Vacuum, *Proc. Phys. Soc. 55*, 222–246.

Van Dooren, A.A., and Müller, B.W. (1983) Effect of Experimental Variables on the Determination of Kinetic Parameters with Differential Scanning Calorimeters. I. Calculation Procedures of Ozawa and Kissinger, *Thermochim. Acta 65*, 257–267.

Vossoughi, S., and El-Shoubary, Y.M. (1990) Kinetics of Liquid-Hydrocarbon Combustion Using the DSC Technique, *Thermochim. Acta 157*, 37–44.

Watson, E.S., O'Neill, M.J., Justin, J., and Brenner N. (1964) A Differential Scanning Calorimeter for Quantitative Differential Thermal Analysis, *Anal. Chem. 36*, 1233–1238.

Wesołowski, M. (1981) Thermal Analysis of Petroleum Products, *Thermochim. Acta 46*, 21–45.

Wesołowski, M. (1985) Thermogravimetric Assessment of Service Performance of M-20 Bp and Ms-20-P Lubricating Oils, *Thermochim. Acta 91*, 265–286.

Wesołowski, M. (1986a) Thermoanalytical Methods in Evaluation of Medicinal Cod-Liver Oils, *Sci. Pharm. 54*, 11–18.

Wesołowski, M. (1986b) Determination of Quality of Fish Oils by Thermal Decomposition, *Seifen Ole Fette Wasche 112*, 231–234.

Wesołowski, M. (1986c) Thermogravimetric Assessment of Service Performance of Marinol CB SAE-30 Lube Oils, *Thermochim. Acta 99*, 333–348.

Wesołowski, M. (1986d) Thermogravimetric Assessment of Service Performance of DS-11 Lube Oils, *Thermochim. Acta 101*, 9–17.

Wesołowski, M. (1987a) Quality Assessment of Edible Fish Oils by Thermal-Analysis Techniques, *Fat Sci. Technol. 89*, 111–116.

Wesołowski, M. (1987b) Thermoanalytical Assessment of the Quality of Edible and Technical Oils, *J. Therm. Anal. 32*, 1781–1784.

Wesołowski, M., and Erecińska, J. (1998) Thermal Analysis in Quality Assessment of Rapeseed Oils, *Thermochim. Acta 323*, 137–143.

Wesołowski, M., Czerwonka, M., Konieczyński, P. (1998) Application of Chemometrically Processed Derivative Thermogravimetric Data for Quality Control of M-20 Bp, MS-20 p, Marinol CB SAE-30 and DS-11 oils, *Thermochim. Acta 323*, 159–168.

Wunderlich, P. (1997) The Basis of Thermal Analysis, in *Thermal Characterization of Polymeric Materials* (Turi, E., ed.), Vol. 1, 2nd ed., pp. 206–483, Academic Press, New York.

Yamazaki, M., Nagao, A., and Komamiya, K. (1980) High Pressure Differential Thermal Analysis (HPDTA) of Fatty Acid Methyl Esters and Triglycerides, *J. Am. Oil Chem. Soc. 57*, 59–60.

Chapter 8

Chemiluminescence in Studying Lipid Oxidation

Lev Zlatkevich[a] and Afaf Kamal-Eldin[b]

[a]Lume, Incorporated, Philadelphia, PA 19115, USA, and [b]Department of Food Science, Swedish University of Agricultural Sciences (SLU) 75007, Uppsala, Sweden

Introduction

Oxidation of organic substances is frequently accompanied by the emission of weak visible light, i.e., chemiluminescence (CL), defined as the emission of light as a direct result of a chemical reaction. Light is emitted when electronically excited species formed during the chemical reaction return to the ground state (Van Dyke *et al.* 1985). CL differs from fluorescence in that the source of the emission is a chemical reaction and not an excitation light. The CL intensity, I_{CL}, is determined by the rate of the chemical reaction R, the quantum efficiency given by the quantum yield, ϕ_{CL}, and the geometric factor G (Kron *et al.* 1997).

$$I_{CL} = \phi_{CL}\,G\,R$$

G is a product of the detection efficiency, the fraction of the emitted photons being captured, and the fraction of photons that leave the sample, rather than being absorbed. The quantum efficiency is the product of the fraction of reactions that produce exited stages, and the fraction of these stages that emit.

The excited species emitting light in CL can be a direct product of the reaction, or can be formed by the reaction of a product in the excited state with a suitable reagent having strong chemiluminescent properties. If a fluorescent molecule is present with a lower-lying electronic energy level than the excited state produced in the chemical reaction, then donor-acceptor energy transfer may occur with the result that the CL spectrum is now that of the foreign fluorescent molecule. This can be used to enhance the yield of CL. The phenomena can be described by the following simple equations, where * denotes excited species

$$A + B \rightarrow C^* + D$$

$$C^* \rightarrow C + h\nu \text{ (direct chemiluminescence), or}$$

$$C^* + E \rightarrow C + E^*$$

$$E^* \rightarrow E + h\nu \text{ (transfer-based chemiluminescence)}$$

Conversely, excited state quenching may decrease the yield of CL. If the emitting state is a triplet, the quantum yield may be only 10^{-6} of that expected; the quantum efficiency of CL ($\phi_{CL} \approx 10^{-9}$) is much lower than that for fluorescence and phosphorescence ($\phi \approx 10^{-2}$–10^{-3}). Quenching by molecular oxygen is believed to be the main reason for the low yield of CL in the solution oxidation of hydrocarbons.

CL can be divided into three types: autonomous CL, enhanced CL, and oxyluminescence (Neeman and Joseph 1985). Autonomous CL measurements are performed in a dynamic mode by increasing the temperature under an inert gas flow to determine the oxidation products caused by an earlier degradation. In this case, integration of the chemiluminescent signal as a function of time has to be performed. Here special care must be paid to the exclusion of oxygen from the sample, reagents and the reaction chamber and to the calibration procedure. Usually the signal is integrated for a fixed period of time or the peak height at a fixed time is used for calculations with reference to a standard treated in the same way. The experimental procedures that allow the evaluation of the kinetic order of decomposition of organic hydroperoxides as well as the temperature dependencies for the decomposition rate constants were described elsewhere (Zlatkevich 1987). The quantum yield of the autonomous CL is usually low, 10^{-11}–10^{-9}, mainly due to quenching of excited species either by molecular oxygen and/or solvent molecules.

Enhanced chemiluminescent measurements were invented as a detection system for peroxidase-related reactions (Whitehead *et al.* 1983). In enhanced CL, the selectivity and sensitivity of chemiluminescent reactions are increased *via* the use of a CL enhancer, e.g., luminol. Because of the complex and often uncertain nature of the CL emission, however, the scope of the studies utilizing CL intensity as the major parameter is limited (Pospisil *et al.* 2003). At best, they can lead to an assay, and their results certainly cannot be generalized.

Oxyluminescence measurements are carried out in an isothermal mode by following the CL response as a function of time. First, the sample is placed in the oven and is equilibrated to the test temperature in an inert atmosphere, usually under nitrogen. Then the gas flow is changed to oxygen and the recording of the CL signal is started. The presence of antioxidants in the samples delays the CL response.

Among the three types of CL mentioned above, the last-mentioned seems to be the most important for our purposes because it is oxyluminescence that is potentially very useful in evaluating the thermal oxidative stability of lipids as well as the quality of antioxidants protecting lipids. Oxyluminescence will be discussed in detail.

Lipid Oxidation and Generation of Light

The oxidation of lipids is one of the oldest examples of a radical chain reaction. Although the number of potential reactions that may occur during the thermal oxidation of organic materials (lipids, in particular) is very large, the list of critical reactions is relatively short. There is a considerable merit in refocusing attention on the simplified, universal oxidation scheme described in terms of the initiation, propagation, and termination reactions. Chain branching that may complicate the oxidation process occurring at high temperatures is often unimportant for lower temperatures and in an excess of oxygen the following reactions are considered:

$$\mathrm{LOOH} \xrightarrow{k_1} \mathrm{LO}^{\bullet} + {}^{\bullet}\mathrm{OH} \xrightarrow[\text{with LH}]{\text{chain transfer}} \mathrm{L}^{\bullet} \qquad \text{INITIATION} \qquad [1]$$

$$\mathrm{L}^{\bullet} + \mathrm{O_2} \xrightarrow{k_2} \mathrm{LO_2}^{\bullet} \qquad \text{PROPAGATION} \qquad [2]$$

$$\mathrm{LO_2}^{\bullet} + \mathrm{LH} \xrightarrow{k_3} \mathrm{LOOH} + \mathrm{L}^{\bullet} \qquad \text{PROPAGATION} \qquad [3]$$

$$\mathrm{LO_2}^{\bullet} + \mathrm{LO_2}^{\bullet} \xrightarrow{k_6} \mathrm{LOOL} + \mathrm{O_2} \qquad \text{TERMINATION} \qquad [4]$$

The question of initiation was the subject of some controversy and, indeed, remains a matter of opinion to this day. Nonetheless, there is considerable indication that practically all lipids contain some minute, often undetectable, quantities of hydroperoxides. In view of the instability of these groups, it seems likely that their decomposition will be a major factor in initiating the oxidative degradation.

As far as the source of emission during oxidation of lipids is concerned, there are several possibilities. Although there is spectral and chemical evidence for emission from excited singlet oxygen (Miyamoto *et al.* 2003), the essential portion of this emission lies at long wavelengths not detectable by photomultipliers. The spectral analysis is in most cases consistent with emission from excited triplet-state carbonyl compounds (Boverus *et al.* 1980, Cadenas and Sies 1982, Shulte-Herbruggen and Sies 1989, Timmins *et al.* 1997).

The reactions that have been considered feasible are either the bimolecular termination reaction of two alkyl peroxy radicals *via* the Russell mechanism (Russell 1957):

$$\mathrm{L\text{-}CH_2\text{-}OO}^{\bullet} + {}^{\bullet}\mathrm{OOL} \longrightarrow \mathrm{L\text{-}\underset{\underset{\displaystyle H}{|}}{C}{=}O^{*}} + \mathrm{LOH} + \mathrm{O_2}$$

or the decomposition of lipid hydroperoxides:

$$\mathrm{L\text{-}CH_2\text{-}OOH} \longrightarrow \mathrm{L\text{-}CH_2\text{-}O}^{\bullet}\ {}^{\bullet}\mathrm{OH} \longrightarrow \mathrm{L\text{-}\underset{\underset{\displaystyle H}{|}}{C}{=}O^{*}} + \mathrm{H_2O}$$

The contribution of the above mechanisms to the actual CL response is not yet resolved. From a practical perspective, however, and as the consequence of the equality of the rates of initiation and termination for a linear steady-state chain reaction (reactions [1]–[4]), the two mechanisms are equivalent (Vassil'ev 1969). In view of the predominately first-order decomposition kinetics of hydroperoxides (Zlatkevich 2004, Zlatkevich and Martella 1995), in the following section the kinetic relationships for CL generated during oxidation of lipids will be developed by utilizing the direct proportionality between the CL intensity and the concentration of hydroperoxides, i.e., I = C [LOOH] where C is a constant.

Instrumentation

An essential feature of chemiluminescent reactions is the very low quantum yield from the reactions involved, necessitating highly sensitive light detecting to measure the very low emission. Instrumentation used in CL studies varies greatly in complexity. The simplest system can consist of a hot plate and chemical reaction flask in a light-tight box with a photomultiplier tube (PMT) and associated electronics. Complex apparatus could include a specially designed cell, sophisticated temperature and gas flow controls, as well as an automated data acquisition and evaluation system. Some instruments use dc circuitry, in which the pulse stream is read as a current; in others, the PMT output is analyzed by pulse-counting circuitry.

In most cases the experimental set-up is rather simple. The sample is placed in a small-volume cell containing a temperature-regulated oven. The cell should be tight but should allow the flexible exchange of gases (often nitrogen and oxygen) and well-regulated gas flow; the oven should have a stable and wide temperature range with variable, rapid heating ability. The sample cell is covered by a lens focusing the emitted light into the PMT, which is protected from extraneous light and heat.

Descriptions of several noncommercial single-cell instruments of various degrees of sophistication are available in the literature (Marino and Ingle 1981, Mendenhall 1977, Schard and Russell 1964, Stieg and Nieman 1978). Independently of the degree of sophistication, however, single-cell instruments are insufficient in evaluating oxyluminescence from highly stabilized materials tested at relatively low temperatures because such experiments may require many hours or even days. The same problem pertains to differential scanning calorimetry (DSC), which is usually limited to testing with timescales of a few hours at most.

From this perspective, the utilization of the multicell CL apparatus with eight completely independent cells (Fig. 8.1) has great potential. This instrument was originally developed for evaluation of solid polymers, but it is equally suitable for studying liquids, for example, automotive oils (Zlatkevich and Martella 1995). The apparatus is computerized, allowing fully automatic operation with

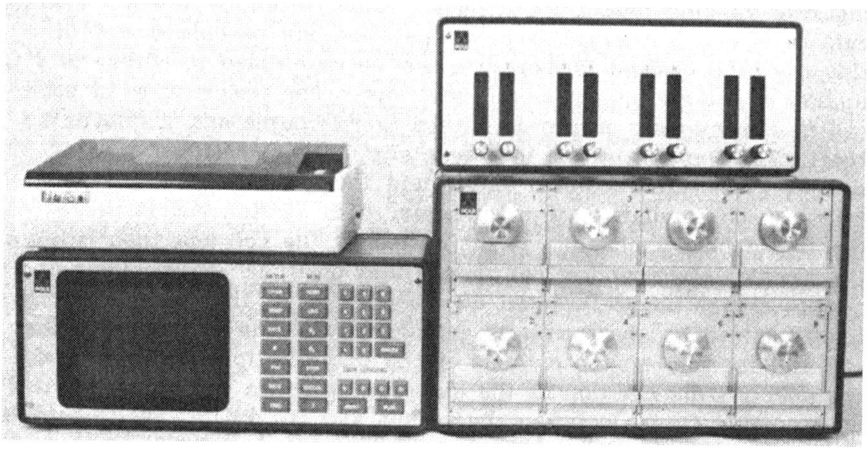

Fig. 8.1. The computerized chemiluminescence apparatus with eight completely independent cells.

the computer fulfilling two major functions: (i) control and monitoring of the temperature and atmosphere experienced by samples; (ii) data storage, retrieval, and analysis.

The computer consists of two separate units, a controller and a host processor. Each unit is controlled by its own microprocessor hardware/software system, and communication between the units is carried out *via* an RS-232 serial interface. The controller is an eight-channel temperature controller/programmer and a data acquisition system. The temperature controllers are capable of independently raising the temperature at a controlled rate to a set-point and maintaining that point. The data acquisition system inputs the individual temperature values *via* thermocouple sensors and light emission intensities *via* photomultiplier tubes. The temperature values are fed back to the temperature control system. They are also linearized, filtered, and sent along with the filtered emission intensity values to the host processor. The host processor receives and stores the incoming data. It provides channel selection and allows the user to enter/modify heating rates and set-points as well as start/stop logging and temperature cycling. It also allows the graphical representation of stored data and the results of calculations on the monitor screen as well as printing of the data.

To perform an experiment, the material is placed in a small-volume cell (Fig. 8.2) which can be heated from room temperature up to 250°C at a constant heating rate varying from 1 to 15°C/min. For liquid samples such as oils, the sample is placed into an aluminum cuvette that is covered by a thin glass, thus restricting the reaction volume to ~0.1 cm^3. Along with the constant heating rate mode, an isothermal mode of operation can be chosen; in this case, the desired temperature can be maintained to within 0.2°C in a flow of either oxygen or an inert gas. The temperature of each cell is continuously surveyed with a thermocouple and dis-

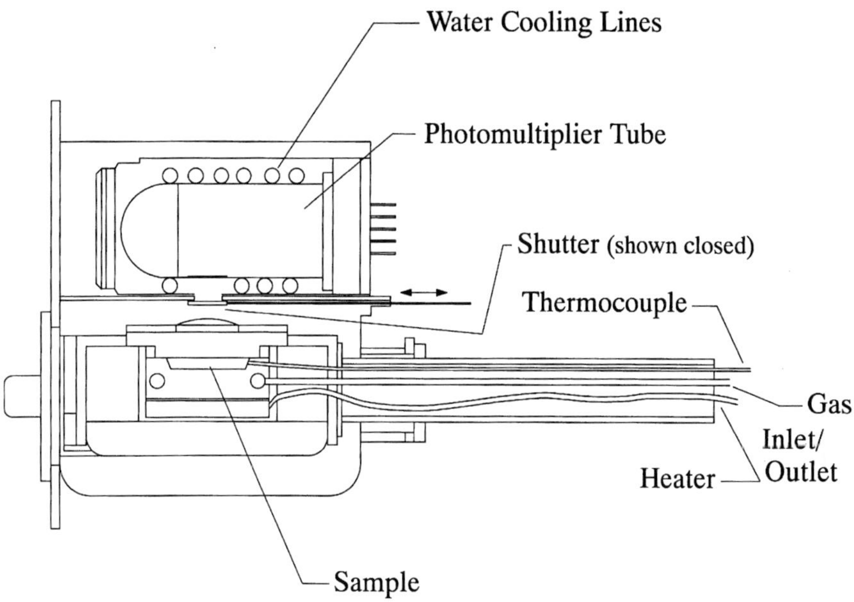

Fig. 8.2. Schematic of a cell.

played on the monitor. Two computer-activated solenoid valves select oxygen or the inert gas supplies, and the gas flow rate in each of the cells can be regulated by a flow meter. The chosen gas is passed through a heat exchanger at the bottom of the cell and then reaches the sample. Symmetrical cell design and positions of gas inlets and outlets ensure no preferential directional flow of the gas in the cell. Light emitted by the sample is gathered by a lens in the lid of the cell and focused on the cathode of the photomultiplier. If desired, the spectral distribution of the CL can be determined by inserting filters between sample and the PMT. Each photomultiplier is placed into an assembly of two coaxial aluminum cylinders. The inner cylinder can be rotated around its major axis, thus opening or closing the light pathway from the sample to the photomultiplier. The outer cylinder is stationary and has a copper coil around it for water circulation, which provides cooling to the entire photomultiplier housing assembly.

The cooling of the PMT helps to reduce the level of thermoelectric emission from the photocathode and is essential to obtain a high signal-to-noise ratio at low levels of emitted light. The sensitivity of different photomultipliers in different cells can be adjusted to the same level by using a standard light intensity source and, if necessary, varying the photomultiplier voltage supply. Because the light flux from the sample has a radial distribution, the sample-to-photocathode distance is kept to a minimum of ~2 cm. The signal output from the anode of the photomultiplier is amplified and passed on to the computer, which functions as a data collection terminal, simultaneously recording the time and the intensity of emitted light.

The two readings thus obtained are placed into an array and a graphical representation of the intensity of light emitted by the sample as a function of time is displayed on the monitor. This information can be obtained for each of eight cells, and the progress of the experiment in each of cells can be followed simply by switching from one cell to another. After the completion of the experiment, the data stored in the computer can be used for analytical evaluations as well as calculation of the area under the curve within any chosen time interval, and the corresponding results are displayed on the monitor. At any time, the experimental data as well as the results of calculations displayed on the monitor can be transferred to the printer to give a hard copy of the monitor image.

Application of Chemiluminescence in the Analysis of Lipid Oxidation

Uninhibited oxidation

Under mild conditions of oxidation, the chain lengths are long and the amount of oxygen participating in the reaction approximately equals the amount of hydroperoxides formed. Under this condition, the amount of hydroperoxide that decomposes to initiate further oxidation is very small and its omission from the expression for oxidation rate is valid. Using the steady-state approximation, the solution of Equations 1 through 4 is

$$-(d[O_2]/dt) = d[LOOH]/dt = k_3(k_1/k_6)\,[LOOH][LH] \qquad [5]$$

If we denote lipid and hydroperoxide concentrations by [A] and [B], respectively, Equation 5 can be rewritten as

$$d[B]/dt = k[A][B] \qquad [6]$$

Equation 6 presents the autocatalytic reaction with regard to substrate and hydroperoxide and, as is typical for autocatalytic reactions, induction and acceleration periods should be expected. Designating the increase in [B] during oxidation as $X = [B] - [B]_0$ and noting that the increase in [B] is equal to the decrease in [A], $[B] - [B]_0 = [A]_0 - [A]$,

$$dX/dt = k([A]_0 - X)([B]_0 + X) \qquad [7]$$

where $[A]_0$ and $[B]_0$ are the initial lipid and hydroperoxide concentrations, respectively, and k is the oxidation rate constant. Integration of [7] gives:

$$k([A]_0 + [B]_0) = \ln\left[\frac{([B]_0 + X)[A]_0}{([A]_0 - X)[B]_0}\right] \qquad [8]$$

Because in all practical cases $[A]_0 >> [B]_0$, the accumulation of hydroperoxides during oxidation can be expressed as

$$X = \frac{[B]_0[\exp(k[A]_0 t) - 1]}{\frac{[B]_0}{[A]_0}\exp(k[A]_0 t) + 1} \quad [9]$$

In view of the direct proportionality between the CL emission intensity and the hydroperoxide concentration

$$I_t = C[B] = C([B]_0 + X) \quad [10]$$

a similar change in the intensity of emitted light with the time of isothermal oxidation is also expected.

As predicted by Equation 9, the hydroperoxide buildup (and, therefore, the buildup in the CL intensity) is slow at the beginning of oxidation (Fig. 8.3a). Then, as the process progresses, it begins to display a characteristic autoacceleration pace with an exponential rise in the intensity of emitted light. Next, the light intensity passes the inflection point (the maximum rate of the hydroperoxide buildup) and

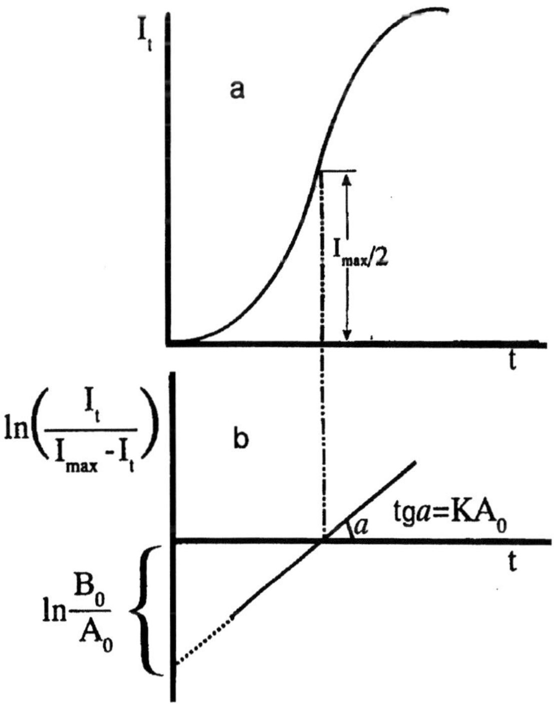

Fig. 8.3. Kinetic data plotted in accordance with Equations 9 (panel a) and 12 (panel b).

its growth gradually slows down, approaching a limiting value. In practice, however, the light intensity often starts to decline after passing through a maximum. The latest descending phase may be associated with secondary reactions and appreciable volatilization at high conversions. In such cases, the limiting light intensity value can be approximated by its maximum value.

The oxidation stages described above lead to a well-recognized sigmoidal change in the accumulation of hydroperoxides and the intensity of emitted light with the time of oxidation. The information gained from a typical CL experiment is shown in Figure 8.4, which represents the intensity of light emitted as a function of

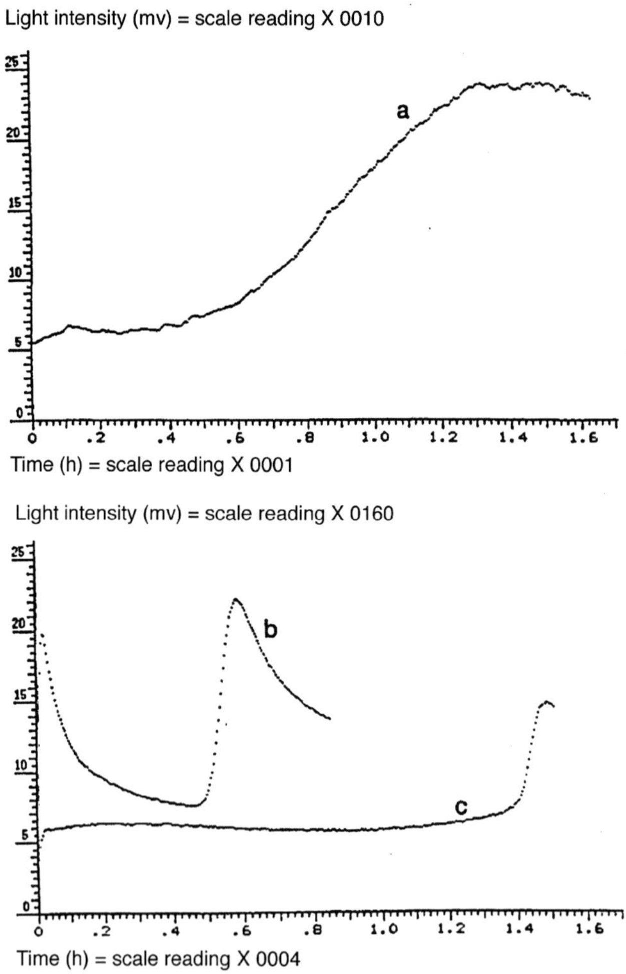

Fig. 8.4. The chemiluminescence curves for three lipid samples: (a) methyl linoleate, 100°C, (b) peanut butter, 130°C, and (c) frying fat, 140°C.

the time the sample is heated at constant temperature in an oxidizing atmosphere.

When CL intensity reaches maximum:

$$I_{max} = C([A]_0 + [B]_0) \text{ or } I_{max} = C([A]_0 - X + [B]_0 + X) \qquad [11]$$

Substituting $([B]_0 + X)$ and $([A]_0 - X)$ from Equations 10 and 11 into Equation 8 and taking into account that $[A]_0 >> [B]_0$,

$$\ln\left(\frac{I_t}{I_{max} - I_t}\right) = \ln\left(\frac{[B]_0}{[A]_0}\right) + k[A]_0 t \qquad [12]$$

The latter equation offers a convenient way of estimating the induction period $\ln([A]_0/[B]_0)$ and the oxidation rate constant $k[A]_0$: a plot of $\ln[I_t/(I_{max} - I_t)]$ vs. t has intersect $\ln([B]_0/[A]_0)$ and slope $k[A]_0$ (Fig. 8.3b). Figure 8.5 demonstrates the evaluation of these parameters for the three lipid samples shown in Figure 8.4. The detailed analysis of [12] and proof that $\ln([A]_0/[B]_0)$ and $k[A]_0$ represent the induc-

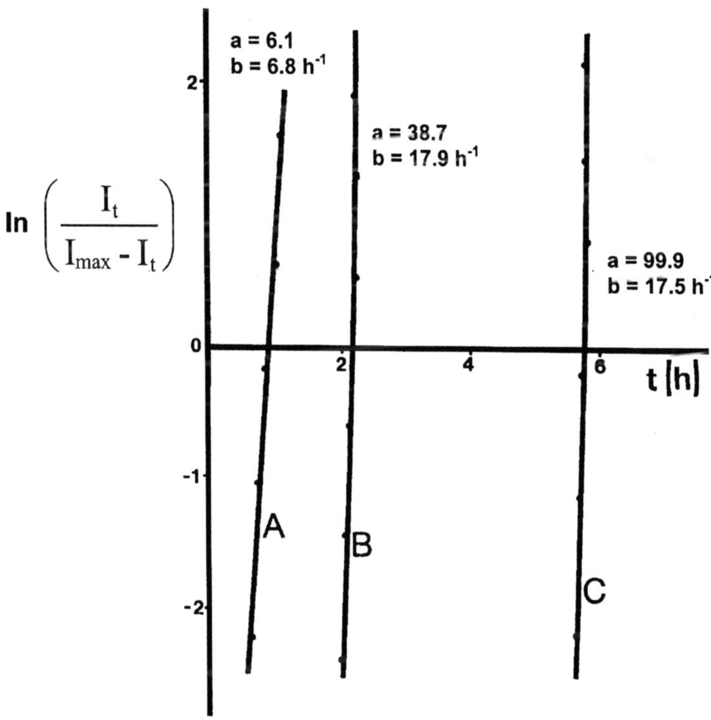

Fig. 8.5. Evaluation of the induction period (a) and the oxidation rate constant (b) values for samples of (A) methyl linoleate, (B) peanut butter, and (C) frying fat.

tion period and the oxidation rate constant were presented elsewhere (Zlatkevich 1989 and 2001).

Inhibited Oxidation

The sequence of the elementary reactions [1] through [4] is applicable to uninhibited autoxidation. If an inhibitor InH is added to the system, then at least two further elementary reactions should be considered:

$$LO_2^{\bullet} + InH \xrightarrow{k_7} LOOH + In^{\bullet} \quad [13]$$

$$LO_2^{\bullet} + In^{\bullet} \xrightarrow{k_8} LOOIn \quad [14]$$

In the presence of a sufficient quantity of inhibitor, the normal chain termination reaction [4] would be completely replaced by the reactions [13] and [14]. Reaction [14], which is the recombination of radicals, does not require any activation energy except for the kinetic energy required for the translational motion. Consequently, the k_8 value is very large and the rates of the consecutive reactions [13] and [14] are determined by the former reaction.

High Inhibitor Reactivity. Initiation is almost immediately followed by chain termination, and no chain propagation can develop. The autoxidation is delayed until all inhibitor is consumed. Hydroperoxides are depleted in the initiation reaction [1] and restored in the termination reaction [13]; it can be assumed that the initial hydroperoxide concentration remains unchanged as long as the inhibitor is present. The steady-state concentration of peroxy radicals is established very quickly:

$$d[LO_2^{\bullet}]/dt = k_1[B]_0 - k_7[LO_2^{\bullet}][InH] \cong 0 \quad [15]$$

or

$$[LO_2^{\bullet}] = k_1[B]_0/k_7[InH] \quad [16]$$

The rate of inhibitor consumption is:

$$-d[InH]/dt = k_7[LO_2^{\bullet}][InH] \quad [17]$$

Substituting [16] into [17], we obtain

$$-d[InH]/dt = k_1[B]_0 \quad [18]$$

Upon integration at the initial condition $t = 0$, $[InH] = [InH]_0$:

$$[\mathrm{InH}] = [\mathrm{InH}]_0 - k_1[\mathrm{B}]_0 t \qquad [19]$$

which means that the consumption of the inhibitor is linear in time and its concentration will be zero at time $t_1 = [\mathrm{InH}]_0/k_1[\mathrm{B}]_0$. The autoxidation is delayed by a period of time that is directly proportional to the initial concentration of the inhibitor and inversely proportional to the rate of initiation. After time t_1, oxidation resumes the autoxidation pace. Thus, the highly active inhibitor prolongs the induction period leaving the oxidation rate unchanged, shifting the oxidation curve as a whole toward longer times. This particular case is shown in Figure 8.6 and the equation that describes strong inhibition is:

$$\ln\left(\frac{I_t}{I_{max} - I_t}\right) = -\left[\ln\frac{[\mathrm{A}]_0}{[\mathrm{B}]_0} + [\ln \mathrm{H}]_0 \frac{k[\mathrm{A}]_0}{k_1[\mathrm{B}]_0}\right] + k[\mathrm{A}]_0 t \qquad [20]$$

Moderate Inhibitor Reactivity. The steady-state approximation for inhibited autoxidation (reactions [1], [2], [3], and [13]) leads to the following equation for the rate of oxidation:

$$\mathrm{dB/dt} = k_3(k_1/k_7)\frac{[\mathrm{A}][\mathrm{B}]}{\ln \mathrm{H}} = k^1\frac{[\mathrm{A}][\mathrm{B}]}{\ln \mathrm{H}} \qquad [21]$$

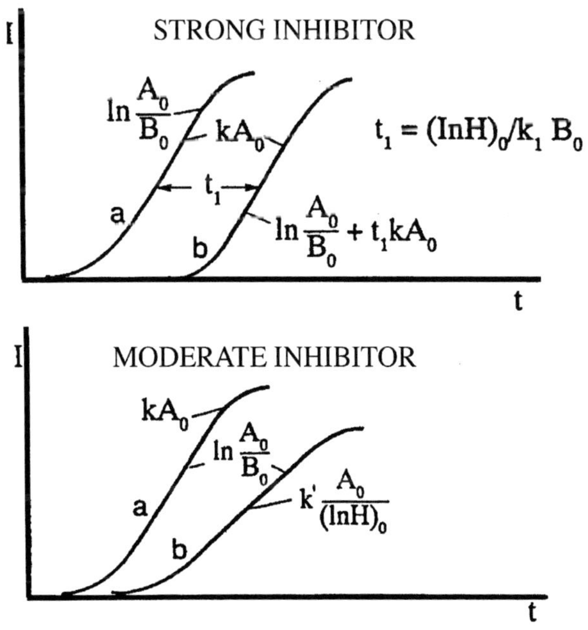

Fig. 8.6. The influence of strong and moderate inhibitors on oxidation: (a) uninhibited oxidation and (b) inhibited oxidation.

and for the early stages of oxidation, the equation analogous to [12] is:

$$\ln\left(\frac{I_t}{I_{max} - I_t}\right) = \ln\left(\frac{[B]_0}{[A]_0}\right) + \frac{k'}{[\ln H]_0}[A]_0 t \qquad [22]$$

In the case of moderate inhibition, the inhibitor does not postpone the autoxidation but rather slows it down. Both uninhibited and inhibited oxidations are characterized by the same induction period. At the same time, moderate inhibition lowers the oxidation rate constant and yields a less steep oxidation curve (Fig. 8.6).

Assessment of the Efficiency of an Antioxidant from the Experimental Data

In the process of oxidation, the degree of conversion Y changes with time according to the relation (see Equation 9):

$$Y = \frac{X}{[A]_0} = \frac{\frac{[B]_0}{[A]_0}[\exp(k[A]_0 t) - 1]}{\frac{[B]_0}{[A]_0}\exp(k[A]_0 t) + 1} \qquad [23]$$

If we define the critical degree of conversion Y_c as the conversion above which lipid quality deteriorates to an unacceptable level, durability t_c of the lipid can be expressed as (Zlatkevich 2002):

$$t_c = \frac{\ln\left[\frac{Y_c \exp(a) + 1}{1 - Y_c}\right]}{b} \qquad [24]$$

where $a = \ln([A]_0/[B]_0)$ and $b = k[A]_0$ are the induction period and the oxidation rate constant, respectively. In the particular case, $Y_c = 0.5$, $t_c' \cong a/b$.

Let us consider two arbitrary CL curves with the parameters a_1 and b_1 (stabilized sample) and a_0 and b_0 (unstabilized sample). To evaluate how many times the antioxidant improves durability of a lipid at a certain degree of conversion, one has to calculate the stabilized to unstabilized lipid durability ratio for this particular degree of conversion. From a practical perspective, utilization of the value Y = 0.12 might be of interest because it corresponds to the graphically defined induction time (Zlatkevich 2002).

In the case in which the critical degree of conversion is 50% ($Y_c = 0.5$), durabilities of the stabilized and unstabilized samples are a_1/b_1 and a_0/b_0, respectively, and the coefficient of improvement represented by their ratio is

$$\frac{\frac{a_1}{b_1}}{\frac{a_0}{b_0}} = \frac{a_1 b_0}{a_0 b_1}$$

The latter expression is similar to the parameter A introduced by Yanishlieva-Maslarova (2001) as a measure of overall antioxidant activity.

Concluding Remarks

A significant advantage of CL over other methods is its exceptional sensitivity. CL detects oxidative changes much earlier than spectral and calorimetric techniques and it has been estimated that a rate of initiation of oxidation as low as 10^{-11}–10^{-12} mol/(L·s) and radical concentrations of only 2×10^{10} radicals/cm^3 can be studied with ease by CL (Emanuel *et al.* 1984). The other important feature that favorably distinguishes CL from other methods, DSC, in particular, is the remarkable long-term baseline stability. The simplicity of the CL experiment can be deceptive, however. Although the CL technique has very high sensitivity, it suffers from the fact that there is still no fully accepted mechanism for the origin of the CL emission and, indeed, more than one mechanism may be involved in some cases. In addition, the CL intensity depends upon the geometry of the sample and the detector system. Thus, the comparison of light intensities between samples is not reliable. It also depends on the thickness and transparency of the sample.

Until recently, the conventional wisdom was that although CL is inherently very sensitive, the generally complex nature of emission and the frequent interference from trace contaminants make interpretation of the CL data difficult. Because of the lack of a general quantitative approach, many CL studies, although of a certain scientific interest, have had little practical value. At present, however, there are certain developments in the field that allow consideration of CL as a valuable technique in the research regimen. Furthermore, this trend is expected to continue as the scientific community becomes more cognizant of the knowledge to be gained by the use of CL because it may yield data that cannot be provided by any other methodology.

References

Boverus, A., Cadenas, E., and Chance, B. (1980) Low-Level Chemiluminescence of the Lipoxygenase Reaction, *Photochem. Photobiophys. 1*, 172–175.

Cadenas, E., and Sies, H. (1982) Low-Level Chemiluminescence of Liver Microsomal Fractions Initiated by *tert*-Butyl Hydroperoxide, *Eur. J. Biochem. 124*, 349–356.

Emanuel, N.M., and Maizus, Z.K. (1984) *Oxidation of Organic Compounds: Medium Effects in Radical Reactions*, pp. 1–625, Pergamon Press, Oxford.

Kron, A., Stenberg, B., and Reitberger, T. (1997) Chemiluminescence Applied to Oxidation of Polyolefins, *Prog. Rubber Plast. Technol. 13*, 81–107.

Marino, D.F., and Ingle, J.D. (1981) Microprocessor-Based Data Acquisition System for Chemiluminescence Measurements, *Anal. Chem. 53*, 1175–1179.

Mendenhall, G.D. (1977) Analytical Applications of Chemiluminescence, *Angew. Chem. Int. Ed. Engl. 16*, 225–232.

Miyamoto, S., Martinez, G.R., Medeiros, M.H.G., and Di Mascio, P. (2003) Singlet Molecular Oxygen Generated from Lipid Hydroperoxides by the Russell Mechanism: Studies Using ^{18}O-Labeled Linoleic Acid Hydroperoxide and Monomol Light Emission Measurements, *J. Am. Chem. Soc. 125*, 6172–6179.

Neeman, I., and Joseph, D. (1985) Induced Chemiluminescence of Oxidized Fatty Acids and Oils, *Lipids 20*, 729–734.

Pospisil, J., Horak, Z., Pilar, J., Billingham, N.C., Zweifel, H., and Nespurek, S. (2003) Influence of Testing Conditions on the Performance and Durability of Polymer Stabilizers in Thermal Oxidation, *Polym. Degrad. Stab. 82*, 145–162.

Russell, G.A. (1957) Deuterium-Isotope Effects in the Autoxidation of Aralkyl Hydrocarbons. Mechanism of the Interaction of Peroxy Radicals, *J. Am. Chem. Soc. 79*, 3871–3877.

Schard, M.P., and Russell, C.A. (1964) Oxyluminescence of Polymers. I. General Behavior of Polymers, *J. Appl. Polym. Sci. 8*, 985–995.

Schulte-Herbruggen, T., and Sies, H. (1989) The Peroxidase/Oxidase Activity of Soybean Lipoxygenase. II: Triplet Carbonyls and Red Photoemission during Polyunsaturated and Glutathione Oxidation, *Photochem. Photobiol. 49*, 705–710.

Stieg, S., and Nieman, T.A. (1978) Experimental and Theoretical Considerations of Flow Cell Design in Analytical Chemiluminescence, *Anal. Chem. 50*, 401–404.

Timmins, G.S., Santos, R.E., Whitewood, A.C., Catalani, L.H., Di Mascio, P., Gilbert, B.C., and Bechara, E.J.H. (1997) Lipid Peroxidation-Dependent Chemiluminescence from the Cyclization of Alkylperoxyl Radicals to Dioxetane Radical Intermediates, *Chem. Res. Toxicol. 10*, 1090–1096.

Van Dyke, K., McCapra, F., and Behesti, I. (1985) *Bioluminescence and Chemiluminescence Instruments and Applications*, Vol. 1, pp. 1–42, CRC Press, Boca Raton, FL.

Vassil'ev, R.F. (1969) On the Mechanism of Chemiluminescence in Oxidation of Organic Substances and Polymers, *Macromol. Chem. 126*, 231–238.

Whitehead, T.P., Thorpe, G.H.G., Carter, T.J.N., Croucutt, C., and Kricka, L. (1983) Enhanced Luminescence Procedure for Sensitive Determination of Peroxidase-Labelled Conjugates in Immunoassay, *Nature 305*, 155–159.

Yanishlieva-Maslarova, N.V. (2001) Inhibiting Oxidation, in *Antioxidants in Food—Practical Applications* (Pokorný, J., Yanishlieva, N., and Gordon, M., eds.), pp. 22–70, CRC Press, Boca Raton, FL.

Zlatkevich, L. (1987) New Chemiluminescence Apparatus and Method for Evaluation of Thermal Oxidative Stability of Lubricants, *Lubr. Eng. 44*, 544–552.

Zlatkevich, L. (1989) Chemiluminescence in Evaluating Thermal Oxidative Stability of Polymers, in *Luminescence Techniques in Solid State Polymer Research* (Zlatkevich, L., ed.), pp. 135–197, Marcel Dekker, New York.

Zlatkevich, L. (2001) Oxidation Induction Period and Its Evaluation, *Proc. Am. Chem. Soc. Div. Polym. Mater. Sci. Eng. 84*, 965–966.

Zlatkevich, L. (2002) Various Procedures in the Assessment of Oxidation Parameters from a Sigmoidal Oxidation Curve, *Polym. Test. 21*, 531–537.

Zlatkevich, L. (2004) On the Kinetic Order of Decomposition of Polymeric Hydroperoxides, *Polym. Degrad. Stab. 83*, 369–371.

Zlatkevich, L., and Martella, D.J. (1995) Chemiluminescence in Evaluating the Thermal Oxidative Stability of Automotive Oils, *Society of Automotive Engineers International Congress and Exposition*, Detroit, Michigan, Technical paper 951025

Chapter 9

Accelerated Stability Tests

Tom Verleyen, Stefaan Van Dyck, and Clifford A. Adams

Nutritional Services Department, Kemin Agrifoods Europe, Toekomstlaan 42, Herentals, Belgium

Introduction

The oxidation of lipids is one of the most fundamental reactions in food chemistry, and the degree of lipid oxidation has several important consequences for food quality and acceptability. Oxidation of various lipid components in food reduces the nutritional value and generates rancidity, causing undesirable odors and flavors. The oxidative stability of a food is therefore an important parameter in determining its shelf life. Consequently, many different procedures have been developed in attempts to assess oxidative stability in various food ingredients and in the final food products.

The assessment of oxidative stability, however, faces two major difficulties. First, the complexity of the reactions involved in lipid oxidation and the wide range of oxidative compounds produced cause great difficulty in evaluating oxidative status as indicated by Márquez-Ruiz *et al.* (2003). Second, oxidative stability determined in foods in the laboratory may not give an indication of the shelf life of the food in practice.

The process of lipid oxidation develops slowly in the initial stages but then accelerates quickly at later stages. In the lipid oxidation process, there is usually an induction period before massive oxidation occurs. This induction period of a fat is theoretically defined as the time required to obtain a continuous oxidation cycle in the oxidation process of the fat (Frankel 1998). In practice, the induction time is measured as the time required for a sudden and rapid change in the rate of the oxidation process to develop. This induction point should be determined by sensitive analytical techniques. Several methods have been developed to identify this induction point in oxidation studies.

Most of the methods used to determine the induction point and therefore the sensitivity of oils and fats to oxidation are based primarily on the determination of oxidized compounds such as the peroxide value (PV), thiobarbituric acid (TBA)-value, *para*-anisidine value, amount of conjugated dienes, or analysis of volatile oxidation products. Sensorial characteristics of the lipid and absorption of molecular oxygen during the oxidation process can also be measured.

These traditional analytical parameters are frequently also used as quality indicators of fats and oils. A PV or a TBA-value may be determined at one time to assess the oxidative status of a particular food or food ingredient. This approach provides information on the oxidative status of the oil and food product at the time of analysis. However, it does not provide information on changes in the oxidative status of the sample in the future, i.e., during its further shelf life.

Generally foods require a long shelf life; consequently, the oxidative status during storage is very important. As a result, classical shelf-life storage tests could last up to 12 mon; obviously, there is a need to obtain information on the oxidative status of lipids in a much shorter period. Ideally, such accelerated tests should allow predictions to be made on the oxidative stability of the product as a function of time.

To make predictions concerning shelf life, it is necessary to subject the fat, oil, or food product to a continuous accelerated oxidation test for a reasonably short time. A suitable end-point must be chosen to determine the extent of oxidation in the sample. The oxidation process can be accelerated in several ways. The most common methods expose the sample to an increased temperature and elevated oxygen pressure. Other methods based on an initiation of the oxidation process by metal contamination have also been developed but are less widely used. New methods based on free radical generation were developed recently and offer interesting opportunities. These tests have the advantage that they can be used at a lower temperature, which more closely resembles oxidation under normal shelf-life conditions.

In practice, accelerated oxidation tests based on an increase in the temperature and consumption of oxygen are most commonly used. Several tests were developed over the past decades, e.g., weight gain or Schaal Oven test, active oxygen method (AOM), Rancimat, Oxidative Stability Instrument (OSI), or oxygen bomb (Hill 1994, Shermer and Giesen 1997). These methods have been used traditionally in scientific and commercial laboratories dealing with lipids and oxidative stability.

Weight Gain and Schaal Oven Test

The weight gain method consists of measuring the increase in the weight of an oil sample. In an oven at 60°C, the sample will gradually oxidize over time as oxygen is inserted into the lipid molecules. Consequently, the weight of the oil sample will increase due to the oxidation. The weight gain of the oil is an indicator of the degree of oxidation. This test is not sensitive and the end-point is questionable because it requires a very high degree of oxidation. This method has a poor correlation with actual shelf-life predictions (Frankel 1998).

The Schaal Oven test involves heating an oil sample to 50–60°C in an oven. The end-point of oxidation can be detected either by sensorial characteristics or by suitable end-point detection (PV, TBA value). Because this test uses relatively low temperatures, the oil is exposed to mild oxidative stress. The Schaal Oven test correlates well with actual shelf-life predictions (Frankel 1998). This method is time- and labor-consuming and thus impractical as a routine method, which requires a short analysis time.

Active Oxygen Method (AOM Test)

The AOM was the first accelerated technique to be introduced into the lipid industry to obtain information on the oxidative stability of oils and fats. It was introduced initially in 1933 (Hill 1994). For many years, the AOM test was the most widely used test in lipid oxidation research. The parameter selected to follow lipid oxidation in the AOM test is the level of peroxides formed in the oil. The method suffers from many drawbacks in terms of operation and application and is now frequently replaced by other accelerated tests.

In the official AOM test, purified air is bubbled through a fat sample held in a heated oil bath at 97.8°C. At various time intervals, aliquots of oil sample are taken out of the flask for determination of the PV. The PV are plotted against time and the time required to reach a PV of 100 mEq/kg fat is reported as the AOM time (AOCS Cd 12–57). Because lipid oxidation is a dynamic process, the PV must be determined at regular time intervals. Estimation of the time required to reach a PV of 100 mEq/kg should be based on two peroxide measurements ranging between 75 and 175 mEq/kg. If a PV > 175 mEq/kg is obtained, the analysis should be restarted. AOM values determined for several oils and fats are shown in Table 9.1.

Several modifications to the official AOM method were made over the years. Some laboratories operate the method at higher temperatures (Laubli and Bruttel 1986) to shorten the analysis time, whereas other laboratories evaluate the time required to reach a PV of 20 mEq/kg (Johnson 1974, Romoser 1982).

The AOM test has several disadvantages; it is a rather time-consuming and labor-intensive test. Problems with the AOM test may also arise from shortcuts introduced by the operator but that are also partly inherent in the procedure. The AOM test requires at least two titrations of the PV on one oil sample. However, in some quality control laboratories, the method is substituted by a pass/fail system in which a sample is analyzed for a predetermined time range and a single PV is determined. If the resulting PV is below a fixed value, the sample passes the test and is deemed to meet the specification. By using this method, no data on the actual

TABLE 9.1
Active Oxygen Method Time Determined for Several Lipid Samples at Three Different Temperatures[a,b]

Lipid type	100°C	110°C	120°C
Peanut oil	14.2 ± 0.05	7.6 ± 0.07	3.9 ± 0.15
Sunflower oil	8.3 ± 0.53	4.4 ± 0.02	2.4 ± 0.00
Olive oil	30.9 ± 0.23	14.2 ± 0.05	7.7 ± 0.00
Lard	4.4 ± 0.10	2.7 ± 0.02	0.5 ± 0.02
Margarine	22.2 ± 0.07	11.7 ± 0.23	6.1 ± 0.07
Butter	22.2 ± 0.20	11.1 ± 0.13	8.8 ± 0.10

[a]Values are means ± SD, n = 3.
[b]*Source:* Laubli and Bruttel (1986).

stability of the oil sample are obtained because the PV often goes through a maximum; thus, an underestimation of its oxidative status might be obtained.

Even when the AOM method is used correctly, it still has several inherent deficiencies. Oil samples are taken at intervals of several hours, whereas the lipid oxidation process is a continuous process. Peroxides are the first and least stable oxidation products and will easily decompose to more stable secondary oxidation products. Indeed, at a temperature of 98°C, peroxides are readily broken down. Consequently, the AOM method is based on a rather unstable parameter. Some laboratories run the AOM test at even higher temperatures to shorten the analysis time for saturated fats (animal fats, hard vegetable oils). This is likely to exacerbate problems associated with the unstable nature of peroxides at elevated temperatures.

Another major deficiency is the determination of the end-point during the rapid oxidation process. During the rapid and accelerated oxidation phase, the reaction is dependent upon the oxygen supply. Variations in oxygen supply can result in poor reproducibility between duplicate samples. Data obtained from an interlaboratory study published an actual coefficient of variation (CV) of 35%. This means that independent laboratories would report an AOM value of 100 ± 35 h for an oil sample with an AOM specification of 100 h (Jebe *et al.* 1993). The combined effect of these problems is a large variability in the AOM time reported for any particular sample. Consequently, alternative methods were developed to replace the AOM test as an accelerated method to study the stability of oils and fats.

Oxidative Stability Index (Rancimat and OSI)

Several alternatives for the AOM test have been developed. First reports on the basic idea of the OSI principle originate from 1970 (Pardun and Kroll 1972). Upon oxidation of lipids, volatile acids will be formed, which can be monitored by a conductivity measurement. Gradually this method was thoroughly validated (Van Oosten *et al.* 1981), resulting in the translation of these ideas into commercial instruments. The first automated version was the Swift test, followed by Brinkmann 617 Rancimat (Brinkmann Instruments, Westbury, NY), which became available in early 1980. This instrument was capable of running six samples at the same time. The end-point had to be determined manually by drawing tangents to the conductivity curve to identify the inflection point. Over time, these instruments underwent several changes to improve their ease of operation. The initial Rancimat 617 instrument was replaced by the Rancimat 679 instrument (Metrohm, Ltd., Switzerland), which was able to determine the end-points automatically. In 1983, Archer Daniels Midland (ADM) Company developed a computer-assisted instrument that was also able to determine the endpoint automatically. Omnion, Inc. (Rockland, MA) now produces a version of this instrument commercially under a license agreement with ADM. The Rancimat and OSI instruments are recognized by AOCS and have been incorporated into the *Official and Recommended Analysis Manual* (AOCS Cd 12b-92).

Operation and Principle of OSI. The same basic principle lies behind the Rancimat and OSI. These instruments differ only slightly in design and operating convenience. OSI operate by a stream of purified air passing through a sample of fat or oil that is held in a thermostated aluminum heating block. The air distribution system does not heat the air before being bubbled into the oil. The incoming air is regulated with a needle valve to control the flow rate. After passing through the oil, the effluent air is passed into a detection cell that contains deionized water. OSI use the formation of volatile oxidation products as a marker to detect the induction point in the lipid oxidation process. The effluent air containing volatile organic acids from oil oxidation increases the conductivity of the water in the detection cell. Initially, a manual integration of the induction point on the conductivity curve was required. Currently, the conductivity measurement is linked to a computer software program, which allows an automated selection of the induction point in the conductivity curve. OSI analyses are highly reproducible with an interlaboratory SD <6% (Jebe *et al.* 1993).

A typical OSI chart is shown in Figure 9.1. At the start of the oxidation experiment, the conductivity of the water in the detection cell is very low. Heating the oil and simultaneously passing air through it will accelerate the oxidation process. Initially, peroxides will be formed which are unstable and break down to secondary oxidation products. Different secondary oxidation products will be formed depending on the type of oil. Many of these secondary oxidation products have a relatively low volatility. They will not be distilled over into the water in the detection cell, but will remain in the oil. Aldehydes will be further oxidized to short-

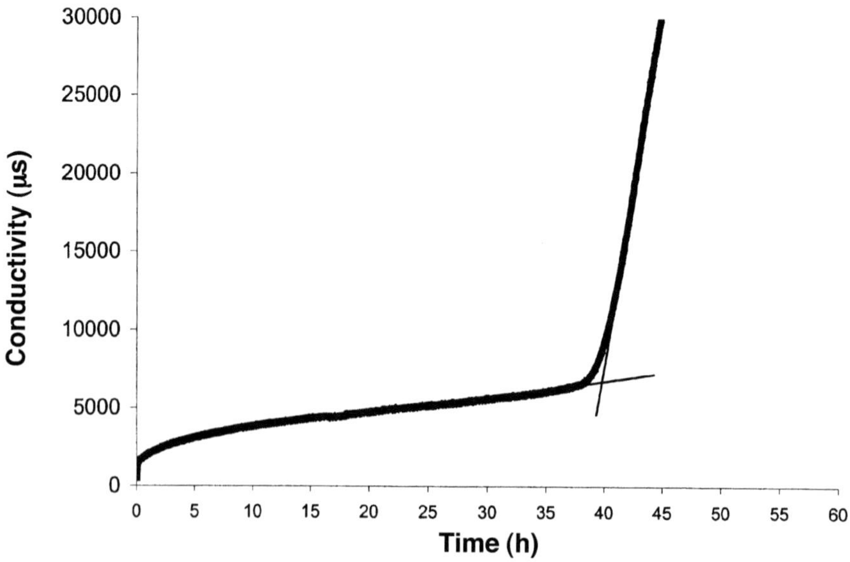

Fig. 9.1. Typical Oxidative Stability Instrument chart.

chain fatty acids. These short-chain acids are volatilized and will condense in the water of the detection cell, increasing its conductivity. Consequently, the conductivity of the water has a direct relation to the degree of oil oxidation. At the induction time of the conductivity curve, several acids will be present in the water. Analysis of the water fraction for its content in short-chain acids indicated that different volatile acids are formed (formic acid, acetic acid, and propionic acid) in varying amounts depending upon the fatty acid distribution of the oil (Table 9.2). For all types of lipids, formic acid is the most important acid formed upon oxidation of aldehydes. Formic acid has a much greater effect on conductivity than acetic acid, and the contribution of other acids to the conductivity is even smaller and can be ignored (De Man *et al.* 1987).

Advantages of OSI Compared with the AOM Test. Methods based on OSI have several advantages compared with the AOM test, which can be summarized as follows:

- The AOM test provides only a single value expressing oxidative stability, which has to be estimated between two measurements. By contrast, OSI provide continuous data, allowing a more accurate detection of the induction point.
- The end-point in the AOM test is measured during the initial oxidation phase, which is strongly dependent upon oxygen availability. This will consequently lead to a higher variability. In OSI, the end-point is determined at the end of the induction period and is less sensitive to the airflow.
- The AOM method relies on the analysis of unstable primary oxidation products, whereas the OSI is based on stable tertiary oxidation products. This has a serious effect on the reproducibility of the two tests. The AOM test has an interlaboratory SD of 35%, which is reduced to 5.7% for the OSI.

TABLE 9.2
Formation of Volatile Acids During Oxidation Expressed as a Percentage of Total Volatile Acids[a]

	Type of lipid (%)				
Organic acid	Butterfat	Canola	Lard	Soybean	Sunflower
Formic	57.8	74.4	79.3	77.4	69
Acetic	15.4	19.4	11.9	13.3	5.4
Propionic	3.8	3.4	0.8	2.1	0.1
Butyric	7.6	ND[b]	ND	0.9	ND
Valeric	5.5	0.3	2.2	0.5	2.2
Caproic	9.7	2.5	7.4	5.8	23.3

[a]*Source:* De Man *et al.* (1987).
[b]ND, not detected.

- In the AOM test, the induction point can be exceeded, requiring a reanalysis of the sample. In OSI, the induction point can never be exceeded because the data are acquired continuously.
- The AOM test is labor- and time-intensive, whereas OSI are fully automated.
- The AOM method is very sensitive when operated at high temperatures because the end-point detection is based upon peroxides, which are unstable oxidation products. The OSI are less sensitive when operated at elevated temperatures because the end-point detection is based on ternary oxidation products, which are heat stable.
- The AOM end-point detection is based on a manually determined end-point, whereas the oxidation stability equipment has instrumental end-point detection.

In general terms, there are clearly several differences in favor of the OSI compared with the AOM method. At present, OSI have become valuable and reliable methods with which to evaluate the stability of oils and fats. Over the past years, a shift from using AOM equipment to a generalized application of OSI was observed.

Influence of Operating Parameters. The induction time reported by OSI depends on several operational parameters. The most important parameter influencing the induction time is clearly the operating temperature because lipid oxidation is a chemical reaction, which is temperature dependent. At higher temperatures, lipid oxidation will proceed more quickly, leading to a lower induction point. The temperature of the heating block in the Rancimat and OSI instrument can be varied between 50 and 160°C. Generally, the oxidative stability of oils and fats is determined at 100 or 110°C. Accelerated tests on saturated fats are often carried out at temperatures of 120–130°C to reduce the analysis time. On the other hand, lipids sensitive to oxidation are often analyzed at lower temperatures in the OSI, ranging between 60 and 80°C. Fish oil, for example, is frequently analyzed at 68°C (Mendez *et al.* 1996).

The effect of temperature on the induction point has been thoroughly evaluated, and linear equations are shown in Table 9.3. For vegetable oils, a strong linear relation between 100 and 140°C was obtained upon plotting the logarithm of the induction point vs. temperature (Hasenhuettl and Wan 1992). Fish oils also had a good linear relation between the oxidation stability induction point and the temperature from 55 to 90°C (Mendez *et al.* 1997). For these temperature ranges, an almost perfect linear response was obtained with linear correlation coefficients >0.99 (Anwar *et al.* 2003, Hasenhuettl and Wan 1992, Mendez *et al.* 1996). Furthermore, information on the kinetics of lipid oxidation can be derived from these equations. An increase in the temperature by 10°C results in a decrease in the induction point by a factor of 1.90–2.14 for all oils (Anwar *et al.* 2003, Hasenhuettl and Wan 1992, Laubli *et al.* 1988, Laubli and Bruttel 1986, Mendez *et al.* 1996). This is in good

TABLE 9.3
Relation Between Oxidative Stability Instrument (OSI) Induction Time and Temperature for Several Vegetable Oils and Fats

Type of oil	Log OSI = a × Temperature (°C) + b[a]	Δ 10°C
Olive[a]	$y = -0.029235x + 3.873735$	1.96
RBD corn[a]	$y = -0.030792x + 4.106362$	2.03
Peanut[a]	$y = -0.031561x + 3.873735$	2.07
Soybean[a]	$y = -0.030099x + 4.238271$	2.00
Safflower[a]	$y = -0.027826x + 3.481556$	1.90
Lard[a]	$y = -0.031009x + 4.389382$	2.04
Anchovy[b]	$y = -0.076x + 7.3$	2.14
Hake liver[b]	$y = -0.073x + 6.2$	2.07
Sardine[b]	$y = -0.075x + 6.4$	2.12

[a]Equations were obtained in a temperature range between 100 and 140°C. *Source:* Hasenhuettl and Wan (1992).
[b]Equations were obtained in a temperature range between 55 and 90°C. *Source:* Mendez *et al.* (1997).

agreement with general data on the kinetics of the lipid oxidation process (Frankel 1998). At temperatures >150°C, the logarithm of the oxidation stability induction time loses its linear response to temperature. At temperatures >120°C, volatilization of synthetic antioxidants might occur, leading to an underestimation of the oxidative stability (Dijkstra *et al.* 1996, Hill 1994). The rate of oxidation may also be limited by the mechanism of degradation because the rate of formation of volatile acids is likely reduced above a certain temperature. The end-point detection in OSI is based on the formation of volatile acids (Reynhout 1991). At high temperatures, the induction time will be too low for an accurate measurement. In general, oxidation times should not be lower than 0.5 h. Ideally induction time should be at least 2 h to minimize deviation between analyses.

The effect of other operating parameters on the induction point determined by OSI was also investigated, but these had a smaller influence than temperature (Hill and Perkins 1995). The size of the oil sample (2.5 or 5 g) influenced the oxidation stability induction time. A small sample size of 2.5 g will oxidize with a much greater variability than a larger sample size. A thorough and uniform distribution of air in the oil sample is crucial to obtain repeatable results. The temperature of the water in the detection cell has no influence on the conductivity and the induction time. This temperature should be as low as possible to limit water loss by evaporation (Hill and Perkins 1995). Operating the oxidation stability tests according to the AOCS standard method is strongly advised to obtain accurate and reproducible results.

Correlation Between Rancimat and OSI. An extensive collaborative study was developed to establish the confidence limits for OSI. In that study, several vegetable oil samples having variable stabilities were analyzed. The samples studied were nonhydrogenated vegetable oils, hydrogenated vegetable oils, margarines, and free fatty acids. They were analyzed by both the automated Rancimat and the

OSI instrument. All samples were analyzed at both 110 and 130°C. Results were obtained by weighting the number of results for each temperature class and instrument type. The interlaboratory CV is shown in Table 9.4. These results represent the variation of the instrument when viewed across several instruments of that type. The OSI instrument had the lowest CV of 5.7%. The automated Rancimat had a significantly higher CV of 15.0%. According to Jebe *et al.* (1993), this difference is probably related to a lower variation in operating temperature of the OSI instrument compared with the Rancimat instrument. Another interlaboratory test performed with the Rancimat method indicated that the automated Rancimat equipment also had CV on the order of 5% (Woestenburg and Zaalberg 1986).

Correlation Between OSI and Other Oxidation Methods. Accelerated shelf-life tests may be very useful in lipid oxidation studies; however, these methods should of necessity correlate closely with the Schaal oven test and other studies representing shelf-life conditions. It is crucial to establish a correlation between accelerated stability tests and classical oxidation tests based on the PV, secondary oxidation products, and most importantly, with sensory analyses (Frankel 1998, Liang and Schwarzer 1998, Warner *et al.* 1989).

Gordon and Mursi (1994) established a correlation between the Rancimat induction time and the PV upon storage at room temperature. Oil samples were stored in a beaker, loosely covered with a foil lid, and stored in the dark at 20°C. Samples were removed periodically and the PV determined. The time for various oil samples to reach PV of 5, 10, and 20 mEq/kg was monitored and correlated with the initial Rancimat induction time operated at 100°C. This study was done on several oils, *viz.*, refined rapeseed oil, soybean oil, corn oil, sunflower oil, safflower oil, and olive oil. The effect of antioxidant stabilization was also monitored. For natural vegetable oils, the time required to obtain a PV of 10 or 20 mEq/kg correlated well with the initial Rancimat induction time. Correlations >0.95 were obtained. Oil samples stabilized with the antioxidants butylated hydroxytoluene (BHT) and butylated hydroxyanisole (BHA) were very efficient in delaying the formation of peroxides at room temperature. It appeared that the induction time of

TABLE 9.4
Interlaboratory Coefficient of Variation (CV) by Instrument Type and Temperature[a]

Instrument	Temperature (°C)	Number of samples analyzed	Number of instruments	Average CV
Automated	110	305	9	14.2
Rancimat	130	231	9	16.1
	Overall	536		15.0
OSI	110	137	4	5.9
	130	107	4	5.4
	Overall	244		5.7

[a]*Source:* Jebe *et al.* (1993).

samples stabilized with antioxidants determined by the Rancimat method gave an underestimation of the predicted oxidative stability compared with the actual shelf life during storage. Rapeseed oil stabilized with 200 ppm BHT had a Rancimat induction time of 20.5 h. Taking into account that the oxidation reaction will be halved upon lowering the temperature by 10°C, it would take 157 d at 20°C to reach a PV of 20 mEq/kg. However, the PV had reached only 2.3 mEq/kg after 185 d of actual storage.

Broadbent and Pike (2003) carried out a correlation between the induction time determined in the OSI and the sensory analysis of canola oil. The canola oil was stored in the dark at 60°C for 1 wk. Samples were taken daily for stability analysis and sensory evaluation. The time required for an oil sample to reach an average sensory score of 5 on a 10-point scale was used as the sensory induction time. During storage, the induction time of canola oil decreased from 9.88 h (d 0) to 5.7 h (d 6), whereas the sensory evaluation decreased from 6–8 (good quality) to 1–4 (strongly rancid). However, a large variation in sensory characteristics was reported among different panelists. This large variation among panelists is frequently observed in sensory tests (Warner and Nelson 1996). Nevertheless, mean sensory scores correlated well with the oxidative stability induction time as a function of storage at 60°C, with correlation coefficients of 0.89.

The relation between the OSI induction time and the sensory evaluation of soybean oil was investigated by Coppin and Pike (2001). Soybean oil was treated with different levels of copper-2-ethylhexanoate as a prooxidant to induce oxidation. Samples were exposed to light and held at ambient temperature for 3 wk to oxidize. The PV of the oil samples treated with this prooxidant increased gradually and consistently with the amount of prooxidant added. The initial oxidation stability induction values correlated very well with the sensory-determined induction values having a correlation coefficient of 0.92.

During the storage period oxidative stability values were monitored as well to correlate the oxidative stability of partially oxidized oil with the sensory induction period. A quite good correlation was obtained between the oxidative stability values and sensory evaluations, ranging between 0.90 and 0.98 (Coppin and Pike 2001). Thus, the oxidative stability value gave a good indication of the oxidative stability of oils, even when partially oxidized.

Barrera-Arellano and Esteves (1992) analyzed the oxidative stability of potato chips with Rancimat. Fried potato chips were stored at elevated temperature and their stability was tested by both Rancimat and sensory analysis. Rancimat analysis involved pulverizing the chips before instrumental analysis. The conductivity curve of the potato chips showed two inflection points. A first inflection point occurred after a few minutes due to volatiles present in the fried product and a second inflection point was observed at later stages. A correlation of 0.86 was obtained between the Rancimat induction time and the sensory evaluation.

The onset of rancidity, as determined by human sensory analysis, is the ultimate test to evaluate the oxidative stability and oxidative status of oils and fats.

Only sensory analysis can detect off-flavor formation by oxidative and nonoxidative degradation reactions. The sensory induction time can be defined as the time required for an oil sample to become slightly rancid as determined by a sensory panel. A sensory test as a function of the shelf life of an oil should have a perfect correlation with the storage conditions applied by the consumer (Frankel 1998).

However, there is considerable evidence to confirm the usefulness of accelerated tests using instruments such as OSI and Rancimat to predict the oxidative stability of oils and fats. In general, correlation coefficients >0.90 are obtained between OSI induction time and chemical or sensory analyses. Correlations between the identification of volatiles and the sensory evaluation also ranged between 0.95 and 0.99 for different vegetable oils (Warner and Nelson 1996). Consequently accelerated oxidation instruments have a good correlation with the oxidation of oils and fats under actual shelf-life conditions. Accelerated data should always be interpreted carefully. It gives a good indication of the current oxidative status and the oxidative stability of the fat as a function of time. Qualitative information on the oxidative status is obtained in a short analysis time, justifying the application of these methods.

The Oxygen Bomb

Another method frequently used to evaluate the oxidative stability of food products is the oxygen bomb method. Initially, the oxygen bomb method was developed in the petroleum industry. Subsequently, it was transferred to the food and feed industries, finding application as a useful method with which to evaluate the stability of oils and final food products containing lipids. The oxygen bomb method has been frequently used to evaluate the oxidative stability of food products (potato chips, crackers, biscuits and nuts) and feed products (grains, animal feed, fish meal) (Gearhart *et al.* 1957, Inglis and Willington 1976, Shermer and Giesen 1997). The major advantage of the oxygen bomb method is that the final food products can be analyzed because unlike the oxidation stability methods, the method is not restricted to pure oils and fats.

The oxygen bomb apparatus consists of a stainless steel container (bomb) connected to a pressure recorder. A sample of the product to be analyzed is weighed into a glass jar and inserted into the bomb, which is partially closed, and the system is purged with pure oxygen. This is done to replace the air in the vessel by pure oxygen. After purging, the bomb is closed tightly and the oxygen pressure is increased up to 5 bar to accelerate the oxidation process and shorten the analysis period. The bomb is heated by placing it in an oil bath or a heating block. Depending upon the product to be analyzed, the bath temperature is varied between 80°C (for products rich in polyunsaturated fatty acids such as linseed oil and fishmeal) to 100°C (for grains, cereals, and biscuits). The pressure in the headspace of the bomb is monitored continuously through a pressure transducer. For ease of interpretation, the amount of oxygen taken up by the product is usually calculated and plotted against time. As the product oxidizes, oxygen from the headspace will be incorporated into the lipid molecules, leading to a reduction in the oxygen pressure inside the bomb. In the initial stages, the product is frequently stable to oxidation and the pressure remains constant. After some time, oxygen will be readily consumed and incorporated into the product. This is the induction point in the pressure chart as a function of time. Products having a fast and large oxygen uptake will be more prone to oxidative degradation (Blankens *et al.* 1973, Gearhart *et al.* 1957). The susceptibility to oxidation of products in the oxygen bomb apparatus is based mainly on their total fat content. The effect of soybean oil mixed into milled wheat on the oxidation rate is shown in Figure 9.2. Increasing the soybean oil level clearly resulted in a faster oxidation of the meal in the oxygen bomb instrument.

Evaluation of Antioxidant Activity by Accelerated Tests

It is important to be able to evaluate the relative efficacy of various antioxidants, and accelerated tests make a useful contribution here. In general, a combination of different accelerated stability tests should be used when assessing the oxidative stability of lipids and evaluating the effectiveness of antioxidant stabilization (Liang and Schwarzer 1998).

In the past, the weight gain and the Schaal oven test were often used in antioxidant research. These methods gave good results in evaluating the activity of antioxidants and they have been used on several lipid substrates. For example, *tert-*

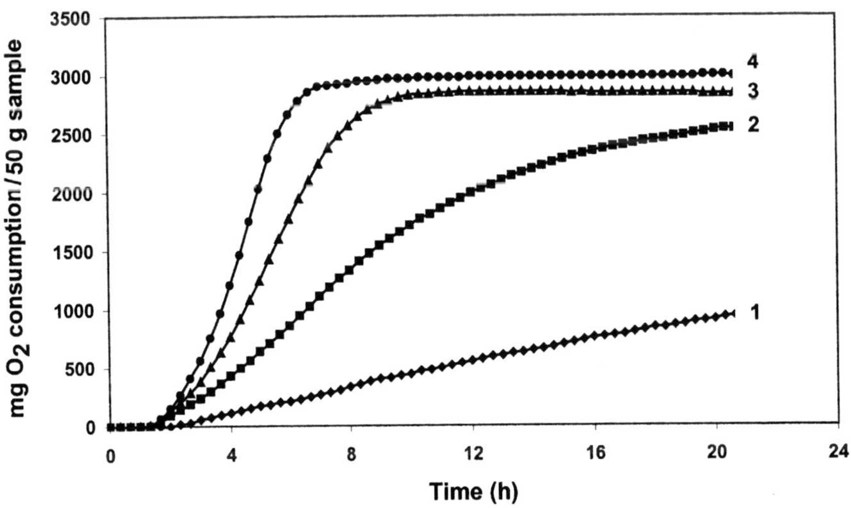

Fig. 9.2. Effect of lipid concentration on oxidative stability of products in the oxygen bomb: (1) 0% soybean oil, (2) 5% soybean oil, (3) 8% soybean oil, and (4) 12% soybean oil in milled wheat.

butylhydroquinone (TBHQ) was the most efficient antioxidant to stabilize crude whale oil, followed by propyl gallate and BHA (Chahine and Macneill 1974). The Schaal oven test was used to evaluate the effect of synthetic antioxidants and of a natural canola extract to stabilize canola oil against oxidation (Wanasundara and Shahidi 1994). The effect of natural antioxidants present in olive oil to protect the oil against oxidation was studied by determining the PV (Gutfinger 1981, Gutierrez *et al.* 2002, Papadopoulos and Boskou 1991) and by following the formation of volatiles in the Schaal oven test (Satue *et al.* 1995). The oxidative stability of corn oils with different fatty acid compositions was also analyzed in the Schaal oven test (Shen *et al.* 1999). The Schaal oven test, however, has the serious disadvantages of being labor-intensive and time-consuming because samples have to be stored for several weeks before results are available.

The AOM method has been used frequently to evaluate the efficacy of antioxidants in the stabilization of oils and fats (Ahmad *et al.* 1983, Kurechi and Kunugi 1983, Liang and Schwarzer 1998, Romoser 1982). The AOM method was also used to evaluate the synergistic activity of a mixture of antioxidants (Kurechi and Kato 1980, Kurechi and Yamaguchi 1980). The ability of antioxidants to stabilize oils and fats in the AOM test is shown in Table 9.5. The antioxidants TBHQ and propyl gallate had the highest efficacy in delaying the formation of peroxides in sunflower oil in the AOM test. In animal fats (yellow grease and poultry fat), ethoxyquin and BHA had the highest efficacy.

The Rancimat and OSI were used frequently to evaluate the efficiency of both synthetic and natural antioxidants (Aparicio *et al.* 1999, Chen and Ho 1997, Liang and Schwarzer 1998, Pongracz 1984) to stabilize oils and fats against oxidation. Natural extracts were also screened for their antioxidant potential by the Rancimat and Oxidation Stability Instrument (Gu and Weng 2001, Wang *et al.* 2000, Weng and Wang 2000). These methods are also very useful for studying the intrinsic sta-

TABLE 9.5
Effect of Antioxidants in Stabilizing Oils and Fats as Determined by the Active Oxygen Method (AOM)

Antioxidant	Sunflower oil[a] (h)	Yellow grease[b] (h)	Poultry fat[b] (h)
Control	6.8	4.5	0.33
TBHQ[c]	27.9	82	5.25
Propyl gallate	15.3	—	—
BHT	9.8	22	4
BHA	7.9	125	15
EQ	—	312	142

[a]AOM time required to reach a peroxide value of 100 mEq/kg antioxidant dosage 200 g/t. *Source:* Ahmad et al. (1983).
[b]AOM time required to reach a peroxide value of 20 mEq/kg, antioxidant dosage 500 g/t. *Source:* Romoser (1982).
[c]Abbreviations: TBHQ, *tert*-butylhydroquinone; BHT, butylated hydroxytoluene; BHA, butylated hydroxyanisole; EQ, ethoxyquin.

bility of oils and fats. Crude oils generally have a higher oxidative stability compared with fully refined vegetable oils due to the presence of natural antioxidants. During the refining process, natural antioxidants are partially lost, resulting in a lower oxidative stability of the refined vegetable oil in the accelerated tests as shown in Figure 9.3. In addition, it was demonstrated that antioxidants are more effective in stabilizing refined vegetable oils than crude vegetable oils (Akoh 1994, Kajimoto and Murakami 1998). At a high concentration, tocopherols, however, might become prooxidant and decrease the oxidative stability of the oil in the accelerated tests (Akoh 1994, Satue *et al.* 1995). Nakatani *et al.* (2001) suggested using a model substrate based on a mixture of methyl linoleate and silicone oil to evaluate the activity of antioxidants by the accelerated oxidation stability test.

The efficiency of different antioxidants in stabilizing lard, soybean oil, and fish oil is illustrated in Table 9.6. All antioxidants were effective in stabilizing lard as demonstrated by an increase in the oxidation stability induction time. Even at a low antioxidant concentration, a significant stabilization of lard was observed for the different antioxidants. The antioxidant ethoxyquin is frequently used in the rendering industry. However, ethoxyquin as the sole antioxidant resulted in only a

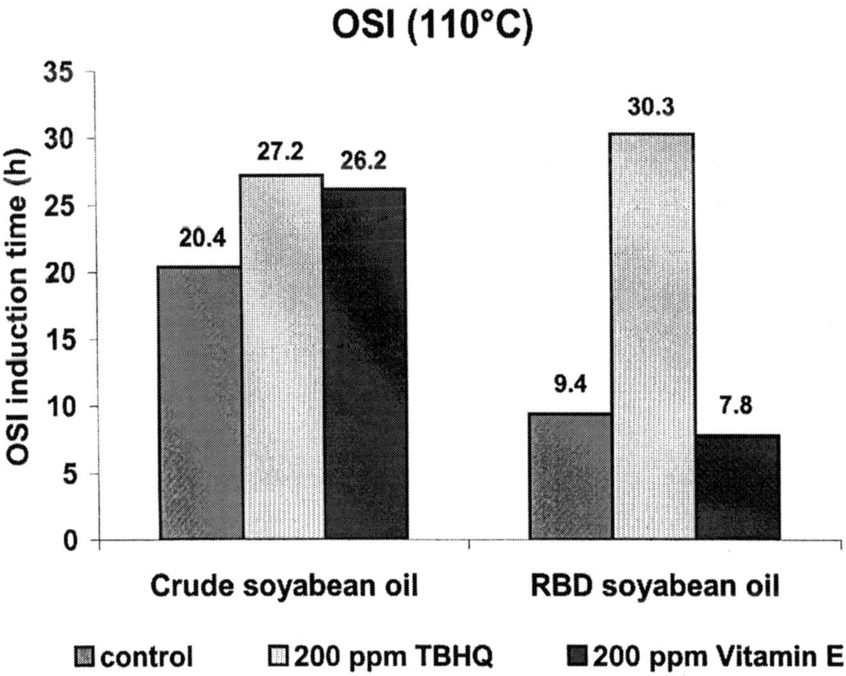

Fig. 9.3. Effect of antioxidants on the stabilization of crude vs. refined soybean oil. OSI, Oxidative Stability Instrument. *Source:* Akoh 1994.

TABLE 9.6
Oxidative Stability Instrument (OSI) Induction Time (h) of Lard, Soybean, and Fish Oil Stabilized with Synthetic Antioxidants at a Concentration of 125 and 250 ppm[a]

Type of oil	Concentration (ppm)	Antioxidant				
		Control	EQ	BHT	BHA	Propyl gallate
Lard (98°C)	125	6.0	12.1	17.9	31.3	33.1
	250	6.0	14.1	27.4	—	—
Soybean oil (98°C)	125	11.9	11.0	15.4	12.7	30.2
	250	11.9	10.9	18.8	13.5	40.6
Fish oil (80°C)	125	4.2	6.3	4.6	4.4	6.9
	250	4.2	7.7	5.0	4.8	9.4

[a]See Table 9.5 for abbreviations.

moderate stabilization of lard in the oxidation stability test. Other antioxidants commonly used (BHA, BHT) had a markedly higher antioxidant activity in lard compared with ethoxyquin.

Soybean oil and fish oil were much more difficult to stabilize by the different antioxidants because the increase in induction time was considerably lower. Because fish oil is very susceptible to oxidation, the OSI was operated at 80°C. Ethoxyquin, which had little antioxidant activity in stabilizing soybean oil, effectively stabilized fish oil. Ethoxyquin has long been recognized as a very effective antioxidant to stabilize fish oils. The antioxidants BHT and BHA gave only a moderate increase in oxidative stability of fish oil. Propyl gallate was the most effective antioxidant for stabilizing soybean oil and fish oil.

Based on these observations, it is clear that the antioxidant efficiency depends strongly on the matrix. Saturated lipids can be stabilized more effectively by antioxidants compared with unsaturated lipids. Therefore, the lipid oxidation mechanism depends on the fatty acid composition of the matrix. This observation was reported in the literature as well (Frankel 1998, Loliger 1991). The antioxidant activity of BHT and BHA clearly depends on the matrix. In a saturated matrix (e.g., lard), BHA was a better antioxidant than BHT, whereas in an unsaturated matrix (e.g., soybean oil, fish oil) BHT was a better antioxidant than BHA. In both saturated and unsaturated oils, propyl gallate and TBHQ have the highest antioxidant activity (Fritsch *et al.* 1975, Luckaddo and Sherwin 1972, Moore and Bickford 1952, Sherwin 1976 and 1978).

The oxygen bomb instrument is also very useful to evaluate the effect of antioxidant stabilization. A typical oxygen bomb chart of nuts stabilized with natural antioxidants is shown in Figure 9.4. The untreated nuts were very prone to oxidation, having an induction point after 9 h. Rosemary-based antioxidants were very effective in stabilizing the nuts and protecting them against oxidation. In conclusion, a combination of different accelerated stability tests should be used when

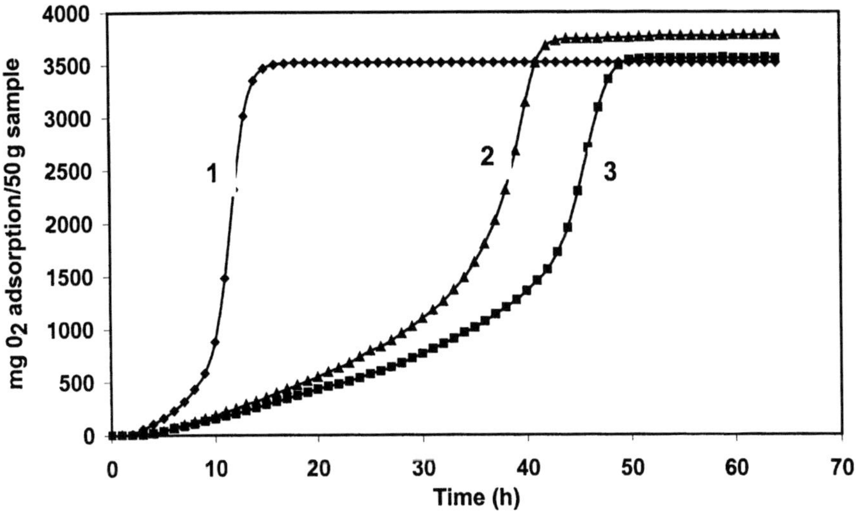

Fig. 9.4. Oxygen bomb chart demonstrating the effect of rosemary extract in stabilizing nuts against oxidation: (1) control nuts (9 h); (2) nuts stabilized with 1000 ppm rosemary extract (35 h); and (3) nuts stabilized with 1500 ppm rosemary extract (44.5 h).

assessing the oxidative stability of lipids and evaluating the effectiveness of antioxidants (Liang and Schwarzer 1998).

Evaluation of Antioxidant Activity by Accelerated Tests at Low Temperatures

The conventional accelerated methods such as OSI, oxygen bomb, and Rancimat operate under forced conditions using high temperature to initiate the oxidation process (Liang and Schwarzer 1998, Rossell 1989). In many cases, very good correlations can be found between these accelerated methods and real shelf life.

In specific cases, the use of higher temperatures limits the correlation with shelf life. One important reason for this is the relation between the rate of oxidation and temperature, which complies with the Arrhenius equation. Starting from the equation, it is possible to calculate that the fraction of the molecules able to react doubles by increasing the temperature by 10°C. That causes the rate of reaction to almost double. This rule-of-thumb is often used in simple rate of reaction work. Like most simple rules, it is only an approximation and therefore should be used with great care, especially for the estimation of shelf life. Essential limitations exist relating to the applicability of Arrhenius's law. The reaction mechanism of the oxidation process typically changes at higher temperature. In most instances, the corresponding activation energy of the new mechanism will be different, and the linear correlation between the test and the actual shelf life may be lost.

Another drawback of high-temperature analysis is the risk of modification of the original structure of the food or feed matrix. Emulsions can break easily, fats will melt, proteins will coagulate, and water may evaporate out of the product. All of these transitions can change the matrix considerably and therefore have a dramatic effect on the analytical results (Frankel and Meyer 2000). In addition, the analysis of more volatile or heat-sensitive antioxidants at high temperatures may cause an unknown amount of antioxidant activity to be lost due to evaporation or decomposition of the active compounds. This effect may result in an underestimation of the real antioxidant efficacy. In addition to increased temperature, there are other ways to increase the oxidation rate. Prooxidant metal ions can be used, but this is not an option for food products (Yoshida and Niki 1992a and 1992b). A high level of metal ions would prohibit the evaluation of preventive antioxidants such as metal chelators.

Apolar diazo-type free radical sources are also known to accelerate the oxidation reaction and have already been used in oxidation studies. However, the most important shortcoming is the use of model systems that are often not relevant to food systems such as free fatty acids or their methyl esters in an emulsion or a liposome (Decker *et al.* 2000, Mei *et al.* 1999). These systems are primarily aqueous and often use charged synthetic emulsifiers such as SDS. The models often neglect important compositional and interfacial phenomena that are pertinent to food matrices and therefore are not useful for antioxidant evaluation in food, or as a tool for the prediction of shelf life. 2,2′-azobis(2,4-dimethylvaleronitrile) (AMVN) is the preferred azo initiator to induce lipid oxidation because its lipophilic properties guarantee the formation of radicals in the lipid phase. The alternative water-soluble 2,2′-azobis(2-amidinopropane) dihydrochloride (AAPH) initiator is not suitable because it will create reactive oxygen species (ROS), which are more typical for the aqueous phase. Consequently, this pathway of radical generation may lead to artifacts in the evaluation of food samples because the oxidation mechanism differs greatly from the nonaccelerated lipid oxidation at normal shelf-life temperature. The free radical generation by AMVN is much more closely related to the natural oxidation process because the radicals generated in the lipid phase will react immediately with both mono- and polyunsaturated fatty acid moieties. This propagation reaction accelerates the generation of the lipid radicals that are of interest in lipid oxidation studies and hence accelerates the lipid oxidation process at temperatures that are much lower than in existing methodologies. One important precaution to take is to avoid a massive flux of radicals into the lipid system because this would not be a realistic model for oxidation. Because the generation of radicals can be controlled by changing the analysis temperature, it is possible to select conditions that give the reactive products enough time to migrate, branch, or proceed to other reactions.

The use of radical initiators can be combined with conventional accelerated techniques to circumvent the disadvantages of high-temperature accelerated oxidation. To distinguish this novel low-temperature method from the traditional accelerated tests, an adapted nomenclature was proposed, i.e., analyses that are accelerated using a free radical source instead of heat could be classified as "Free Radical Generation assays," abbreviated as "FRG assays." Depending on the method used to detect oxidation, one could use combinations comparable to the nomenclature of hyphenated analytical techniques. Possibilities are FRG-OSI in which oxidation is measured through an increase of conductivity as in the OSI, or FRG-OB in which the pressure in the headspace is measured as in the oxygen bomb. Combinations with other techniques were assigned accordingly (Van Dyck *et al.* 2005).

FRG-OSI and FRG-OB. The AMVN-induced oxidation of soybean oil was evaluated in combination with OSI. In the FRG-OSI, the temperature could be reduced easily from 98°C to at least 40°C. A total of three different concentrations of AMVN (0.4, 0.6, and 0.8%) were used to accelerate the oxidation of soybean oil at 50°C. The results (Table 9.7) show that the oxidation is remarkably accelerated. Normally the induction point in OSI for soybean oil at 50°C is expected to be ~2–3 wk. The time of analysis for FRG-OSI could be reduced to <1 d.

In addition, the FRG-OB could be used successfully to accelerate the oxidation of soybean oil at 50°C. Compared with the FRG-OSI, the oxidation rate was even higher due to the high oxygen pressure in the bomb. This accelerates the formation of AMVN peroxyl radicals, which in turn have a higher reactivity toward lipids than the corresponding carbon-centered radicals. For the lower concentrations, only a gradual decrease in the oxygen pressure was observed without a clear induction point; this phenomenon is frequently observed for very slow oxidation reactions or specific food matrices. In that case, the slope of the curves can be used to quantify the rate of oxidation. Also FRG-OB proved to be a reproducible method with a standard deviation comparable to standard OB measurements.

TABLE 9.7
Reproducibility of the Accelerated Oxidation of Soybean Oil at 50°C

Repetition #	Amount AMVN (%)	OSI[a] (h)	Mean ± SD (h)	Relative error (%)
1	0.8	22.66		
2	0.8	22.46	22.84 ± 0.49	2.14
3	0.8	23.39		
1	0.6	33.76		
2	0.6	33.63	33.82 ± 0.14	0.41
3	0.6	33.48		
1	0.4	67.30		
2	0.4	66.00	66.42 ± 0.78	1.17
3	0.4	65.90		

[a]Abbreviations: OSI, Oxidative Stability Instrument; AMVN, 2,2′-azobis(2,4-dimethylvaleronitrile).

TABLE 9.8
Induction Times of Mayonnaise Treated with Rosemary Extract (RE) Antioxidant Measured with Free Radical Generation-Oxygen Bomb (FRG-OB)[a]

	FRG-OB induction time	
Sample	Induction time (h)	Relative error (%)
Control	57.3 ± 1.4	2.4
750 ppm RE	73.8 ± 2.2	2.9
75 ppm EDTA	74.3 ± 2.9	3.9

[a]Values are means ± SD, n = 3.

Applications in Food Systems

Shelf Life Prediction of Antioxidant-Treated Mayonnaise. Most emulsions do not withstand the increased temperatures during conventional accelerated oxidation studies. The emulsions can break and liberate an oil layer. The information obtained consequently relates more to the oxidative stability of bulk oil, rather than an emulsion. Reduction of the temperature of analysis to ~40°C in the FRG assays avoids the destruction of many emulsions.

An attempt was made to predict the possibility of replacing the synthetic antioxidant EDTA, which is commonly used in dressings with a natural rosemary extract (RE). For this purpose, three groups of mayonnaise samples were evaluated with FRG-OB: (i) negative control without antioxidant, (ii) positive control with 75 ppm EDTA, and (iii) treatment with 750 ppm RE. Identical samples were produced without initiator and stored at 22°C for measurement of the peroxide over a 2-mo period.

A comparison of Tables 9.8 and Table 9.9 shows that the FRG-OB method could be used successfully to predict that the antioxidant EDTA could be replaced with RE. The two treatments showed equal improvement in the induction times compared with the control sample. In the nonaccelerated follow-up of the shelf life quantified with PV, the comparable activity of the two antioxidants is reflected.

TABLE 9.9
Evolution of the Peroxide Values of Mayonnaise Treated with Rosemary Extract (RE) and EDTA Antioxidants as a Function of Time

	Treatment (mmol/kg)	
Day	750 ppm RE	75 ppm EDTA
13	6.49	7.23
33	9.40	9.99
58	21.74	19.43

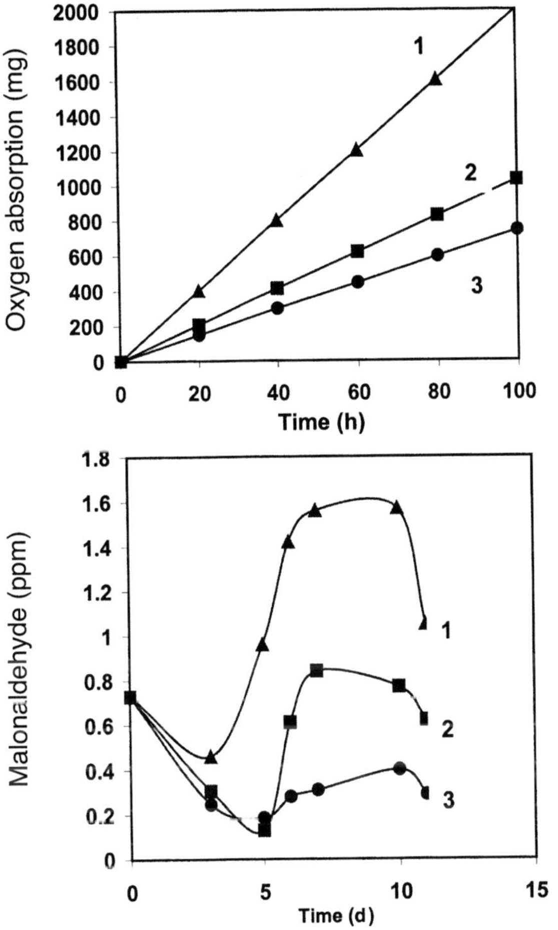

Fig. 9.5. Shelf-life prediction of minced pork: (1) control; (2) 1000 ppm tocopherol formulation; and (3) 1000 ppm rosemary formulation.

The initiator AMVN generates radicals in the lipid phase only (Massaeli *et al.* 1999, Noguchi *et al.* 1998). Krainev and Bigelow (1996) were able to prove with an electron paramagnetic resonance experiment that none of the AMVN-derived radical species can escape from the hydrophobic lipid environment. This avoids the formation of ROS that can be considered to be artificial. The good correlation between the accelerated test and the real shelf life of the emulsion may be due to the close mechanistic relation between the natural oxidation process and the oxidation mechanism in the FRG-assays.

Shelf Life Prediction of Antioxidant-Treated Minced Pork. During high-temperature analysis, raw meat will start to cook, and the original food matrix will be modified dramatically due to coagulation of the proteins, denaturation of the iron-containing methmyoglobin (prooxidant), and destruction of enzyme activity. These problems will not occur during FRG-assays at 40°C.

The correlation between FRG-OB data of antioxidant-treated minced meat of pork and the nonaccelerated shelf life evaluated with thiobarbituric acid-reactive substances (TBARS) was investigated (Fig. 9.5). For this purpose three groups of meat samples were evaluated: (i) negative control without antioxidant, (ii) treatment with 1000 ppm mixed tocopherol formulation and (iii) treatment with 1000 ppm RE. Identical samples were also produced without initiator and stored at 6°C. The TBARS content was measured over a period of 8 d. The TBA values, which reflect shelf life, were compared with the predicted shelf life calculated from the results obtained with FRG-OB.

The main asset of the FRG assays is the improved ability to compare rapidly the qualitative performance of several antioxidants in an accelerated test. Several important weaknesses of traditional accelerated methods are overcome. First, the original status of the food matrix is retained easily because of the low temperatures used. The new assays also allow studying antioxidants with low thermal stability without the risk of degradation.

The FRG-assays are a valuable tool in antioxidant research for the rapid selection of promising antioxidant formulations. The activity of the selected antioxidant can be examined more extensively in real-time shelf-life studies. The entire initial screening process will be faster and more accurate and will drastically increase the speed of the antioxidant screening process in food products.

References

Ahmad, M.M., Hakim, S.A., and Sehata, A.A. (1983) The Behavior of Phenolic Antioxidants, Synergists and Their Mixture in Two Vegetable Oils, *Fette Seifen Anstrichm. 85*, 479–483.

Akoh, C.C. (1994) Oxidative Stability of Fat Substitutes and Vegetable Oils by the Oxidative Stability Index Method, *J. Am. Oil Chem. Soc. 71*, 211–216.

Anwar, F., Bhanger, M.I., and Kazi, T.G. (2003) Relationship Between Rancimat and Active Oxygen Method Values at Varying Temperatures for Several Oils and Fats, *J. Am. Oil Chem. Soc. 80*, 151–155.

Aparicio, R., Roda, L., Albi, M.A., and Gutierrez, F. (1999) Effect of Various Compounds on Virgin Olive Oil Stability Measured by Rancimat, *J. Agric. Food Chem. 47*, 4150–4155.

Barrera-Arellano, D., and Esteves, W. (1992) Oxidative Stability of Potato Chips Determined by Rancimat, *J. Am. Oil Chem. Soc. 69*, 335–337.

Blankens, B.R., Holaday, C.E., Barnes, P.C., and Pearson, J.L. (1973) Comparison of Oxygen Bomb Method to Other Methods for Measuring Oxidative Stability of Peanuts and Peanut Products, *J. Am. Oil Chem. Soc. 50*, 377–380.

Broadbent, C.J., and Pike, O.A. (2003) Oil Stability Index Correlated with Sensorial Determination of Oxidative Stability in Canola Oil, *J. Am. Oil Chem. Soc. 80*, 59–63.

Chahine, M.H., and Macneill, R.F. (1974) Effect of Stabilization of Crude Whale Oil with Tertiary Butylhydroquinone and Other Antioxidants upon Keeping Quality of Resultant Deodorized Oil: A Feasibility Study, *J. Am. Oil Chem. Soc. 51*, 37–41.

Chen, J.H., and Ho, C.T. (1997) Antioxidant Activities of Caffeic Acid and Its Related Hydroxycinnamic Acid Compounds, *J. Agric. Food Chem. 45*, 2374–2378.

Coppin, E.A., and Pike, O.A. (2001) Oil Stability Index Correlated with Sensory Determination of Oxidative Stability in Light-Exposed Soybean Oil, *J. Am. Oil Chem. Soc. 78*, 13–18.

De Man, J.M., Tie, F., and De Man, L. (1987) Formation of Short Chain Volatile Acids in the Automated AOM Method, *J. Am. Oil Chem. Soc. 64*, 993–996

Decker, E.A., Livisay, S.A., and Zhou, S. (2000) A Re-evaluation of the Antioxidant Activity of Purified Carnosine, *Biochemistry* 766–770.

Dijkstra, A.J., Maes, P.J., Meert, D., and Meeussen, W. (1996) Interpreting the Oil Stability Index, *Fondamental 5*, 378–386.

Frankel, E.N. (1998) *Lipid Oxidation*, p. 303, The Oily Press Limited, Dundee, Scotland.

Frankel, E.N., and Meyer, A. (2000) The Problems Using One-Dimensional Methods to Evaluate Multifunctional Food and Biological Antioxidants, *J. Sci. Food. Agric. 48*, 1925–1941.

Fritsch, C.A., Weisss, V.E., and Anderson, R.H. (1975) Effect of Antioxidants on Refined Palm Oil, *J. Am. Oil Chem. Soc. 52*, 517–521.

Gearhart, W.M., Stuckey, B.N., and Austin J.J. (1957) Comparison of Methods for Testing the Stability of Fats and Oils, and of Foods Containing Them, *J. Am. Oil Chem. Soc. 34*, 427–430.

Gordon, M.H., and Mursi, E. (1994) A Comparison of Oil Stability Based on the Metrohm Rancimat with Storage at 20°C, *J. Am. Oil Chem. Soc. 71*, 649–651.

Gu, L., and Weng, X. (2001) Antioxidant Activity and Components of *Salvia plebeia* R. Br.—A Chinese Herb, *Food Chem. 73*, 299 305.

Gutfinger, T. (1981) Polyphenols in Olive Oils, *J. Am. Oil Chem. Soc. 58*, 966–968.

Gutierrez, F., Villafranca, M.J., and Castellano, J.M. (2002) Changes in the Main Components and Quality Indices of Virgin Olive Oil During Oxidation, *J. Am. Oil Chem. Soc. 79*, 669–676.

Hasenhuettl, G.L., and Wan, P.J. (1992) Temperature Effects on the Determination of Oxidative Stability with the Metrohm Rancimat, *J. Am. Oil Chem. Soc. 69*, 525–527.

Hill, E. (1994) Comparisons: Measuring Oxidative Stability, *INFORM 5*, 104–109.

Hill, S.E., and Perkins, E.G. (1995) Determination of Oxidation Stability of Soybean Oil with the Oxidative Stability Instrument: Operation Parameter Effects, *J. Am. Oil Chem. Soc. 72*, 741–743.

Inglis, D.B., and Willington, D.J. (1976) Improved Oxygen Bomb Method for Measurement of Oxidative Stability of Lard, *Chem. Ind. 20*, 905–910.

Jebe, T.A., Matlock, M.G., and Sleeter, R.T. (1993) Collaborative Study of the Oil Stability Index Analysis, *J. Am. Oil Chem. Soc. 70*, 1055–1061.

Johnson, C.W. (1974) A Practical Test for Fat Stability, *NRA Newsletter*, January, pp. 2–3.

Kajimoto, G., and Murakami, C. (1998) Changes in Organic Acid Formation in Volatile Degradation Products During Oxidation of Oils Treated with Antioxidant, *J. Jpn. Soc. Nutr. Food Sci. 51*, 207–212.

Krainev, A.G., and Bigelow, D.J (1996) Comparison of 2,2′-Azobis(2-amidinopropane) Hydrochloride (AAPH) and 2,2′-Azobis(2,4-dimethylvaleronitrile) (AMVN) as Free Radical Initiators: A Spin-Trapping Study, *J. Chem. Soc. Perkin Trans. 2*, 747–754.

Kurechi, T., and Kato, T. (1980) Studies on the Antioxidants: XI. Oxidation Products of Concomitantly used BHA and BHT, *J. Am. Oil Chem. Soc. 57*, 220–223.

Kurechi, T., and Yamaguchi, T. (1980) Studies on the Antioxidants. Oxidation Products of Concomitantly used BHA and Ethyl Protocatechuate, *J. Am. Oil Chem. Soc. 57*, 216–219.

Kurechi, T., and Kunugi, A. (1983) Studies on the Antioxidants XVII: Photo-Oxidation Products of Concomitantly Used BHA and PG, *J. Am. Oil Chem. Soc. 60*, 109–113.

Laubli, M.W., and Bruttel, P.A. (1986) Determination of the Oxidative Stability of Fats and Oils: Comparison Between the Active Oxygen Method and the Rancimat Method, *J. Am. Oil Chem. Soc. 63*, 792–795.

Laubli, M.W., Bruttel, P.A., and Schalch, E. (1988) Bestimmung der Oxidationsstabilität von Fetten und Ölen—Vergleich Zwischen der Active Oxygen Method und der Ranimat Methode, *Fat Sci. Technol. 90*, 56–58.

Liang, C., and Schwarzer, K. (1998) Comparison of Four Accelerated Stability Methods for Lard and Tallow with and Without Antioxidants, *J. Am. Oil Chem. Soc. 75*, 1441–1443.

Löliger, L. (1991) The Use of Antioxidants in Food, in *Free Radicals and Food Additives* (Aruoma, O.I., and Halliwel, B., eds.), pp. 121–151, Taylor and Francis, London.

Luckadoo, B.M., and Sherwin, E.R. (1972) Tertiary Butylhydroquinone as Antioxidant for Crude Sunflower Oil, *J. Am. Oil Chem. Soc. 49*, 95–99.

Márquez-Ruis, G., Martín-Polvillo, M., and Dobarganes, C. (2003) Effect of Temperature and Addition of α-Tocopherol on the Oxidation of Trilinolein Model Systems, *Lipids 38*, 233–240.

Massaeli, H., Sobrattee, S., and Pierce, G. N. (1999) The Importance of Lipid Solubility in Antioxidants and Free Radical Generating Systems for Determining Lipoprotein Peroxidation, *Free Radic. Biol. Med. 27*, 1524–1530.

Mei, L., McClements, D.J., and Decker, E.A. (1999) Lipid Oxidation in Emulsions as Affected by Charge Status of Antioxidants and Emulsion Droplets, *J. Agric. Food. Chem. 47*, 2267–2273.

Mendez, E., Sanhueza, J., Speisky, H., and Valenzuela, A. (1996) Validation of the Rancimat Test for the Assessment of the Relative Stability of Fish Oils, *J. Am. Oil Chem. Soc. 73*, 1033–1037.

Mendez, E., Sanhueza, J., Speisky, H., and Valenzuela, A. (1997) Comparison of Rancimat Evaluation Modes to Assess Oxidative Stability of Fish Oils, *J. Am. Oil Chem. Soc. 74*, 331–332.

Moore, R.N., and Bickford, W.G. (1952) A Comparative Evaluation of Several Antioxidants in Edible Fats, *J. Am. Oil Chem. Soc. 29*, 1–5.

Nakatani, N., Tachibana, Y., and Kikuzaki, H. (2001) Establishment of a Model Substrate Oil for Antioxidant Activity by Oil Stability Index Method, *J. Am. Oil Chem. Soc. 78*, 19–23.

Noguchi, N., Yamashita, H., Gotoh, N., Yamamoto, Y., Numano, R., and Niki, E. (1998) 2,2′-Azobis (4-Methoxy-2,4-dimethylvaleronitrile), a New Lipid-Soluble Azo Initiator: Application to Oxidations of Lipids and Low-Density Lipoprotein in Solution and Aqueous Dispersions, *Free Radic. Biol. Med. 24*, 259–268.

Papadopoulos, G., and Boskou, D. (1991) Antioxidant Effect of Natural Phenols on Olive Oil, *J. Am. Oil Chem. Soc. 68*, 669–671.

Pardun, H., and Kroll, E. (1972) Bestimmung der Oxydationstabilität von Ölen und Fetten mit Hilfe einer Automatischen Version des SWIFT-tests, *Fette Seifen Anstrichm. 74*, 366–375.

Pongracz, G. (1984) Gamma Tocopherol als natürliches Antioxidans, *Fette Seifen Anstrichm. 86*, 455–460.

Reynhout, G. (1991) The Effect of Temperature on the Induction Time of a Stabilized Oil, *J. Am. Oil Chem. Soc. 68*, 983–984.

Romoser, G.L. (1982) *AOM Test Reveals Superiority of Santoqin over Seven Other Commercial Antioxidants in Fats and Greases*, Brochure, Nutrition Chemical Division, Monsanto.

Rossell, J.B. (1989) Measurement of Rancidity, in *Rancidity in Foods*, 2nd ed. (Allen, J.C., and Hamilton, R.J., eds.), pp. 23–52, Elsevier Applied Science, London.

Satue, T.M., Huang, S.W., and Frankel, E.N. (1995) Effect of Natural Antioxidants in Virgin Olive Oil on Oxidative Stability of Refined, Bleached, and Deodorized Olive Oil, *J. Am. Oil Chem. Soc. 72*, 1131–1137.

Shen, N., Duvick, S., White, P., and Pollak, L. (1999) Oxidative Stability and AromaScan Analyses of Corn Oils with Altered Fatty Acid Content, *J. Am. Oil Chem. Soc. 76*, 1425–1429.

Shermer, W.D., and Giesen, A.F. (1997) Quality Control Methods to Monitor Oxidative Status of Fats: What Do Fat Tests Tell You? *Feed Management 48*, 55–58.

Sherwin, E.R. (1976) Antioxidants for Vegetable Oils, *J. Am. Oil Chem. Soc. 53*, 430–436.

Sherwin, E.R. (1978) Oxidation and Antioxidants in Fat and Oil Processing, *J. Am. Oil Chem. Soc. 55*, 809–814.

Van Dyck, M.O., Verleyen, T., Dooghe, W., Teunckens, A., and Adams, C.A. (2005) Free Radical Generation Assays: New Methodology for Accelerated Oxidation Studies at Low Temperature in Complex Food Matrics, *J. Agric. Food Chem. 53*, 887–892.

Van Oosten, C.W., Poot, C., and Hensen, A.C. (1981) The Precision of the Swift Stability Test, *Fette Seifen Anstrichm. 83*, 133–135.

Wanasundara, U.N., and Shahidi, F. (1994) Canola Extract as an Alternative Natural Antioxidant for Canola Oil, *J. Am. Oil Chem. Soc. 71*, 817–822.

Wang, W., Weng, X., and Cheng, D. (2000) Antioxidant Activities of Natural Phenolic Components from *Dalbergia odorifera*, T. Chen, *Food Chem. 71*, 45–49.

Warner, K., and Nelson, T. (1996) AOCS Collaborative Study on Sensory and Volatile Compounds Analyses of Vegetable Oils, *J. Am. Oil Chem. Soc. 73*, 157–165.

Warner, K., Frankel, E.N., and Mounts, T.L (1989) Flavor and Oxidative Stability of Soybean, Sunflower and Low Erucic Acid Rapeseed Oils, *J. Am. Oil Chem. Soc. 66*, 558–564.

Weng, X.C., and Wang, W. (2000) Antioxidant Activity of Compounds Isolated from *Salvia pebeia, Food Chem. 71*, 489–493.

Woestenburg, W.J., and Zaalberg, J. (1986) Determination of the Oxidative Stability of Edible Oils—Interlaboratory Test with the Automated Rancimat Method, *Fette Seifen Anstrichm. 88*, 53–56.

Yoshida, Y., and Niki, E. (1992a) Oxidation of Methyl Linoleate in Aqueous Dispersions Induced by Copper and Iron, *Arch. Biochem. Biophys. 1*, 107–114.

Yoshida, Y., and Niki, E. (1992b) Oxidation of Phosphatidylcholine Liposomes in Aqueous Dispersions Induced by Copper and Iron, *Bull. Chem. Soc. Jpn. 65*, 1849–1854.

Chapter 10

Kinetic Analysis of Lipid Oxidation Data

Afaf Kamal-Eldin[a] and Nedyalka Yanishlieva[b]

[a]Department of Food Science, Swedish University of Agricultural Sciences, 750 07 Uppsala, Sweden, and [b]Institute of Organic Chemistry, Bulgarian Academy of Sciences, 1113 Sofia, Bulgaria

Introduction

Lipid oxidation is one of the important reactions in biology. It has deleterious effects on polyunsaturated fatty acids and other lipid substrates, causing significant losses in our food quality, health, and well-being. To design strategies to inhibit the progression of oxidative reactions in foods and biological systems, it is important to understand the nature of these reactions and how they are influenced by controllable chemical and physical factors. This goal may be achieved through a better understanding of the reaction kinetics (Greek, *kinein* "set in motion or move") and, whenever possible, the energetic and mechanistic aspects of these reactions. Chemical reaction kinetics considers two aspects: (i) the rate (i.e., the speed) at which the reaction takes place, and (ii) the effective factors (mainly temperature, concentration of reactants and products, and presence of catalysts), and how these two are related. If the kinetic data are appropriately collected and analyzed, one might be able to propose a reasonable mechanism for the reaction and to further develop models able to simulate the oxidation of lipid substrates under given experimental physical conditions.

Understanding reaction kinetics is an essential prerequisite for modeling the lipid oxidation cascade, the shelf life of stored foods, durability of functional low density lipoproteins, and so on. To date, only scattered literature is available for trials analyzing lipid oxidation kinetic data. This chapter provides a review of some of these with the aim of highlighting the knowledge achieved in these studies and the remaining gaps awaiting further scientific investigations. The chapter also reviews knowledge about the mechanism of lipid oxidation and points to gray areas of understanding. Finally, the application of this knowledge to the stability of foods and of lipids in biological compartments is discussed briefly.

The Stoichiometry of the Reaction

Establishing the stoichiometry of the reaction and identifying side reactions are generally the first steps in kinetic analysis. The first products of the oxidation of unsaturated fatty substrates are often hydroperoxides (LOOH), which are generated by the following general equation:

$$LH + O_2 \rightarrow LOOH$$

However, the direct reaction of triplet-state molecular oxygen (↑↑) with singlet-state organic compounds (↑↓) is spin forbidden. At the early stage(s) of oxidation, the rates of oxygen uptake, the disappearance of the substrate, and the formation of hydroperoxides are very slow and they all generally agree with the stoichiometric ratio of 1 mol of oxygen/mol of substrate (Chan and Levett 1977, Porter *et al.* 1980, Yamamoto *et al.* 1982a and 1982b). It was shown, however, that the initial rate of inhibited oxidation of tetralin W_{inh} can be expressed by the following formula where AH represents an antioxidant (or an inhibitor) (George *et al.* 1946, George and Robertson 1946).

$$W_{inh} = k\,[LH]^2\,[O_2]^0/1 + k_{AH}\,[AH] \qquad [1]$$

This mathematical equation seems to correspond to a chemical equation of the type

$$2\,LH + O_2 \rightarrow 2L^{\bullet} + H_2O_2$$

During the exponential stage, the situation is different, with hydroperoxides taking a lead role in autoxidation. The stoichiometry during this phase is 2 mol oxygen/mol substrate in the case of linoleate (Brimberg 1993a), and 1 mol oxygen/mol substrate in the case of conjugated linoleate (Brimberg and Kamal-Eldin 2003a). The maximum oxygen absorption in the case of oleate can reach 2 mol oxygen/mol substrate but only part (60–90%) of the absorbed oxygen seems to be converted to hydroperoxides (Brimberg 1993b). These differences agree with the differences in the types of oxidation products formed. The overall stoichiometric ratio of 2 in the case of linoleate is consistent with the fact that the formation of secondary products from linoleate hydroperoxides, such as hydroxy-, keto-, and epoxy-hydroperoxides (Schieberle and Grosch 1981a and 1981b) or volatile scission products (Grosch 1987), are all produced by reactions with 2 mol of oxygen. The stoichiometric ratio of 1 in the case of conjugated linoleate agrees with the fact that most of the oxidation products of this substrate are cyclic or oligomeric peroxides. The secondary products from oleate oxidation have not been well studied because this fatty acyl substrate is often considered of little importance for oxidation. Oleic acid reacts with peroxyl radicals by hydrogen abstraction (Frankel 1998) as well as by addition to form epoxides (Walther and Spiteller 1993).

However, it should be emphasized that a balanced chemical equation, even if established to indicate the stoichiometry of the reaction and possibly the nature of the reactants and the products, cannot give a description of how the lipid oxidation reaction occurs. This requires detailed and sophisticated analyses of the kinetics of formation and degradation of the various reactants and products involved, a mission complicated by the nature of the process involving innumerable numbers of parallel and consecutive reactions.

Graphical Representation of Lipid Oxidation Data

A typical graphical representation of the kinetics lipid oxidation with respect to the consumption of the lipid substrate (typically linoleate) and oxygen and to the formation and degradation of hydroperoxides is shown in Figure 10.1. The kinetic curve of oxidation, measured as oxygen consumption, is usually sigmoid and can be described by three phases or periods:

1. The induction period (or lag phase): During this period the rate of oxidation [measured as oxygen consumption or peroxide value (PV)] is extremely small (Cadenas and Sies 1998) and, in many cases, it cannot be measured by following the change in the concentration of the substrate.
2. The active phase of peroxidation (the exponential phase): Here the rate of oxidation increases dramatically up to a certain maximal PV. The rate of hydroperoxide decomposition is significantly low, and very small amounts of secondary oxidation products are formed. Yet, these small amounts are important with respect to their organoleptic, toxic, and other deleterious effects.

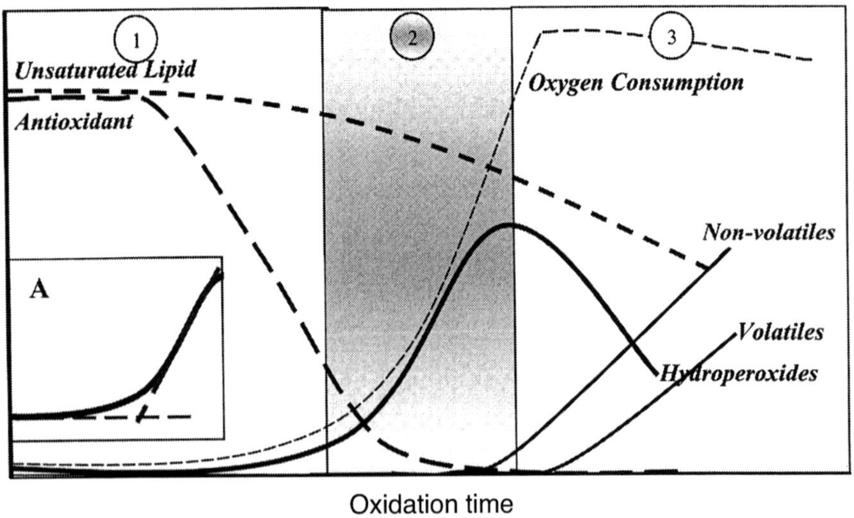

Fig. 10.1. The kinetic curve of autoxidation of polyunsaturated fatty acids is divided into (1) an induction period, (2) a phase of active hydroperoxide formation, and (3) a phase of active hydroperoxide decomposition. The insert (A) shows how the induction period can be determined by the tangent method. *Source:* modified from Labuza (1971).

3. The active phase of hydroperoxide decomposition (or the stationary phase of oxygen consumption): In this phase, the rate of hydroperoxide degradation is greater than the rate of their formation. This phase is marked by an exponential increase in the formation of secondary oxidation products resulting from the hydroperoxides.

The graphic representation of lipid oxidation kinetics, often used in the literature, provides only an arbitrary description of the reaction, especially because only a limited number of reaction products can practically be measured. In actual or simulated shelf-life studies, the induction period provides a rough estimate of the stability of the sample, especially in the presence of antioxidants, which is often the case in biological materials. The end of the induction period is determined graphically by the offset method, i.e., the point at which the oxidation curve departs from the baseline, or better, by the tangent method, i.e., the intersection of the tangents corresponding to the initial oxidation curve and the exponential curve (Zlatkevich 2002). The induction period is decreased in the presence of prooxidants and is increased in the presence of antioxidants, but in many cases, e.g., at high concentrations of antioxidants, the induction period cannot be used for evaluation of the antioxidant effects as will be discussed below.

The Kinetic Analysis of Lipid Oxidation Data

Oxygen consumption is the best parameter with which to follow the kinetics of lipid oxidation (Allen *et al.* 1949, Wewala 1997). The activation energy for the oxidation of different fatty acids was found to range from 10 to 20 kcal/mol and to correlate well with the carbon-hydrogen bond dissociation energy (BDE) according to the following equation (Korcek *et al.* 1972):

$$E_a = 0.55\ [\mathrm{BDE(R–H)} - 62.5]$$

The bond dissociation energy for the bisallylic hydrogen in linoleate is ~83 kcal/mol and that of the allylic hydrogens in oleate is ~10 kcal/mol higher (Reich and Stivala 1969). This gives E_a of ~11 and 17 kcal/mol for linoleate and oleate, respectively. These values are in the range given by Kohen and Klinman (1998) and are comparable to the values of 14.3 and 18 kcal/mol obtained by Brimberg (1991 and 1993b) for linoleate and oleate, respectively. The above equation might, however, provide only an estimation because of the involvement of nonselective initiators, mainly epoxy alkoxyl radicals generated by decomposition of the hydroperoxides (Gardner 1991, Wilcox and Marnett 1993).

There is a critical hydroperoxide concentration that marks the shift in the oxidation kinetics from the initiation to the exponential phase. This was found by Knorre *et al.* (1957) to be as low as 1 mM (PV of 1–2 mEq/kg) and by Crapiste *et al.* (1999) to be ~20 mEq/kg. It was previously suggested that the change in oxidation stage depends on the LOOH/antioxidant ratio, e.g. [LOOH]/[α-tocopherol] ≈

160 (Witting 1969) and it is also reasonable to assume that the critical hydroperoxide concentration is dependent on the substrate and the experimental conditions.

One of the important steps in studying the kinetics of a chemical reaction is to determine its rate law. This can be done experimentally by measuring how the concentration of a product, e.g., hydroperoxides, varies with time and then make characteristic kinetics plots that produce straight lines. The characteristic kinetic plots of zero-, first-, and second-order lipid oxidation reactions, with respect to hydroperoxides, are depicted in Table 10.1. During the induction period, the reaction is zero order overall as well as with respect to hydroperoxides. During the exponential oxidation stage, the reaction is first order with respect to hydroperoxides, but the overall order is not established because the change in substrate might still be too small to allow a reasonable prediction of the reaction order (see below). During the decomposition phase, the reaction is second order (bimolecular) with respect to hydroperoxides. In the case of bulk lipids, this last phase is perhaps termolecular overall (unimolecular with respect to substrate).

Reactions taking place during the induction period, the active phase of hydroperoxide formation, and the active phase of hydroperoxide decomposition are many (although a single reaction might dominate), and cannot be regarded to be of absolutely zero, first, and second order. Studying the oxidation of trilinolein at 25, 60, and 100°C in the absence and presence of β-tocopherol, Marquez-Ruiz *et al.* (2003) found the order of reaction during the induction period to vary between 0.0 and 0.57 depending on temperature and tocopherol concentration. The reactions are perhaps better described as pseudo-zero-, pseudo-first-, and pseudo-second-order reactions, respectively. Considering Equation 1, the reaction during the early stage might even be second order with respect to the substrate and overall, but this might be difficult to confirm directly because of the negligible change in the substrate concentration. Thus, the analysis of the kinetics of lipid oxidation is very complicated and not straightforward. Moreover, the order of the reaction, defined as the sum of all of the exponents of the reactants involved in the rate equation, does not necessarily comply with the stoichiometry of the reaction. This is because reaction order represents the molecules taking part in the reaction, i.e., undergoing collision, but not necessarily undergoing the final change.

Oxidation kinetics can be described by empirical as well as mechanistic models. Empirical models are particularly concerned with the practical consequences of the physicochemical change and often seek to simply describe the data by convenient mathematical relations (McDonald and Sun 1999). Empirical kinetic models can generally be divided into polynomial probability models and logistic kinetic models. The polynomial models are nonlinear, only weakly adhere to theoretical foundations, and might fit only part of the experimental data. In contrast, the empirical kinetic models try to fit the data to parameters related to the reaction kinetics, such as concentrations of reactants, temperature, or catalyst. The latter models can later be developed into mechanistic models. Some of the published models are discussed below.

TABLE 10.1
Basic Equations Describing Zero-, First-, and Second-Order Reactions with Respect to Hydroperoxides

Reaction order	Differential rate law	Integrated rate law	Characteristic kinetic plot	Slope of kinetic plot	Units of rate constant (k)[a]
Zero	d [LOOH]/dt = k	$[LOOH] = [LOOH]_o - kt$	[LOOH] vs. t	–k	mol L^{-1} s^{-1}
First	d [LOOH]/dt = k [LOOH]	$[LOOH] = [LOOH]_o\, e^{-kt}$	ln [LOOH] vs. t	–k	s^{-1}
Second	d [LOOH]/dt = k $[LOOH]^2$	$[LOOH] = [LOOH]/(1 + kt\,[LOOH]_o)$	1/[LOOH] vs. t	k	L mol^{-1} s^{-1}

[a]k is the rate constant or rate coefficient, a value dependent on temperature and other physicochemical constants of the reaction.

Semenov Polynomial Probability Models. Semenov (1939) deduced the first logistic equation to describe autocatalytic processes as

$$X = C\,(e^{\Phi t} - 1) \approx C \cdot e^{\Phi t} \qquad [2]$$

where X is the amount of oxidized substrate, t is the oxidation time, and C and Φ are constants depending on the type and initial concentration of the substrate as well as on other parameters such as temperature, pressure, or surface area.

This equation is comparable to the parameterized equation of Pagliarini *et al.* (2000) who found that the concentration of antioxidants (carotenoids and tocopherols) in autoxidized virgin olive oil and the Rancimat stability decreased with time following pseudo-first-order kinetics, i.e.,

$$[AH] = [AH]_o \qquad [3]$$

The Semenov equation is comparable to the Monod (1949) model for population growth derived from bacterial growth,

$$N_t = N_L \qquad [4]$$

where N_t is the number of "organisms" at time t, N_L is the initial number at the end of the lag time (t_L), and k is the specific growth rate.

The two equations presented above account only for the active phase of the reaction (the exponential phase) but do not account for the lag phase or the stationary phase, and by no means for the death of bacteria or breakdown of hydroperoxides. For bacteria, this problem was treated by Buchanan *et al.* (1997) who developed a three-phase linear model, with

$$\text{Lag phase, } (t \le t_L),\ \log N_t = \log N_L \qquad [5]$$

$$\text{Exponential phase, } (t_L < t < t_{max}),\ \log N_t = \log N_L + k(t - t_L) \qquad [6]$$

$$\text{Stationary phase, } (t \le t_{max}),\ \log N_t = \log N_{max} \qquad [7]$$

where t_{max} is the time for maximum attainable population density (N_{max}).

Sigmoidal curves are typically described by the continuous Gompertz model (Gibson *et al.* 1987) or a modification thereof (Zwietering *et al.* 1990). The Gompertz equation can be written as:

$$\log N = A + C \cdot \exp\{-\exp[-B\,(t - t_M)]\} \qquad [8]$$

where log N is the decimal logarithm of the population density at time t, A is the asymptotic log of population density as time decreases indefinitely (approximately equivalent to the log of the initial population density), C is the log of population density increment as time increases indefinitely (i.e., the number of growth cycles), and B is the relative maximum growth rate at time t_M, i.e., the time required to reach the maximum growth rate. Using these parameters, the specific growth rate is equal to B·C/e where e = 2.7182, the lag phase duration is equal to $[t_M - 1/B)]$, and the maximum population density is equal to A + C.

The Gompertz equation was applied successfully to the kinetics of bacterial growth (Buchanan *et al.* 1997), as well as to the kinetics of crystallization of fats (Foubert *et al.* 2003, Kloek *et al.* 2000) but not yet to lipid oxidation. In essence, these processes are similar. On the one hand, the production of bacteria is comparable to nucleation and crystal growth, and to the initiation and propagation of lipid oxidation reaction chains. On the other hand, consumption of nutrients by bacteria, leading to the stationary phase, is comparable to decreased supersaturation and decreased substrate concentration.

Brimberg Empirical Kinetic Model. Brimberg (1991) empirically found the following general formula to describe lipid oxidation

$$F(x) = K\, f(t) + a \qquad [9]$$

During the course of autoxidation, F(x) was found to assume three different forms corresponding to the three stages of oxidation mentioned above. This equation is different from all other equations because it contains a time function, f(t), which was said to depend on the situation of the catalyst(s). In the beginning of the oxidation, the event starts with $f(t) = (t - t_o)^2$ and is then followed, usually to the end of the experiment, by a steady state with $f(t) = t - t_o$. The functions F(x) and f(t) change independently.

The basic rate equation derived from the above equation is

$$dx/dt = k\,[O_2]\,(1 - x/n)\,f'(t) \qquad [10]$$

where x is the number of moles of O_2 consumed at time t per initial mole of substrate, $[O_2]$ is the oxygen concentration in the substrate, $(1 - x/n)$ is the amount of unreacted substrate at time t, and n is the number of O_2 molecules that can react with 1 mol of the substrate (n = 1 for oleate and conjugated linoleate and n = 2 for linoleate).

In the initial oxidation stage, the value of x is small and $(1 - x/n) \approx 1$. The integration of [10] at constant $[O_2]$ and with $f(t) = (t - t_o)^2$ gives $x = k_1\,(t - t_o)^2$ or

$$\sqrt{x} = \sqrt{k_1}\,(t - t_o) \qquad [11]$$

In some cases, branch [11] may be over before the first measurement. In these cases, the integration of [10] with $f(t) = t - t_o$ gives

$$x = k_1\,(t - t_o) + a \qquad [12]$$

After some time of oxidation, the so-called exponential stage starts. Brimberg (1991, 1993a, and 1993b) assumed that the concentration of oxygen in the substrate increases due to solubilization of the hydroperoxides formed in micelles, resulting in an extrapolated oxidation rate (v_o) at $x = 0$. Thus,

$$dx/dt = k'(x + v_o/k') = k'\ (x + A)\ f'(t) \qquad [13]$$

With correction for the consumed substrate and with $f'(t) = 1$, we obtain

$$dx/dt = k'\ (A + x)(1 - x/n) \qquad [14]$$

where k' is a constant and the value of A can be obtained as $A = v_o/k'$ by plotting $dx/dt(1 - x/n)^{-1}$ vs. x, where $n = 2$ for linoleate and $n = 1$ for oleate and conjugated linoleate. A is expressed in the same units as the degree of oxidation (i.e., in mol oxygen/mol substrate when oxygen consumption is followed, in mEq oxygen/kg substrate when PV is followed, and in g/g substrate when weight increase is followed).

After integration, [14] becomes

$$\ln\ [(A + x)/(1 - x/n)] = k_2t + b \qquad [15]$$

where k_2 is a constant.

In the stationary oxidation stage, $[O_2]$ is constant again but higher than $[O_2]_o$. Thus

$$dx/dt = K\ (1 - x/n)\ f'(t) \qquad [16]$$

with $K = k\ [O_2]$ and $f(t) = t$, and after integration

$$-\ln\ (1 - x/n) = k_3t + c \qquad [17]$$

where k_3 is a constant.

The Brimberg empirical kinetic formula could satisfactorily fit data on the oxidation of linoleate under different oxidation conditions and in the presence of different pro- and antioxidants. The presence of peroxides in lipid substrates enhanced their rate of oxidation especially during the induction period, i.e., affecting mainly k_1 and A, but not k_2 (Table 10.2), in agreement with the early observation that a hydroperoxide concentration as low as 1 mM is able to increase the rate of initiation above that of the direct reaction of hydrocarbons with oxygen (Knorre *et al.* 1957). Other organic molecules were found to enhance (e.g., metal ions and sterols) or inhibit [e.g., α-tocopherol and butylated hydroxytoluene (BHT)] the oxidation of linoleate during the induction period (Brimberg and Kamal-Eldin

TABLE 10.2
Effect of Added Peroxides on the Oxidation of Methyl Linoleate (ML) in the Dark at 50°C[a]

Hydroperoxide added	Concentration mol/mol ML	$k_1 \cdot 10^2$	$A \cdot 10^2$	k_2
Effect of different hydroperoxides (oxidation at 50°C)				
None	0	0.407	2.0	0.175
Cyclohexenyl-*tert*-butylperoxide	0.01	0.422	2.4	0.166
tert-Butylperbenzoate	0.01	0.571	4.5	0.117
tert-Butylperlaurinate	0.01	0.666	3.0	0.152
tert-Butyl-hydroperoxide	0.01	0.636	3.5	0.179
Cumene hydroperoxide	0.01	0.694	4.0	0.158
Tetralin hydroperoxide	0.01	0.775	4.0	0.171
Cyclohexene hydroperoxide	0.01	0.739	4.0	0.169
2,2′-bis-(*tert*-Butylperoxy)-butane	0.01	0.845	6.0	0.115
Perlauric acid	0.01	1.38	8.0	0.159
1,1′-bis-Hydroperoxide-dicyclohexylperoxide	0.01	4.22	26	0.190
Effect of hydroperoxide concentration (oxidation at 60°C)				
Cumene hydroperoxide	0.000	0.802	1.35	0.388
Cumene hydroperoxide	0.0053	0.839	1.60	0.408
Cumene hydroperoxide	0.0200	1.103	2.50	0.383
Cumene hydroperoxide	0.0497	1.684	4.00	0.383
Cumene hydroperoxide	0.1056	2.830	6.75	0.373
Cumene hydroperoxide	0.1237	3.310	8.00	0.373

[a]*Source:* Brimberg and Kamal-Eldin, unpublished.

2003b). Antioxidant synergists (such as the amino acids tryptophan and histidine) do not affect the rate of oxidation during the induction period but inhibit the rate of oxidation during the exponential stage. However, these compounds can synergize the effect of primary antioxidants, such as α-tocopherol and BHT, by a mechanism not yet completely understood. When the general formula [10] was applied to data describing the oxidation of oleate, and conjugated linoleate, small modifications of the equation were found necessary, suggesting some differences in reaction pathway(s) as will be discussed later.

Kinetic Models Based on the Free Radical Mechanism. The mechanistic kinetic models published to date tried to explain the change-time relations pertaining to types and concentrations of reactants and products as well as to physicochemical parameters (such as temperature, pressure, or surface area) with reference to the basic autoxidation mechanism. The reaction is generally recognized as a chain reaction propagated by free radicals and is often described by the well-known lipid oxidation scheme suggested by Bolland (1949) and later extended by Emanuel *et al.* (1965).

Initiation	$LH + O_2 \rightarrow L^\bullet + HOO^\bullet$	(1)
Propagation	$L^\bullet + O_2 \rightarrow ROO^\bullet$	(2)
	$LOO^\bullet + LH \rightarrow LOOH + L^\bullet$	(3)
Branching	$LOOH + LH \rightarrow LO^\bullet + L^\bullet + H_2O$	(3a)
	$2\ LOOH \rightarrow LOO^\bullet + LO^\bullet + H_2O$	(3b)
Repropagation	$LH + X^\bullet \rightarrow L^\bullet + XH$	(3c)
	where $X^\bullet = LO^\bullet$, $^\bullet OH$, or $LOO^\bullet$ formed in branching	
Termination	$2\ LOO^\bullet \rightarrow$ nonradical products	(4)
	$LOO^\bullet + L^\bullet \rightarrow$ nonradical products	(5)
	$2\ L^\bullet \rightarrow$ nonradical products	(6)

From basic chemical kinetics, the overall rate of oxidation, defined as the quantity of reactants (lipids or oxygen) consumed or the quantity of oxidation products formed per unit time, can be given by the following expression:

$$d[LH]/dt = -d[O_2]/dt = d[LOOH]/dt \quad [18]$$

As shown in Figure 10.1, the change in the concentration of substrate can scarcely be measured, in contrast to the change in the concentration of oxygen and hydroperoxides. The change in oxygen concentration provides the best kinetic parameter for the evaluation of the overall rate of oxidation.

The Steady-State Approximation. The steady-state approximation is based upon the above lipid oxidation scheme. From reactions (1)–(6),

$$d[LOOH]/dt = k_3\ [LOO^\bullet][LH] \quad [19]$$

For the solution of [18], the following three assumptions are necessary (Karel 1992, Reich and Stivala 1969)

1. The three rate termination steps are related by $k_5 = (k_4k_6)^{0.5}$. Thus

$$d[L^\bullet]/dt = r_{i1} + k_3\ [LOO^\bullet]\ [LH] - k_2\ [L^\bullet][\ O_2] - (k_4k_6)^{0.5}\ [L^\bullet][\ LOO^\bullet] - k_4[L^\bullet]^2 \quad [20]$$

and,

$$d[LOO^\bullet]/dt = r_{i2} + k_2\ [L^\bullet][\ O_2] - k_3\ [LOO^\bullet]\ [LH] - (k_4k_6)^{0.5}\ [L^\bullet][\ LOO^\bullet] - k_6[LOO^\bullet]^2 \quad [21]$$

where r_{i1}, and r_{i2} are the initial rates of production of alkyl ($L^\bullet$) and peroxyl ($LOO^\bullet$) radicals, respectively.

2. The Bodenstein or "steady-state" assumption holds, i.e., d[radical]/dt ≈ 0. By assuming that the concentrations of alkyl ($L^\bullet$) and peroxyl ($LOO^\bullet$) radicals do not change much with time (pseudo-steady state), then

$$R_i = r_{i1} + r_{i2} = ((k_4)^{0.5}[L^\bullet] + (k_6)^{0.5}[LOO^\bullet]) \quad [22]$$

3. When the reaction chains are long enough, the rate of reactions (2) and (3) are equal; thus,

$$[L^\bullet] = k_3\,[LOO^\bullet]\,[LH]/k_2\,[O_2] \quad [23]$$

By simultaneously solving the last two equations, one arrives at

$$[LOO^\bullet] = k_2\,R_i^{0.5}\,[O_2]/(k_3(k_4)^{0.5}[LH] + (k_2\cdot(k_6)^{0.5}[O_2]) \quad [24]$$

from which, the rate of the overall reaction can be obtained as

$$-d[LH]/dt = \frac{k_3 R_i^{0.5}[LH][O_2]}{(k_6)^{0.5}([O_2] + (k_3(k_4)^{0.5}[LH])/(k_2\cdot(k_6)^{0.5}))} \quad [25]$$

For low oxygen concentrations, $[O_2] \lll (k_3(k_4)^{0.5}[LH])/[k_2\cdot(k_6)^{0.5}]$, the rate is dependent on the oxygen concentration but not on the substrate concentration

$$-d[LH]/dt = k_2\,(R_i/k_4)^{0.5}\,[O_2] \quad [26]$$

At high oxygen concentrations, $[O_2] \ggg (k_3(k_4)^{0.5}[LH])/(k_2\cdot(k_6)^{0.5})$, the rate of oxidation is dependent on the substrate concentration and independent of the oxygen concentration

$$-d[LH]/dt = k_3\,(R_i/k_6)^{0.5}\,[LH] \quad [27]$$

A major problem of the steady-state approximation is that it focuses only on the active phase of hydroperoxide formation; it does not consider the induction period and the decomposition phase and does not take them into account. It also does not address the participation of hydroperoxides in the reactions (3a)–(3b) and considers them to be stable end products. Because of these limitations of the steady-state approximation, many authors have tried to find empirical formulae that would be able to provide a kinetic description of the lipid oxidation reactions.

Primary Kinetic Models. Bolland (1949) proposed the following general equation to account for the rate of lipid oxidation

$$d[LOOH]/dt = k\,([O_2]/K + [O_2])\,[LH][LOOH] \qquad [28]$$

where k is the rate constant and K is the saturation constant.

Özilgen and Özilgen (1990) provided another equation, based on PV determination, i.e.,

$$dC/dt = kC\,(1 - C/C_{max}) \qquad [29]$$

where C is the concentration of the total oxidation products, C_{max} is the maximum attainable concentration of oxidation products (in this case hydroperoxides), k is the rate constant, and t is time. In the early stages of the lipid oxidation process, when $C << C_{max}$, the term $1\text{-}C/C_{max} \approx 1$, and

$$dC/dt = kC \qquad [30]$$

and because of the small change during the induction period, C is almost constant, i.e.,

$$dC/dt = k' \qquad [31]$$

By integration,

$$C = k't + a \qquad [32]$$

where a is a constant equal to $-k't_o$.

In the exponential stage of the reaction, integration of [29] gives

$$C = \frac{C_o.e^{kt}}{1 - C/C_{max}(1 - e^{kt})} \qquad [33]$$

The linearized form of this equation is

$$kt = \ln\,(C_{max}/C_o - 1) + \ln\,(X/(1 - X)) \qquad [34]$$

or

$$\ln\,(C_{max}/C_o - 1) = kt - \ln\,(X/(1 - X)) \qquad [35]$$

where $X = C/C_{max}$. This equation provides a straight line of $\ln\,(X/(1 - X))$ vs. t with the slope k and intercept $-\ln\,(C_{max}/C_o - 1)$. The same equation was used previously for simulating microbial and cellular processes (Özadali and Özilgen 1988).

Adachi *et al.* (1995) modified the original Bolland equation [18] by assuming that [LOOH] is proportional to the consumption of substrate, i.e.,

$$\text{Rate} = k_\alpha\, C_x/(K + C_x)\, C_{LH}\, C_{LOOH} \qquad [36]$$

where C_x, C_{LH}, C_{LOOH} are the concentrations of oxygen, unoxidized substrate, and hydroperoxides, respectively, and k_α is the rate constant. If we can assume that C_{LOOH} is proportional to the consumption of substrate, then

$$dY/dt = -k_x\,(C_x/(K + C_x))\,Y\,(1 - Y) \qquad [37]$$

where Y is the fraction of unoxidized substrate, and t is the time.

When C_x is in equilibrium with the oxygen partial pressure in the atmosphere, $k_x\, C_x/(K + C_x)$ approaches a constant (k_1), and Equation 37 can be rewritten as

$$d[LH]/dt = -k_1\,Y\,(1 - Y) \qquad [38]$$

The integration of this equation under the conditions of $Y = Y_o$ at $t = 0$, gives

$$\ln\,(1 - Y)/Y = k_1 t + \ln\,((1 - Y_o)/Y_o) \qquad [39]$$

This equation described the change in the amount of unoxidized substrate for the entire oxidation period for n-6 fatty acids but only for the first half, $Y \le 0.5$, for n-3 fatty acids. In the latter case, a first-order kinetic formula was found for the second half

$$dY/dt = -k_2\,Y\ (Y \le 0.5) \qquad [40]$$

Integrating this equation with $Y = 0.5$ at $t = t_{0.5}$ gives

$$\ln\,(2Y) = -k_2\,(t - t_{0.5}) \qquad [41]$$

Equations 39 and 41 were used to study the kinetics of autoxidation of ethyl esters of linoleate, linolenate, arachidonate, eicosapentaenoate, and docosahexaenoate at 50°C and 74.5% room humidity (Adachi *et al.* 1995). Selected results on the Arrhenius parameters (frequency factors and activation energies) for the oxidation of different fatty acid substrates are given in Table 10.3.

The Monomolecular and Bimolecular Mechanistic Reaction Models. On the basis of previous work by Semenov (1959), Maloney *et al.* (1966), and Labuza (1971), Børquez *et al.* (1997) derived a two-phase equation to describe the induction period and the exponential stage of lipid oxidation. The derivation of these equations was performed as follows:

TABLE 10.3
The Frequency Factors (k_1, k_2) and Apparent Activation Energies (E_1 and E_2) of the Rate Constants for Autoxidation of Fatty Acid Esters[a]

Substrate	ln k_1	E_1 (kcal/mol)	ln k_2	E_2 (kcal/mol)
Methyl linoleate	12.1	14.3	—	—
Ethyl-γ-linolenate	12.1	14.0	—	—
Ethyl-α-linolenate	10.8	13.2	9.0	12.5
Ethyldocosahexaenoate	13.2	14.2	10.9	13.3

[a]*Source:* Adachi *et al.* 1995.

$$\text{Rate of propagation} = d[LOOH]/dt = -d[LH]/dt = k_3\,[LOO^\bullet][LH] \quad [42]$$

$$\text{Rate of termination } (R_t) = k_4\,[LOO^\bullet]^2 \quad [43]$$

Assuming that the rate of initiation is n times faster than the rate of termination, i.e., $R_i = nR_4$ where $n \neq 0$), then

$$[LOO^\bullet] = \sqrt{}(R_i/nk_4) \quad [44]$$

In this study, the oxidation was followed by measuring the consumption of the n-3 polyunsaturated fatty acid substrates, which might be possible only for these highly oxidizable substrates and under the oxidation conditions used (100°C). The rate of substrate consumption was expressed as:

$$-d[LH]/dt = k_3\,\sqrt{}(R_i/nk_4)\,[LH] \quad [45]$$

where $R_i = k_i$ [LH], and the mole fraction of the oxidized substrate (Y) can be expressed as $[LH]_t/[LH]_o$. The kinetics of the initial stage of oxidation (the induction period) can be expressed as

$$-dY/dt = k_m\,Y^{3/2} \quad [46]$$

where k_m is a combined constant given as $k_m = k_3\,\sqrt{}(k_i/nk_4) = k_o e^{-Bo/T}$
Integration of Equation 40 yields

$$1/\sqrt{}Y = 1 + (k_m/2)t \quad [47]$$

In the exponential stage, the rate of reaction is bimolecular, i.e., dependent on the concentrations of unoxidized lipids and hydroperoxides:

$$-d[LH]/dt = k_b\,[LH]\,[LOOH] \quad [48]$$

or

$$-dY/dt = k_b [Y] [1 - Y] \qquad [49]$$

Integration of this equation yields

$$\ln (Y/(1 - Y)) = k_b t \qquad [50]$$

where $k_b = k_1 e^{-B1/T}$

In this study of Børquez *et al.* (1997), these equations were applied to investigate the fractional losses of n-3 fatty acids (18:4, 20:5, and 22:6) in fresh mackerel lipids oxidized at 100°C in the dark. The rate constants k_m and k_b were dependent on temperature, in accordance with the Arrhenius equation (see below). For the n-3 fatty acids, k_o = 21630.4 s^{-1}, B_o = 7560 K, k_1 = 0.011 s^{-1}, and B_1 = 1822 K (in both cases T was the temperature in kelvin, K). These values correspond to activation energies of 15 and 3.6 kcal/mol for the initiation and exponential phases, respectively.

Børquez *et al.* (1997) proposed that the rate of oxidation in the presence of antioxidant is given by

$$-d[LH]/dt = (k_p \cdot k_i/k_a) [LH]^2/[AH] \qquad [51]$$

where k_p, k_i, and k_a are the constants for the propagation, initiation, and antioxidation, respectively. Integration of this equation as above gives

$$1/Y = 1 + k_A t \qquad [52]$$

where $k_A = k_p \cdot k_i/k_a$.

Crapiste *et al.* (1999) studied the kinetics of oxidation during the active phase of hydroperoxide formation (the exponential phase), and the active phase of hydroperoxide decomposition (the stationary phase). They modeled the rate of peroxide formation in sunflower triacylglycerols by a kinetic model composed of a first-order formation reaction and a second-order decomposition reaction with respect to hydroperoxides, i.e.,

$$\frac{d[LOOH]}{dt} = k_1[LOOH] - k_2[LOOH]^2 \qquad [53]$$

the reaction rate constants (k_1 and k_2) were temperature dependent in accordance with the Arrhenius equation

$$k_i = k_{oi} \cdot e^{-\Delta E_i/RT} \qquad [54]$$

where k_{oi} and ΔE_i represent the frequency factor and the activation energy for the rate constant [k_i (i = 1,2), respectively, R is the universal gas constant (= 8.3×10^{-3} kJ mol^{-1} K^{-1}) and T is the absolute temperature in Kelvin. The values of k_{01} and k_{02} for sunflower triacylglycerols were $1.04 \times 10^5/d^{-1}$ and 0.139×10^5 $(mEq/kg\ d)^{-1}$ and ΔE_i and ΔE_2 values were 38.35 and 48.47 kJ/mol K. The exponential phase started at a critical PV of 18.8 mEq/kg and ended at maximal values that were temperature dependent (PV = 414, 334, and 267 mEq/kg at 30, 47, and 67°C, respectively). After these maximal values, the reaction enters the active decomposition phase in which autoxidation reactions involved mainly hydroperoxides rather than PUFA.

The apparent zero-, first-, and second-order reaction kinetics during the induction period, the exponential, and the oxygen absorption stationary phases, respectively, might not describe the actual molecular interactions. In fact, Bateman *et al.* (1953) described the lipid oxidation reaction as first order during the induction period (peroxides <20 mM) and as second order during the exponential phase (peroxides >20 mM). This scenario agrees with the assumption that reactions (1) and (3a) are the main limiting reactions responsible for the formation of radicals during the initiation and propagation stages of the reaction, respectively. Because the change in polyunsaturated substrate concentration is negligible in most cases during the induction period and very small during the exponential phase, these reactions might appear as pseudo-zero and pseudo-first order, respectively, overall.

Toro-Vazquez *et al.* (1993) used a multiple variable approach to study the oxidation of refined corn oil and how it is affected by initial peroxides and by the presence of antioxidants. A discriminatory multiple regression analysis of oxidation data provided the following relations in the absence of antioxidants:

$$\ln IP = 2.62 - (15.43 \times 10^{-2})\,T + (1.22 \times 10^{-2})\,T \times PV_o - (2.34 \times 10^{-3})\,T \times PVO_2 + (2.06 \times 10^{-2})\,C_o \qquad [55]$$

$$k_3 = (8.9 \times 10^{-3})\,T + (4.2 \times 10^{-5})\,T^2 - (4.96 \times 10^{-2})\,PV_o \qquad [56]$$

and in the presence of antioxidant the following relations:

$$\ln IP = 2.62 - (8.01 \times 10^{-2})\,T + (5.34 \times 10^{-2})\,T \times PV_o - (1.06 \times 10^{-3})\,T \times PVO_2 \qquad [57]$$

$$k_3 = (3.97 \times 10^{-3})\,T - (4.96 \times 10^{-2})\,PV_o \qquad [58]$$

where IP is the induction period, k_3 is the rate of oxidation during the exponential phase, T is temperature, PV_o is the initial PV, and C_o is the initial carotene concentration in the oil.

This model showed that the induction period is a function of linear and quadratic interactions between temperature, and initial peroxides present at a concentration >2.7 mEq/kg oil ($P < 0.02$). The study also showed that although the addition of the antioxidant (*tert*-butyl hydroxyquinoline 0.0014% and citric acid 0.0012%) to the refined corn oil prolonged the induction period (i.e., inhibited the propagation reactions), it did enhance peroxide decomposition and the rate of oxidation during the induction period. Peroxide decomposition was evident at $T \geq 80°C$ and after the induction period in the absence of antioxidants, whereas it was detected at $T > 50°C$ and during the induction period in the presence of antioxidants. These results are in agreement with Cash *et al.* (1987 and 1988) and the loss of efficacy in antioxidant action discussed above.

As mentioned in the introduction, the study of oxidation reaction kinetics is complicated by the fact that hydroperoxides are unstable and that they decompose into secondary oxidation products, such as alcohols, aldehydes, ketones, or acids (Emanuel and Gal, 1986). Hérberger *et al.* (1999) studied the degradation of sunflower oil hydroperoxides under strictly oxygen-free conditions. Using principal component analysis, they found that the degradation can be described by three sets of parameters carrying independent information, i.e., (i) absorbance values at 232 and 268 nm, (ii) *para*-anisidine value (corresponding to α,β-unsaturated aldehydes), and (iii) amount of volatiles, namely, hexanal, 2-*trans*-heptenal, *cis*, *trans*, Δ-2,4-decadienal, and *trans*,*trans*, Δ-2,4-decadienal. The kinetics of the formation of the secondary oxidation products of polyunsaturated fatty acids, formed primarily during the hydroperoxide decomposition phase, is not well investigated.

Compared with polyunsaturated fatty acids, fewer and more stable oxidation products are formed by the oxidation of cholesterol. Chien *et al.* (1998) found that cholesterol oxidizes to form 7-hydroperoxy cholesterol and 5,6-epoxy cholesterol

$$d[A']/dt = k_1 [A'] \times (1 - ([A']/ [A']_{max}) \qquad k_1 - 1587 \pm 1 \text{ h}^{-1} \qquad [59]$$

$$d[A'']/dt = k_2 [A] \times [A'] \qquad k_2 = 1357 \pm 358 \text{ h}^{-1} \qquad [60]$$

where [A] is the concentration of cholesterol, [A′] and $[A']_{max}$ are the concentration and the maximum attainable concentration of cholesterol hydroperoxides before degradation, [A″] is the concentration of epoxides, and k_1 and k_2 are the rate constants for the two reactions. Both reactions are bimolecular (second-order overall). The hydroperoxides are unstable and degrade as soon as they are formed to 7-hydroxy ($k = 781 \pm 107 \text{ h}^{-1}$) and 7-keto ($k = 805 \pm 2 \text{ h}^{-1}$) derivatives. As with polyunsaturated fatty acid oxidation, there is a maximum attainable hydroperoxide concentration ($[A']_{max} = 7.9\%$ in the study).

The Mechanism of Lipid Oxidation from Kinetics

It is understood that the lipid oxidation reaction is autocatalyzed by hydroperoxides, but it is not yet established how the first hydroperoxides are formed and how their formation is efficiently inhibited by primary antioxidants of the phenolic type.

The lipid oxidation reaction follows rigorous rule(s), and the exponential phase can be simulated to a satisfactory extent by the empirical and mechanistic kinetic models presented above. Unfortunately, none of these models can be used satisfactorily to estimate the end of the induction period or the rate of oxidation during that period from the compositional data and physical conditions of the model. The main problem with modeling lipid oxidation is, perhaps, the lack of a theoretical model for the mechanism involved in the initiation of oxidation chains. Brimberg (1991, 1993a, and 1993b) proposed that oxidation is initiated by the formation of hydrogen peroxide from trace amounts of water by catalytic trace metal ions, i.e., the mechanism of Wieland (1912 and 1913). Other possible mechanisms include the formation of catalytic amounts of hydroperoxides by the direct reaction of lipids with singlet oxygen (Rawls and Van Santen 1970) and the branching of chain reactions by the biomolecular decomposition of preformed hydroperoxides (Emanuel and Gagarina 1966).

Another major problem with modeling oxidation of unsaturated fatty acids is the vast multitude of primary and secondary oxidation pathways involving peroxyl and alkoxyl radicals (Kamal-Eldin *et al.* 2003). The results of Allen *et al.* (1949) who studied the oxidation of methyl linoleate at 30°C using oxygen absorption, PV, and conjugated dienes, suggest that the methyl linoleate 9- and 13-hydroperoxides, with conjugated diene structures, represent only a part of the peroxides formed. The other, not yet characterized hydroperoxides should then be epoxy-hydroperoxides resulting from the cyclization of alkoxyl radicals formed by the decomposition of the conjugated hydroperoxides as discussed by Gardner (1987). The study of Brimberg (1993a) confirms that the stoichiometry of the reaction of linoleate with oxygen is two, even in the early stages of the oxidation. Cyclization of peroxyl radicals is also a very important part of the oxidation pathway of linolenate and higher n-3 eicosapentaenoate and docosahexaenoate (Frankel 1998). In the case of oleate and cholesterol, with a single double bond, peroxyl radicals prefer addition to the double bond rather than hydrogen abstraction, resulting in epoxides as the major oxidation products (Chien *et al.* 1998, Dutta 1997, Kim and Nawar 1993, Koelewijn 1972, Ozawa *et al.* 1986, Smith *et al.* 1982, Sugiyama *et al.* 1987, Walther and Spiteller 1993). Substrates with conjugated double bonds prefer oxidation by the addition mechanism rather than by the hydrogen abstraction mechanism (Mayo 1968). Thus, the general assumption that hydroperoxides are the main lipid oxidation products must be revised. At present, oxygen consumption is the preferred method for generating data for kinetic analysis (Brimberg 1993a and 1993b). There is some evidence scattered in literature (see Brimberg and Kamal-Eldin 2003b) that active spots on the surface of the reaction vessel may catalyze the lipid oxidation reaction in the same way as trace metal ions (Davies *et al.* 1956). It is thus also important that lipid oxidation reactions be performed in containers with standardized composition and dimensions.

The stability of vegetable oils is determined by their fatty acid composition and their tocopherol levels (Przybylski and Zambiazi 1998 and 2000). The presence of antioxidants together with polyunsaturated lipid substrates leads to changes in the mechanism and kinetics of the reaction (Denisov and Khudyakov 1987, Roginiskii 1990). It was often discussed that antioxidants (AH) inhibit the rate of lipid oxidation, due mainly to scavenging of propagating peroxyl radicals

$$LOO^{\bullet} + AH \rightarrow LOOH + A^{\bullet} \quad (7)$$

$$LOO^{\bullet} + A^{\bullet} \rightarrow A\text{-}OOL \quad (8)$$

However, these reactions do not really explain the inhibitory effect during the induction time, in which an antioxidant such α-tocopherol would scavenge chain-initiating species (or radicals); however, it is not yet established whether these species are really peroxyl radicals. Brimberg (1991, 1993a, and 1993b) provides an alternative explanation that the glass wall of the reaction vessel does, like metal catalysts, split trace amounts of adsorbed water into $H^{\bullet}$ absorbed in the glass, and $^{\bullet}OH$ adsorbed on the glass. According to that hypothesis, the latter provides the major catalyst to the initiation of lipid oxidation and when antioxidants are present, they adsorb on the glass wall and inhibit the formation of $^{\bullet}OH$.

Moreover, antioxidants with high hydrogen donation ability, like tocopherols (TOH), lose inhibition efficacy when present at high concentrations, due mainly to their decomposing effects on hydroperoxides (9) (Naumov and Vasli'ev 2003, Tavadyan *et al.* 2003, Yanishlieva *et al.* 2002). Because of the very low $[TO^{\bullet}]$ compared with [TOH], the loss of efficacy is also due to a smaller extent to the reactions of the tocopheroxyl radical [e.g., (10) and (11)]

$$LOOH + TOH \rightarrow LOO^{\bullet} + TO^{\bullet} + H_2O \quad (9)$$

$$TO^{\bullet} + LH \rightarrow TOH + L^{\bullet} \quad (10)$$

$$TO^{\bullet} + LOOH \rightarrow TOH + LOO^{\bullet} \quad (11)$$

Rate constants for the different reactions taking place during the oxidation of methyl linoleate in the absence and presence of α-tocopherol were established (Table 10.4). Due to these opposing effects, the inhibitory effect of different antioxidants during the induction period decreases with increasing concentration in ways dependent on the structures of the antioxidants and their hydrogen-donating powers (Naumov and Vasli'ev 2003, Yanishlieva and Marinova 2003). This aspect is of tremendous importance when considering the shelf life of lipids and lipid-containing foods.

The rate of inhibited oxidation (determined as the slope of the kinetic curve during the induction period (Yanishlieva and Marinova 2003) can be given by the expression

$$d[LOOH]/dt = \frac{R_i k_3 [LH]}{k_7 f [AH]} \quad [61]$$

TABLE 10.4
The Approximate Rate Constants of the Different Reactions Involved in the Autoxidation of Methyl Linoleate in the Absence or Presence of α-Tocopherol (at 60°C)

No.	Reaction	k/L mol^{-1} s^{-1}	Reference
1	$LH + O_2 \rightarrow L^\bullet + HOO^\bullet$	5.8×10^{-11}	Kasaikina *et al.* 1999
2	$L^\bullet + O_2 \rightarrow LOO^\bullet\Sigma$	3×10^{8}	Hasegawa and Patterson 1987
		5×10^{6}	Kasaikina *et al.* 1999
		1×10^{8}	Naumov and Vasli'ev 2003
3	$LOO^\bullet + LH \rightarrow LOOH + L^\bullet$	31	Babbs and Steiner 1990
		90	Kasaikina *et al.* 1999
		100	Naumov and Vasli'ev 2003
3a	$LOOH + LH \rightarrow LOO^\bullet + L^\bullet + H_2O$	2.3×10^{-7}	Kasaikina *et al.* 1999
3b	$2\ LOOH \rightarrow LOO^\bullet + LO^\bullet + H_2O$	2.4×10^{-6}	Kasaikina *et al.* 1999
3c′	$LO^\bullet + LH \rightarrow LOH + L^\bullet$	1×10^{7}	Small *et al.* 1979
3c″	$LH + {}^\bullet OH \rightarrow L^\bullet + H_2O$	1×10^{9}	Andar *et al.* 1966
4	$2\ LOO^\bullet \rightarrow 2\ (LO) + {}^1O_2$	1×10^{5}	Barclay *et al.* 1989
		4.4×10^{6}	Kasaikina *et al.* 1999
		1×10^{7}	Naumov and Vasli'ev 2003
5	$LOO^\bullet + L^\bullet \rightarrow LOOL$	5×10^{7}	Naumov and Vasli'ev 2003
6	$2\ L^\bullet \rightarrow L\text{-}L$	1×10^{8}	Naumov and Vasli'ev 2003
7	$LOO^\bullet + TOH \rightarrow LOOH + TO^\bullet$	1×10^{6}	Niki *et al.* 1984
		2×10^{6}	Naumov and Vasli'ev 2003
8	$LOO^\bullet + TO^\bullet \rightarrow TO\text{-}OOL$	2.5×10^{6}	Kaouadji *et al.* 1987
9	$LOOH + TOH \rightarrow LOO^\bullet + TO^\bullet + H_2O$	4×10^{-6}	Naumov and Vasli'ev 2003
10	$TO^\bullet + LH \rightarrow TOH + L^\bullet$	0.02	Mukai and Okauchi 1989
		0.5	Naumov and Vasli'ev 2003
		0.07	Remorova and Roginsky 1991
11	$TO^\bullet + LOOH \rightarrow TOH + LOO^\bullet$	0.1–0.5	Mukai *et al.* 1993
		10	Naumov and Vasli'ev 2003
12	$2\ TO^\bullet \rightarrow (TO)_2$	3×10^{3}	Burton *et al.* 1985
13	$TO^\bullet + O_2 \rightarrow TQ + HOO^\bullet$ or $OTOO^\bullet$	1	Naumov and Vasli'ev 2003

where R_i is the mean rate of initiation, k_3 and k_7 are the rate constants for reactions (3) and (7), respectively, [LH] and [AH] are the concentrations of the lipid substrate(s) and antioxidant, and f is the stoichiometric coefficient of inhibition by the antioxidant ($f = 2$ in case of tocopherols).

Typical kinetic curves describing the antioxidant effect of α-tocopherol, for example during the oxidation of triacylglycerols of lard at four different concentrations, are shown in Figure 10.2. Because of this anomalous behavior of such type of antioxidants, it is often not possible to evaluate the antioxidant performance solely on the grounds of the length of induction period. Yanishlieva and Marinova (1992 and 2003) suggested a new kinetic parameter to evaluate and compare the antioxidant activity (A), which can be determined as

$$A = \frac{IP_{inh} \times W_o}{W_{inh} \times IP_o} \qquad [62]$$

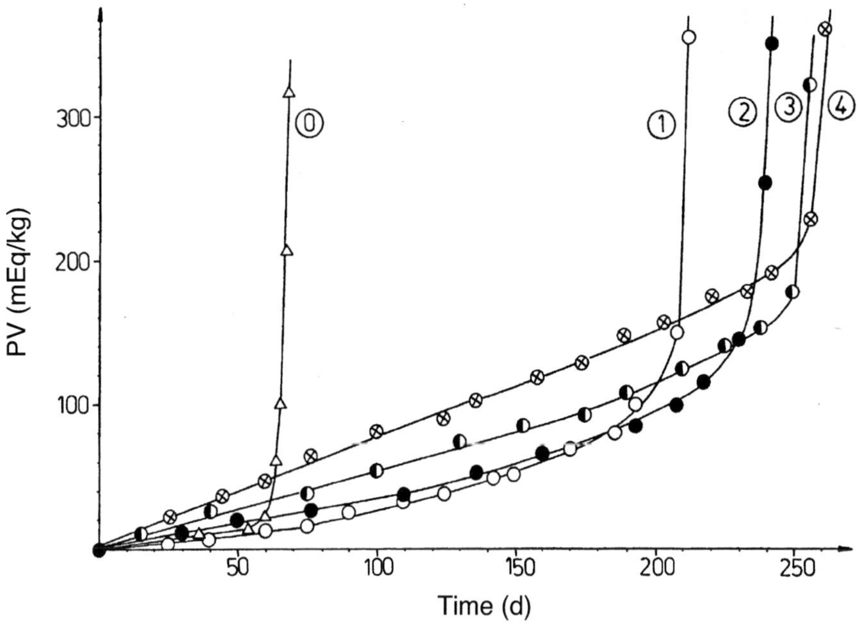

Fig. 10.2. Kinetic curves of peroxide accumulation during oxidation of triacylglycerols of lard at 25°C in the presence of α-tocopherol: 0) 0%; 1) 0.02%; 2) 0.05%; 3) 0.1%; 4) 0.02%. *Source:* Marinova *et al.* 2004.

where IP_o and IP_{inh} are the induction period for the uninhibited and inhibited oxidations, and W_o and W_{inh} are the rates for the uninhibited and inhibited oxidations, respectively. Using this approach, the antioxidant activity of α-tocopherol decreases with increasing concentration as shown in Table 10.5. Despite the lower induction periods, the antioxidant activity is higher at higher oxidation temperatures (Yanishlieva and Marinova 2003), suggesting less participation in side reactions under more aggressive oxidation conditions.

The Effect of Physicochemical Factors on the Rate of Oxidation

According to the collision theory, molecules of reactants must collide with each other before a reaction can occur. The number of collisions in a given time, the collision frequency, positively controls the rate of reaction. The reaction will take place during a collision only if the molecules hit each other at the right angle and if they have enough energy. Therefore, reaction rates are increased by increases in the concentration of reactants and temperature because both of these parameters increase the collision frequency. The rate of the reaction is also affected by the medium in which the reaction occurs (e.g., whether a medium is aqueous or organ-

TABLE 10.5
The Antioxidant Activity of Different Concentrations of α-Tocopherol During the Oxidation of Triacylglycerols Purified from Sunflower and Soybean Oils[a]

	α-Tocopherol concentration (ppm)						
Lipid system	10	50	100	200	500	1000	2000
At 25°C							
Sunflower	3.9	14.4	25.9	35.7	53.3	52.0	44.6
Soybean	3.6	13.4	20.4	29.6	24.5	22.7	16.8
At 100°C							
Sunflower	21.4	106	147	222	222	220	169
Soybean	19.4	130	191	310	267	257	228

[a]The antioxidant activity (A) is a measure of the effect of antioxidant on the length of the induction period as well as the rate of oxidation during the induction period (see original references: Yanishlieva and Marinova 2003).

ic, polar or nonpolar). The collision frequency also increases with the increase in the surface area of the reactants and in the presence of catalysts, which provide an alternative route so that the reactant molecules require less activation energy to start reacting without being consumed in the process. Catalysts increase the collision frequency by altering the orientation of reactants so that more collisions are effective. They also reduce the intramolecular bonding within reacting molecules and provide electron-dense environments to the reactants, thus helping the reaction to proceed more quickly to equilibrium. Other chemical species in the medium may decrease the rate of a reaction by competing for a reactant or altering its orientation by hydrogen bonding or dipole interactions, for example.

The quantitative relationship between the rate constant of the reaction and temperature is described by the Arrhenius equation

$$k = A \times e^{-Ea/RT} \qquad [63]$$

where A, the so-called preexponential factor, is a constant related to the geometry needed, e is a constant (approximately 2.7281), E_a is the activation energy, R is the universal gas constant (8.314×10 $Jmol^{-1} \cdot K^{-1}$), and T is the absolute temperature (in degrees Kelvin). By integration

$$\ln k = \ln A - E_a/RT \qquad [64]$$

The value of E_a was 41.8, 46.1, and 51.1 kJ/mol for trilinolein containing 0, 250, and 500 ppm α-tocopherol and the induction period was directly proportional to E_a/RT (Márquez-Ruiz *et al.* 2003).

The same relation can also be described by the transition-state (Eyring) equation

$$K = k_B T/h \exp(-\Delta G^{\neq}/RT) = k_B T/h \exp(-\Delta H^{\neq}/RT) \exp(\Delta S^{\neq}/R) \qquad [65]$$

where k_B is Boltzmann's constant (1.4×10^{-23} J/K); h is Plank's constant (6.6×10^{-34} J·s), $\Delta G^{\neq}$ is the activation Gibbs free energy (J/mol); $\Delta H^{\neq}$ is the activation enthalpy (J/mol); and $\Delta S^{\neq}$ is the activation entropy ($J \cdot mol^{-1} \cdot K^{-1}$).

Concluding Remarks

This chapter reviewed the importance of chemical kinetics in predicting how fast the lipid oxidation can proceed and how it is affected by compositional and environmental parameters, some of which can be controlled to provide a better stability to lipids vulnerable to oxidative degradation. To identify the mechanism of a reaction, it is important to determine how the rate of reaction varies as the reaction progresses. It was readily shown in the above discussion that our understanding of the kinetics and mechanism is hampered by a lack of knowledge about the initial events leading to autocatalytic, peroxyl radical-driven reactions. As mentioned in the introduction, it should always be remembered that kinetic analysis of rate data does not provide an unambiguous evidence for a mechanism because more than one mechanism can be consistent with the kinetic analysis. A better approach to defining the mechanism involves combining kinetic data with extra information about the reaction products and the thermodynamics of the elementary steps involved in the reaction. However, it is not obligatory to have a mechanistically based kinetic model to be able to predict stability because successful empirical models can also achieve this mission.

References

Adachi, S., Ishiguro, T., and Ryuichi, M. (1995) Autoxidation Kinetics for Fatty Acids and Their Esters, *J. Am. Oil Chem. Soc. 72*, 547–551.

Allen, R.R., Jackson, A., and Kummerow, F.A. (1949) Factors Which Affect the Stability of Highly Unsaturated Fatty Acids.1. Differences in the Oxidation of Conjugated and Non-Conjugated Linoleic Acid, *J. Am. Oil Chem. Soc. 26*, 395–399.

Anbar, M., Meyerstain, D., and Neta, P. (1966) Reactivities of Aliphatic Compounds Toward Bromine and Hydrogen Atoms in Aqueous Solution, *J. Chem. Soc. A*, 572–575.

Babbs, C.F., and Steiner, M.G. (1990) Simulation of Free Radical Reactions in Biology and Medicine, *Free Radic. Biol. Med. 8*, 471–485.

Barclay, L.R.C., Baskin, K.A., Locke, S.J., and Vinquist, M.R. (1989) Absolute Rate Constants for Lipid Peroxidation and Inhibition in Model Biomembranes, *Can. J. Chem. 68*, 2258–2269.

Bateman, L., Hughes, H., Morris, A.L. (1953) Hydroperoxide Decomposition in Relation to the Initiation of Radical Chain Reactions, *Faraday Soc. 14*, 190–199.

Bolland, J.L. (1949) Kinetics of Olefin Oxidation, *Q. Rev. (London) 3*, 1–21.

Børquez, R., Koller, W.-D., Wolf, W., and Spieβ, W.E.L. (1997) A Rapid Method to Determine the Oxidation Kinetics of n-3 Fatty Acids in Fish Oil, *Lebensm.-Wiss. Technol. 30*, 502–507.

Brimberg, U.I. (1991) Über die Kinetik der Autoxidation von Fetten, *Fat Sci. Technol. 93*, 298–303.

Brimberg, U.I. (1993a) On the Kinetics of the Autoxidation of Fats, *J. Am. Oil Chem. Soc. 70*, 249–254.
Brimberg, U.I. (1993b) On the Kinetics of the Autoxidation of Fats: Monounsaturated Substrates, *J. Am. Oil Chem. Soc. 70*, 1063–1067.
Brimberg, U.I., and Kamal-Eldin, A. (2003a) On the Kinetics of the Autoxidation of Fats: Substrates with Conjugated Double Bonds, *Eur. J. Lipid Sci. Technol. 105*, 17–22
Brimberg, U.I., and Kamal-Eldin, A. (2003b) On the Kinetics of the Autoxidation of Fats: Influence of Prooxidants, Antioxidants and Synergists, *Eur. J. Lipid Sci. Technol. 105*, 83–91.
Buchanan, R.L., Whiting, R.C., and Damert, W.C. (1997) When Is Simple Good Enough: A Comparison of the Gompertz, Baranyi, and Three-Phase Linear Models for Fitting Bacterial Growth Curves, *Food Microbiol. 14*, 313–326.
Burton, G.W., Doba, T., Gabe, E.J., Hughes, L., Lee, F.L., Prasad, L., and Ingold, K.U. (1985) Autoxidation of Biological Molecules. 4. Maximizing the Antioxidant Activity of Phenols, *J. Am. Chem. Soc. 107*, 7053–7065.
Cadenas, E., and Sies, H. (1998) The Lag Phase, *Free Radic. Res. 28*, 601–609.
Cash, G.A., George, G.A., and Bartley, J.P. (1987) A Chemiluminescence Study of the Decomposition of Methyl Linoleate Hydroperoxides on Active Substrates, *Chem. Phys. Lipids 43*, 265–282.
Cash, G.A., George, G.A., and Bartley, J.P. (1988) A Chemiluminescence Study of the Oxidation of Vegetable Oils and Model Compounds and the Effects of Antioxidants, *J. Sci. Food Agric. 43*, 277–287.
Chan, H.W.-S., and Levett, G. (1977) Autoxidation of Methyl Linoleate. Separation and Analysis of Isomeric Mixtures of Methyl Linoleate Hydroperoxides and Methyl Hydroxy Linoleates, *Lipids 12*, 99–104.
Chien, J.T., Wang, H.C., and Chen, B.H. (1998) Kinetic Model of the Cholesterol Oxidation During Heating, *J. Agric. Food Chem. 46*, 2572–2577.
Crapiste, G.H., Brevedan, M.I.V., and Carelli, A.A. (1999) Oxidation of Sunflower Oil During Storage, *J. Am. Oil Chem. Soc. 76*, 1437–1443.
Davies, D.S., Goldsmith, H.L., Gupta, A.K., and Lester, G.R. (1956) Radical Capture Agents in Tetralin: Measurement of Relative Efficiencies and Correlation with Structure, *J. Chem. Soc. (London)*, 4926–4933.
Denisov, E.T., and Khudyakov, I.V. (1987) Mechanism of Action and Reactivities of the Free Radical Inhibitors, *Chem. Rev. 87*, 1313–1357.
Dutta, P.C. (1997) Studies on Phytosterol Oxides. 2. Content in Some Vegetable Oils and in French Fries Prepared in These Oils, *J. Am. Oil Chem. Soc. 74*, 659–666.
Emanuel, N.M., and Gagarina, A.B. (1966) Critical Phenomena in Chain Reaction with Degenerate Branching, *Usp. Khim. 35*, 619–655
Emanuel, N.M., and Gal, D. (1986) Modelling of Oxidation Processes, *Akad. Kiadó, Budapest*, 173–186.
Emanuel, N.M., Denisov, E.T., and Maizus, Z.K. (1965) *Chain Reactions of Hydrocarbon Oxidation in Liquid Phase*, Nauka, Moscow.
Foubert, I., Dewettinck, K., and Vanrolleghem, P.A. (2003) Modelling of the Crystallization Kinetics of Fats, *Trends Food Sci. Technol. 14*, 79–92.
Frankel, E.N. (1998) *Lipid Oxidation*, The Oily Press, Dundee, Scotland.
Gardner, H.W. (1987) Reactions of Hydroperoxides: Products of High Molecular Weight, in *Autoxidation of Unsaturated Lipids* (Chan, H.W.-S., ed.), pp. 51–93, Academic Press, London.

Gardner, H.W. (1991) Recent Investigations into the Lipoxygenase Pathway of Plants, *Biochim. Biophys. Acta 1084*, 221–239.

George, P., and Robertson, A. (1946) The Oxidation of Liquid Hydrocarbons. II. The Energy-Chain Mechanism for the Thermal Oxidation of 1,2,3,4-Tetrahydronaphthalene, *Proc. R. Soc. Lond. A185*, 309–336.

George, P., Rideal, E.K., and Robertson, A. (1946) The Oxidation of Liquid Hydrocarbons. I. The Chain Formation of Hydroperoxides and Their Decomposition, *Proc. Roy. Soc. Lond. A185*, 288–309.

Gibson, A.M., Bratchell, N. and Roberts, T.A. (1987) The Effect of Sodium Chloride and Temperature on Rate and Extent of Growth of *Clostridium botulinum* Type A in Pasteurized Pork Slurry, *J. Appl. Bacteriol. 62*, 479–490.

Grosch, W. (1987) Reactions of Hydroperoxides: Products of Low Molecular Weight, Chapter 4, in *Autoxidation of Unsaturated Lipids* (Chan, H.W.-S., ed.), pp. 95–140, Academic Press, London.

Hasegawa, K., and Patterson, L.K. (1987) Pulse Radiolysis in Model Lipid Systems: Formation and Behaviour of Peroxyl Radicals in Fatty Acids, *Photochem. Photobiol. 28*, 817–823.

Hérberger, K., Keszler, Á., and Gude, M. (1999) Principal Component Analysis of Measured Quantities During Degradation of Hydroperoxides in Oxidized Vegetable Oils, *Lipids 34*, 83–92.

Kamal-Eldin, A., Mäkinen, M., and Lampi, A.-M. (2003) The Challenging Contribution of Hydroperoxides to the Lipid Oxidation Mechanism, in *Lipid Oxidation Pathways* (Kamal-Eldin, A., ed.), Chapter 1, pp. 1–36, AOCS Press, Champaign, IL.

Kaouadji, M.N., Jore, D., Ferradini, C., and Patterson, L.K. (1987) Radiolytic Scanning of Vitamin E/Vitamin C Oxidation Reduction Mechanisms, *Bioelectrochem. Bioenerg. 18*, 59–70.

Karel, M. (1992) Kinetics of Lipid Oxidation, in *Physical Chemistry of Foods* (Schwartzberg, H.G., and Hartel, R.W., eds.), Chapter 15, pp. 651–668, Marcel Dekker, New York.

Kasaikina, O.T., Kortenska, V.D., and Yanishlieva, N.V. (1999) Effect of Chain Transfer and Recombination/Disproportionation of Inhibitor Radicals on Inhibited Oxidation of Lipids, *Russ. Chem. Bull. 48*, 1891–1896.

Kim, S.K., and Nawar, W.W. (1993) Parameters Influencing Cholesterol Oxidation, *Lipids 28*, 917–922.

Kloek, W., Walstra, P., and Van Vliet, T. (2000) Crystallization Kinetics of Fully Hydrogenated Palm Oil in Sunflower Oil Mixtures, *J. Am. Oil Chem. Soc. 77*, 389–398.

Knorre, D.G., Maizus, Z.K., Obukhova, L.K., and Emanuel, N.M. (1957) Modern Concepts of the Mechanism of Oxidation of Hydrocarbons in Liquid Phase, *Usp. Khim. 26*, 416–458.

Koelewijn, P. (1972) Epoxidation of Olefins by Alkylperoxy Radicals, *Rec. Trav. Chim. Pay-Bas 91*, 759–779.

Kohen, A., and Klinman, J.P. (1998) Enzyme Catalysis: Beyond Classical Paradigms, *Acc. Chem. Res. 31*, 397–404.

Koreck, S., Cheiner, J.H.B., Howard, J.A., and Ingold, K.U. (1972) Absolute Rate Constants for Hydrocarbon Autoxidation, *Can. J. Chem. 50*, 2285–2288.

Labuza, T.P. (1971) Kinetics of Lipid Oxidation in Foods, *CRC Crit. Rev. Food Technol. 2*, 355–405.

Maloney, J.F., Labuza, T.P., Wallace, D.H., and Karel, M. (1966) Autoxidation of Methyl Linoleate in Freeze-Dried Model Systems. 1. Effect of Water on the Autocatalyzed Oxidation, *J. Food Sci. 31*, 878–884.

Marinova, E.M., Kamal-Eldin, A., Yanishlieva, N. V., and Toneva, A. G. (2004) Antioxidant Activity of α- and γ-Tocopherols in Vegetable Oil Triacylglycerols, *Riv. Ital. Sostanze Grasse LXXXI*, 98–106.

Márquez-Ruiz, G., Martin-Polvillo, M., and Dobarganes, C. (2003) Effect of Temperature and Addition of α-Tocopherol on the Oxidation of Trilinolein Model Systems, *Lipids 38*, 233–240.

Mayo, F.R. (1968) Free Radical Autoxidation of Hydrocarbons, *Acc. Chem. Res. 1*, 193–201.

McDonald, K., and Sun, D-W. (1999) Predictive Food Microbiology for the Meat Industry: A Review, *Int. J. Food Microbiol. 52*, 1–27.

Monod, J. (1949) The Growth of Bacterial Cultures, *Annu. Rev. Microbiol. 3*, 371–394.

Mukai, K., and Okauchi, Y. (1989) Kinetic Study of the Reaction Between Tocopheroxyl Radicals and Unsaturated Fatty Acid Esters in Benzene, *Lipids 24*, 936–939.

Mukai, K., Sawada, K., Kohno, Y., and Terao, J. (1993) Kinetic Study of the Prooxidant Effect of Tocopherol. Hydrogen Abstraction from Lipid Hydroperoxides by Tocopheroxyls in Solution, *Lipids 28*, 747–752.

Naumov, V.V., and Vasli'ev, R.F. (2003) Antioxidant and Prooxidant Effect of Tocopherol, *Kinet. Catal. 44*, 101–105.

Niki, E., Saito, T., Kawakami, A., and Kamiya, Y. (1984) Inhibition of Oxidation of Methyl Linoleate in Solution by Vitamin E and Vitamin C, *J. Biol. Chem. 259*, 4177–4182.

Özadali, F., and Özilgen, M. (1988) Microbial Growth Kinetics of Fed-Batch Fermentation, *Appl. Microbiol. Biotechnol. 29*, 203–207.

Ozawa, T., Hayakawa, M., Takamura, T., Sugiyama, S., Suzuki, K., Iwata, M., Taki, F., and Tomita, T. (1986) Biosynthesis of Leukotoxin, 9,10-Epoxy-12-Octadecenoate by Leukocytes in Lung Lavages of Rat After Exposure to Hyperoxia, *Biochem. Biophys. Res. Commun. 134*, 1071–1078.

Özilgen, S., and Özilgen, M. (1990) Kinetic Model of Lipid Oxidation in Foods, *J. Food Sci. 55*, 498–501, 536.

Pagliarini, E., Zanoni, B., Giovanelli, G. (2000) Predictive Study on Tuscan Extra Virgin Olive Oil Stability Under Several Commercial Conditions, *J. Agric. Food Chem. 48*, 1345–1351.

Porter, N.A., Weber, B.A., Weenen, H., and Khan, J.A. (1980) Autoxidation of Polyunsaturated Lipids. Factors Controlling the Stereochemistry of Product Hydroperoxides, *J. Am. Chem. Soc. 102*, 5597–5601.

Przybylski, R., and Zambiazi, R.C. (1998) Effects of Endogenous Minor Components on the Oxidative Stability of Vegetable Oils, *Lipid Technol. 10*, 58–62.

Przybylski, R., and Zambiazi, R.C. (2000) Predicting Oxidative Stability of Vegetable Oils Using Neural Network Systems and Endogenous Oil Components, *J. Am. Oil Chem. Soc. 77*, 925–931.

Rawls, H.R., and Van Santen, P.J. (1970) A Possible Role for Singlet Oxygen in the Initiation of Fatty Acid Autoxidation, *J. Am. Oil Chem. Soc. 47*, 121–125.

Reich, L., and Stivala, S.S. (1969) *Autoxidation of Hydrocarbons and Polyolefins*, Marcel Dekker, New York.

Remorova, A.A., and Roginsky, V.A. (1991) Rate Constants for the Reaction of α-Tocopherol Phenoxy Radicals with Unsaturated Fatty Acid Esters and the Contribution of this Reaction to the Kinetics of Inhibition of Lipid Peroxidation, *Kinet. Catal. 32*, 726–731.

Roginiskii, V.A. (1990) Kinetics of Polyunsaturated Fatty Acid Esters Oxidation Inhibited by Substituted Phenols, *Kinet. Catal. 31*, 546–552.

Schieberle, P., and Grosch, W. (1981a) Detection of Monohydroperoxides with Unconjugated Diene Systems as Minor Products of the Autoxidation of Methyl Linoleate, *Z. Lebensm.-Unters.-Forsch 173*, 199–203.

Schieberle, P., and Grosch, W. (1981b) Decomposition of Linoleic Acid Hydroperoxides by Radicals. II. Breakdown of Methyl 13-Hydroperoxy-*cis,trans*-9,11-octadecadienoate by Radicals of Copper II Ions, *Z. Lebensm.-Unters.-Forsch 173*, 192–198.

Semenov, N.N. (1939) The Kinetics of Complex Reactions, *J. Chem. Phys. 7*, 683–699.

Semenov, N.N. (1959) *Some Problems in Chemical Kinetics and Reactivity*, Vol. 2, Princeton University Press, Princeton.

Small, R.D. Jr., Scaiano, J.C., and Patterson, L.K. (1979) Radical Processes in Lipids: A Laser Photolysis Study of *t*-Butoxy Radical Reactivity Toward Fatty Acids, *Photochem. Photobiol. 29*, 49–51.

Smith, L.L., Teng, J.I., Lin, Y.Y., Seitz, P.K. (1982) Lipid Peroxidation of Cholesterol, in *Proceedings of the International Conference on Lipid Peroxides in Biology and Medicine 1980* (Yagi, K., ed.), pp. 89–105, Academic Press, New York.

Sugiyama, S., Hayakawa, M., Nagai, S., Ajioka, M., and Ozawa, T. (1987) Leukotoxin (9,10-Epoxy-12-Octadecenoate) Causes Cardiac Failure in Dogs, *Life Sci. 40*, 225–231.

Tavadyan, L.A., Martoyan, G.A., and Minasyan, S.H. (2003) Determination of the Kinetic Significance of Elementary Steps in the Reaction of Ethylbenzene Oxidation Inhibited by *para*-Substituted Phenols: Choice of an Effective Antioxidant, *Kinet. Catal. 44*, 91–100.

Toro-Vazquez, J.F., Castillo, M.A.A., and Hermàndez, C.R. (1993) A Multiple-Variable Approach to Study Corn Oil Oxidation, *J. Am. Oil Chem. Soc. 70*, 261–267.

Walther, U., and Spiteller, G. (1993) Formation of Oleic Acid Epoxide During the Storage of Technical Quality Oleic Acid, *Fat Sci. Technol. 95*, 472–474.

Wewala, A.R. (1997) Prediction of Oxidative Stability of Lipids Based on the Early Stage Oxygen Consumption Rate, in *Natural Antioxidants: Chemistry, Health Effects, and Applications* (Shahidi, F., ed.), Chapter 21, pp. 331–345, AOCS Press, Champaign, IL.

Wieland, H. (1912) [Über Hydrierung und Dehydrierung], *Ber. 45*, 484–493.

Wieland, H. (1913) [Über den Mechanismus der Oxydationsvorgänge], *Ber. 46*, 3327–3342.

Wilcox, A. L., and Marnett, L. J. (1993) Polyunsaturated Fatty Acid Alkoxyl Radicals Exist as Carbon-Centered Epoxyallylic Radicals: A Key Step in Hydroperoxide-Amplified Lipid Peroxidation, *Chem. Res. Toxicol. 6*, 413–416.

Witting, I.A. (1969) The Oxidation of α-Tocopherol During the Autoxidation of Ethyl Oleate, Linoleate, Linolenate, and Arachidonate, *Arch. Biochem. Biophys. 129*, 142–151.

Yamamoto, Y., Niki, E., and Kamiya, Y. (1982a) Oxidation of Lipids. I. Quantitative Determination of the Oxidation of Methyl Linoleate and Methyl Linolenate, *Bull. Chem. Soc. Jpn. 55*, 1548–1550.

Yamamoto, Y., Niki, E., and Kamiya, Y. (1982b) Oxidation of Lipids. III. Oxidation of Methyl Linoleate in Solution, *Lipids 17*, 870–877.

Yanishlieva, N.V., and Marinova, E.M. (1992) Inhibited Oxidation of Lipids. I. Complex Estimation and Comparison of the Antioxidative Properties of Some Natural and Synthetic Antioxidants, *Fat Sci. Technol. 94*, 374–379.

Yanishlieva, N.V., and Marinova, E.M. (2003) Kinetic Evaluation of the Antioxidant Activity in Lipid Oxidation, in *Lipid Oxidation Pathways*, (Kamal-Eldin, A., ed.), pp. 85–110, AOCS Press, Champaign, IL.

Yanishlieva, N.V., Kamal-Eldin, A., Marinova, E.M., and Toneva, A.G. (2002) Kinetic Study of the Antioxidant Activity of α- and γ-Tocopherols in Sunflower and Soybean Triacylglycerols, *Eur. J. Lipid Sci. Technol. 104*, 262–270.

Zlatkevich, L. (2002) Various Procedures in the Assessment of Oxidation Parameters from a Sigmoidal Oxidation Curve, *Polym. Testing 21*, 531–537.

Zwietering, M.H., Jongerburger, I., Roumbouts, F.M. and van't Riet, K. (1990) Modeling of the Bacterial Growth Curve, *Appl. Environ. Microbiol. 56*, 1875–1881.

Chapter 11

Analysis of Interaction Products of Oxidized Lipids with Amino Acids, Proteins, and Carbohydrates

Jan Pokorný[a], Anna Kolakowska[b], and Grzegorz Bienkiewicz[b]

[a]Institute of Chemical Technology, Faculty of Food and Biochemical Technology, Prague, Czech Republic, and [b]University of Agriculture, Faculty of Marine Fisheries and Food Technology, Szczecin, Poland

Introduction

Oxidized fats, oils, and other lipids are analyzed using numerous physicochemical and chemical methods. Their analysis becomes difficult if lipids are in mixture with amino acids, proteins, or carbohydrates because oxidized fatty acids react with them, and are bound in the reaction product to hydrophilic food components not only by physical bonds, but also by covalent bonds. The isolation of the oxidized lipid fractions before the analysis is often necessary to obtain correct results; otherwise, the content of oxidation products would be found to be lower than it is in reality.

Interaction of Lipids with Proteins

Interaction of Native (Fresh) Lipids with Proteins

Native lipids are bound in foods to protein, carbohydrates, and other polar components by relatively weak hydrogen or hydrophobic bonds; these bonds are easily broken by solvents. Complexes of lipids with proteins are called lipoproteins. In the presence of water, they form either oil-in-water dispersions, membranes, or liposomes. Lipoproteins are a topic of intensive research in association with human or animal tissues, but much less information is available on lipoproteins present in food.

The structure of lipoproteins is very complex; therefore, the lipidic moiety and the protein moiety are often studied separately. Lipids can be extracted from dry food materials using nonpolar or medium polar solvents, such as hexane or diethyl ether. In the presence of water, it is necessary to use more polar solvent systems that are able to dissociate hydrogen bonds. Combinations of chloroform or diethyl ether with methanol or ethanol are often used. Special procedures are used for the extraction of lipids from milk and dairy products, meat and meat products, or cereal products. These procedures are readily found in collections of the respective standard analytical methods.

Interactions of Oxidized Lipids with Amino Acids and Proteins

Interactions between oxidized lipids and proteins are very complex because both moieties possess several reactive functional groups (Pokorný and Kolakowska 2002). The oxidation of both lipids and proteins starts with the formation of free radicals, which can be detected by electron spin resonance. Various polymers and copolymers are probable reaction products, but very little is known about their chemical structure. Protein products corresponding to dimers or trimers were detected among the reaction products. Lipid dimers or trimers were identified also. Copolymers, however, are difficult to study because the red or yellow primary copolymers are very unstable and rapidly polymerize to form brown macromolecular pigments.

The primary amino groups of amino acids and proteins react easily with lipid hydroperoxides, which are the most commonly occurring primary lipid oxidation products. The reaction mechanism was studied in model systems using lysine and other amino acids. The reaction chain of a hydroperoxide is cleaved with the formation of an imine and a volatile product (Fig. 11.1A). The decomposition of lipid hydroperoxides is accompanied by the formation of oxidized lipid-amino acid reaction products in systems containing proteins (Hidalgo and Zamora 2002). The covalently bound lipid moiety increases the hydrophobicity of the protein molecule. Lysine reacts not only as a free amino acid, but also if it is bound on a protein

A – Reaction of an amine group of protein with a lipid hydroperoxide

$$R^1\text{-}\underset{\underset{\text{OOH}}{|}}{\text{CH}}\text{-CH=CH-}R^2 + H_2N\text{-}R^3 \rightarrow R^1\text{-CH=N-}R^3 + \text{HO-CH=CH-}R^2 + H_2O$$

B – Reaction of an amine group of protein with an aldehyde produced by decomposition of lipid hydroperoxide

$$R^1\text{-C=O} + H_2N\text{-}R^2 \rightarrow R^1\text{-CH=N-}R^2 + H_2O$$

C – Reaction of an amine group of protein with a lipid epoxide

$$R^1\text{-}\underset{\diagdown\ \text{O}\ \diagup}{\text{CH—CH}}\text{-}R^2 + H_2N\text{-}R^3 \rightarrow R^1\text{-}\underset{\underset{\text{HO}}{|}}{\text{CH}}\text{-}\underset{\underset{\text{NH-}R^3}{|}}{\text{CH}}\text{-}R^2$$

D – Reaction of a lipid aldehyde with phosphatidylethanolamine

$$R^1\text{-C=O} + H_2N\text{-}CH_2\text{-}CH_2\text{-O-}\underset{\underset{\text{OH}}{|}}{\text{P(O)}}\text{-O-}R^2 \rightarrow R^1\text{-CH=N-}CH_2\text{-}CH_2\text{-O-}\underset{\underset{\text{OH}}{|}}{\text{P(O)}}\text{-O }R^2 + H_2O$$

Fig. 11.1. Reaction of oxidized lipids with the amine groups of proteins.

molecule because the 6-amino group remains free. Free amino acids, containing primary amine groups, react by analogous mechanisms. Because they are almost exclusively α-amino acids, they are also cleaved following the mechanism of the Strecker degradation, and ammonium salts or amines and aldehydes are formed. They interact with other lipid or protein molecules present.

Lipid hydroperoxides react with some amino acids, either free or bound in protein or peptides as well, but according to different mechanisms than in the case of lysine. Thiol groups of cysteine are oxidized into disulfide bonds of cystine, and methionine is oxidized into methionine sulfoxide. However, the lipidic reaction product does not remain bound to the protein molecule in the case of sulfur-containing amino acids.

Secondary lipid oxidation products are also reactive. The most important products are aldehydes and ketones. They react with primary amine groups, again with the formation of imines (Fig. 11.1B). Unsaturated aldehydes are more reactive than saturated aldehydes. Interactions of β-lactoglobulin with saturated aldehydes were studied in dry milk. Protein polymers were formed, and the fluorescence changed appreciably (Stapelfeldt and Skibsted 1994). Soy proteins were modified by reaction with 4-hydroxynon-2-enal, and the modified proteins were detected by immunoblotting. The protein molecule becomes more hydrophobic because the lipid residue remains bound to protein. The lysine residue is not available for human nutrition so that the nutritional value of food products decreases in the course of these reactions.

Another important group of oxidation products is that of epoxy (oxirane) derivatives, such as *trans*-4,5-(*E*)-2-heptenal. They react with primary amine groups, and the oxirane cycle is cleaved during this reaction (Fig. 11.1C). The aminolysis of epoxides by lysine is an important mode of oxidized lipid-protein interactions (Lederer *et al.* 1999). The reaction products are less easily hydrolyzed by digestive enzymes than the original proteins (Zamora and Hidalgo 2001). Monomeric and dimeric albumins were identified by Sephacryl S-200-HR chromatography; the content of lysine residues decreased, and that of ε-*N*-pyrrolylnorleucine increased during incubation at 37°C overnight (Alaiz *et al.* 1997). Other oxidized hydrophobic compounds, such as oxidized sterols or terpenes, react with proteins similarly to lipids.

Naturally, other compounds containing amine groups react similarly to amino acids; alkyl amines, such as histamine, are typical examples. Phospholipids often contain bound ethanolamine, serine, sialic acid, or sphingosine, which are amines, and can react with oxidized lipids (Fig. 11.1D). However, the reaction mechanism of oxidized phospholipids with amine groups of proteins is different from the reaction mechanism in the case of oxidized lipids and proteins (Nielsen 1981) when the effects of ultraviolet spectra and fluorescence spectra are compared. The phosphate groups of phospholipids can bind to protein as well. Phosphatidylcholine can participate in interactions by different mechanisms, even when it contains no primary amine group.

Changes in Lipid and Protein Solubility by Reactions with Amino Acids and Proteins

Effect of Bound Lipids on the Solubility of Proteins

The lipophilic moiety of lipoproteins is surrounded in aqueous solution by a hydrophilic layer of hydrated protein so that lipoproteins are relatively stable in an aqueous medium. Changes in their solubility can be used for an estimation of interactions with oxidized lipids. The solubility of the protein moiety decreases with such interactions because they become more hydrophobic by the addition of lipids or even form dimers or higher oligomers. Lipids are easily soluble in organic solvents; however, during the oxidation, they are converted into higher oligomers, which are nearly insoluble (Pokorný *et al.* 1992). The insoluble proteins resulting from interactions with oxidized lipids are cleaved with proteases only slowly and/or incompletely so that the digestibility decreases.

The solubility changes are accompanied by changes in protein structure and functional properties as was demonstrated in the case of muscle proteins of Atlantic mackerel (Suhur and Howell 2002). Various functional properties were modified by interaction with oxidized lipids during cold or frozen storage. In addition to the decreased protein solubility, other protein properties changed by reactions with oxidized lipids, such as emulsifying capacity, relative viscosity of the slurry, and water binding activity (Sarma *et al.* 2000). The older literature on the effect of lipid oxidation on functional properties of proteins was reviewed (Xiong and Decker 1995).

Effect of Water on the Formation and Solubility of Lipid-Protein Products

Oxidized lipids react only slowly and only under heating in dry mixtures of lipids with proteins with a water content <10%, and heating is necessary to detect any reaction. However, the reaction is fast in the presence of water, i.e., under conditions similar to those of most foods, and proceeds even at ambient temperature. In mixtures of esters of sunflower oil fatty acids with dry casein, the content of nonextractable lipids was equal to 31%, whereas 52% was found in the case of casein swollen in water (Pokorný *et al.* 1975). Another example observed in a model experiment is shown in Table 11.1. The amount of polar lipid fraction, insoluble in chloroform, but soluble in a solvent system containing methanol, is very low in dry systems. The extraction with a solvent mixture of medium or high polarity, such as methanol, is thus not necessary in dried samples.

In model systems containing water, not only lipid polymers, but also protein polymers were formed easily so that the method could be used for monitoring oxidation. The fraction of soluble monomeric ovalbumin decreased in a mixture with lipids by the formation of albumin polymers or albumin-lipid copolymers. The process could be conveniently followed by size exclusion chromatography (El-Zeany *et al.*

TABLE 11.1
Fractionation of Mixtures of Sunflower Oil with Casein (1:1 wt/wt) Stored at 60°C for 6 d[a,b]

Solvent used	Dry casein (%)	Casein swollen in water (%)	Swollen casein after drying (%)
Hexane	93	55	56
Chloroform	4	5	12
Methanol	1	8	2
After hydrolysis	2	32	30

[a]*Source:* J. Pokorný and E. Morávková, unpublished results.
[b]Subsequent extractions were used, e.g., hydrolysis with 2M HCL for 12 h.

1975). The kinetics of the polymerization was studied in ethyl linoleate-collagen mixtures, and the formation of higher oligomers was a more useful method for the study of oxidized lipid-protein interactions than the loss of monomers or the formation of dimers (Pokorný *et al.* 1990).

Fractionation of Lipoproteins by Centrifugation and Ultrafiltration

Separation of Native Lipoproteins

The density (weight of 1 mL) of lipoproteins depends on the ratio of the lipid and protein moieties. Therefore, various classes of natural lipoproteins are very often separated by centrifugation into groups of similar composition. The most detailed separation of lipoproteins was achieved for human serum lipoproteins. An example is shown in Table 11.2. It is evident that not only the proportion of lipids but also the composition of the lipid fraction changes. More sophisticated separation is obtained using modern instrumentation with more efficient centrifuges. The separation of the lipid and protein phases is not sharp because an intermediary layer is obtained at the interface, containing both lipid (preferentially, polar lipids) and proteins. The sediment contains not only proteins, but also lipids covalently bound to proteins and even lipids bound by weak physical forces.

The degree of separation of physically bound lipids from proteins depends on the procedure used for centrifugation. When comminuted tissue of frozen-stored mackerel was centrifuged at the speed of 83 Hz (5000 rot/min) for 6 min, lipids were separated only partially. A part of the more polar lipids bound by hydrogen bonds to protein was not separated from protein so that only a third of the phospholipids appeared in the separated lipid phase (Kolakowska and Kwiatkowska-Zienkowicz 1990). Soy beverage (incorrectly called soymilk) was fractionated into particulate, soluble, and floating fractions by centrifugation. The floating fraction contained all neutral lipids (Guo *et al.* 1997). Centrifugation may be used for the study of the type and strength of lipid-protein interactions, and for research into

TABLE 11.2
Characteristics of the Fractions Obtained by Centrifugation of Lipoproteins[a,b]

Parameter	Chylomicrons	VLDL	IDL	LDL	HDL
Density (g/cm^3)	0.93	0.93–1.006	1.006–1.019	1.019–1.063	1.063–1.210
Flotation constant (S_f)	400	400–20	20–10	11	—
Mean (mm)	500	80	30	28	13
Composition			(%)		
Triacylglycerols	80–90	50–70	20–25	5–10	3–5
Free cholesterol	1–3	7–10	7–10	5–8	3–5
Cholesterol esters	2–5	4–13	1–12	40–45	15–20
Phospholipids	3–7	15–20	15–20	20–22	20–30
Apoproteins	1–2	8–12	18–20	20–25	45–55

[a] *Source:* adapted from Michajlik and Bartnikowska (1999).
[b] Abbreviations: VLDL, very low density lipoproteins; IDL, intermediate density lipoproteins; LDL, low density lipoproteins; HDL, high density lipoproteins.

various extraneous effects influencing the interactions. Protein found in the lipid fraction and lipids found in the protein fractions prove the occurrence of lipid-protein interactions. Separation of milk is a typical example because of the relatively high protein concentration in the fat fraction and the presence of fat globules in the pellet fraction (Guo *et al.* 1998). The centrifugation method is advantageous for measuring the capacity of proteins to bind lipids in an aqueous medium (Ludwig *et al.* 1989).

In addition to the above low-speed centrifugation, high-speed procedures may be used for the more efficient separation of lipoprotein fractions of different densities. Shenouda and Pigott (1977) used sucrose gradient to separate complexes formed during interaction of fish myosin and both polar and neutral lipids, marked with ^{14}C. They separated the mixture at 583 Hz (35000 rot/min) for 20 h. The mixture was studied by polyacrylamide electrophoresis and electron paramagnetic resonance (EPR) at the same time.

Ultrafiltration is frequently used for the separation of lipoproteins in blood. Lipoprotein classes have slightly higher densities than water. Therefore, they decompose in the respective apertures of the filter so that the particles of lower density flow to the surface, whereas the particles of higher density fall as sediment to the bottom. The sodium chloride solution of different concentrations is used for the separation. The rate of separation is expressed in units of flotation S_f (Svedberg) in relation to the rate in a solution of density 1.063 g/cm^3 at a temperature of 26°C. The fraction is separated by centrifugation. A sodium chloride solution of higher density is then used, and the lipoprotein fraction is separated by centrifugation.

The ultracentrifugation separates lipoproteins into the following fractions (Michajlik and Bartnikowska 1999): (i) The fraction of chylomicrons (density of ~0.98 g/cm^3) is obtained during the first step of centrifugation at the force of 10^6 g. After the separation of this fraction, the following fractions are obtained by applying the respective sodium chloride solutions: (ii) very low density lipoproteins (VLDL) in solutions with a density of 1.006 g/cm^3; (iii) intermediary density lipoproteins (IDL) with a density of 1.006–1.019 g/cm^3; (iv) low density lipoproteins (LDL) with a density of 1.019–1.063 g/cm^3; (v) high density lipoproteins (HDL) with a density of 1.063–1.21 g/cm^3. All of these fractions consist of lipid globules covered by polar lipids and protein, called apoproteins. The precipitate obtained after centrifugation still contained a small amount of bound oil, which could be measured on the basis of fluorescence intensity (Tsutsui *et al.* 1986).

Separation of Lipoproteins Containing Oxidized Lipids

The LDL fraction contains high amounts of unsaturated fatty acids, especially linoleic and arachidonic acids, which are easily oxidized. LDL containing oxidized lipids have higher density. Modifications of apoprotein accompany the oxidation of lipids. They cause fragmentation of apoprotein B and polymerization of polypeptides. Aldehydes formed from oxidized lipids contribute to these reactions.

Malonaldehyde and 4-hydroxynonenal are particularly active in these interactions. The LDL particles with the blocked 6-amino group of lysine are not recognized by cellular receptors except for the receptor on the surface of macrophages. The oxidized LDL can thus be determined on the basis of their reception by macrophages or the determination of unavailable lysine in apoprotein B. Other suitable methods are the determination of electrophoretical mobility (Coni *et al.* 2000) compared with native LDL (Esterbauer 1991), the use of EPR, measurement of the fluorescence of apoprotein B-100 (Esterbauer *et al.* 1987), and determination of the degree of oxidation using classical methods. Either the extracted lipids or the original LDL may be analyzed. For the determination of the oxidized fraction in the triacylglycerol moiety, lipoproteins are separated by centrifugation, triacylglycerols are extracted with a mixture of chloroform and methanol, and the extract is purified on prepacked silica columns (Suomela *et al.* 2004a). The peroxide content in oxidized LDL is generally low; this is attributed to the presence of metals catalyzing their decomposition. A reaction with amine groups could be another cause. The degree of oxidation of the triacylglycerol fraction, losses of polyunsaturated fatty acids, and diene conjugation of the HPLC profile of polar triacylglycerols can be used to characterize the process (Suomela *et al.* 2004b). The Fourier transform infrared (FTIR) spectra also give good information on the amount of oxidation products in the range of 4000–650 cm^{-1}, particularly between 1180 and 935, at 1745, 1710, and 1621 cm^{-1}.

Fractionation of Lipoproteins by Solvent Extraction

Isolation of Free and Physically Bonded Lipids

Solvent extraction is a convenient procedure because no expensive equipment is required. Native lipids are extracted quantitatively with nonpolar hydrocarbons, but oxidized lipids are extracted only partially. Some oxidation products are bound by multiple hydrogen bonds to the protein molecule so that they resist the action of nonpolar solvents such as hexane. If a stored or heated food product is extracted with hexane, only the nonoxidized lipids or slightly polar oxidation products are extracted. Another fraction of oxidized lipids can be extracted using only mixtures of nonpolar and polar solvents, e.g., a mixture of chloroform and methanol, as proposed by Folch. The methanol-soluble lipid fraction rapidly increases in the early stages of oxidation, but decreases again in the later stages by conversion into a fraction that is extractable only after hydrolysis (Pokorný *et al.* 1988).

Lipids covalently bound to proteins are not extracted from the reaction mixture even by polar solvents so that the yield of extracted lipids is usually lower than the original amount of lipids before the oxidation. In advanced oxidation stages, the extractable lipid fraction may decrease by as much as ~20–40% (Pokorný 1963). In the case of hide tanning with fish oil, the reaction is very intensive so that the resulting tanned product contains only traces of extractable lipids.

Most lipids are bound to hide proteins with multiple covalent and physical bonds. The hide is then very resistant to microorganisms and insects. The lipid content in the lipid-protein complex left after removal of the extractable lipids can be characterized by electrophoresis on a polyacrylamide gel (Janitz 1987), and staining with liposoluble dyes, such as Rhodamine 6 G or Oil Red O (St. Angelo and Ory, 1975).

Isolation of Lipoproteins from Insoluble Protein Complexes

The unextractable lipid residues can be obtained only if the material was previously hydrolyzed by acids, and liberated lipids are then extracted. Alkaline hydrolysis is possible, too, with comparable results. It was found suitable, especially for model experiments, but a small amount of bound lipids remained in the insoluble material left after the alkaline hydrolysis, which could be recovered only after a subsequent acid hydrolysis. Examples of distribution of lipid fractions in model experiments were discussed earlier (Pokorný 1963), and other results are shown in Table 11.2. The hydrolysis can change the composition of oxidized lipids, especially, if oxygen had access to the reaction mixture during the operation. The mass balance between the original lipids and the sum of the fractions is still not completely satisfactory. Some volatile oxidation products may be lost during the extraction. In contrast, the weight of the lipids increases by absorption of oxygen during the oxidation. The extract obtained after the hydrolysis sometimes contains nitrogen because condensation products of oxidized lipids with amino acid residues, left bound to lipids after the hydrolysis, are partially soluble in organic solvents. Solvent fractionation is expensive, time consuming, and has a negative effect on the environment. Because of the above disadvantages, solvent fractionation is used only rarely for the determination of the degree of protein-oxidized lipid interactions.

Use of Markers for the Characterization of Nonextractable Lipids

An alternative procedure to hydrolysis and the subsequent solvent fractionation is the analysis of the degree of oxidation in the fraction of soluble lipids, proteins, and interaction products. Selected indicators of either lipid oxidation or protein oxidation are then used for the estimation of total oxidized lipids. Products of oxidized lipid-protein interactions can also be used as markers. The most important examples are given in Table 11.3.

In the case of the lipid fraction, the determination of the thiobarbituric acid (TBA) value or the content of TBA-reactive substances (TBARS) is the most widely used procedure. The reason is that the procedure is very simple and suitable for the analysis of extensive series of samples, but the results are often not very reliable. Nevertheless, satisfactory correlations were observed in several experiments, e.g., between the loss of free amino acids and increase in the TBA values and sensory bitterness of an autolytic extract from frying of mackerel wastes. The

TABLE 11.3
Markers Used for Detecting Oxidized Lipid-Protein Interactions[a]

Markers of the degree of oxidation of the lipid moiety	Markers of the degree of oxidation of the protein moiety	Markers of the degree of formation of protein-lipid interaction products
Peroxide value	Carbonyls (DNPH, TBA)	Browning
TBA, TBARS	Available lysine, free amino acids, free amine groups, degradation of tryptophan, cysteine, methionine	Formation of hexanal-lysine, lysinoalanine, pyrrolization
Anisidine value	Protein-bound fluorescence, tryptophan fluorescence	Polymerization, SDS-PAGE, electrophoresis
DNPH value	Decrease of thiol groups	Fluorescence of the interaction products
Volatile fraction, hexanal, malonaldehyde, GC	Protein solubility, protein hydrophobicity, HPSEC	Formation of lipofuscin
Free fatty acids	Changes of heme pigments	Circular dichroism
Conjugated dienes	Functional properties: viscosity, solubility, emulsifying capacity	Immunoreactions (formation or decrease)
Essential fatty acids, polyenoic fatty acids,		Sensory analysis
DHA and EPA, saturated acids		
HPLC, HPSEC		
Fluorometry		
Oxidized sterol formation		
Tocopherol decrease		
Oxygen uptake		

[a]Abbreviations: DNPH, 2,4-dinitrophenylhydrazine; TBA, thiobarbituric acid; TBARS, TBA-reactive substances; GC, gas chromatography; HPSEC, high-pressure size exclusion chromatography; DHA, docosahexaenoic acid; EPA, eicosapentaenoic acid; HPLC, high-performance liquid chromatography.

peroxide value is also used, but hydroperoxides are very unstable under the conditions of sample preparation. Another important marker is the determination of fatty acid composition. Essential fatty acids and other polyunsaturated fatty acids are sensitive to oxidation so that the content of polyunsaturated fatty acids decreases during the storage and heating of foods. In contrast, saturated fatty acids are relatively stable so that their relative content in residual lipids increases. During the storage of smoked tuna, the ratio of docosahexaenoic acid:palmitic acid decreased by 20%, and the process was accompanied by intensive browning (Zotos *et al.* 2001). The disadvantage of this method is that it is necessary to know the fatty acid composition of the original sample. Modern instrumental methods were proposed more recently, such as FTIR or proton nuclear magnetic resonance.

Various methods were proposed for markers of protein transformation by oxidized lipids. The measurement of protein carbonyl was most often used because the method is very simple. Fluorescence, e.g., the loss of the tryptophan fluorescence (Viljanen *et al.* 2004) is also a frequently used method because it is simple and suitable for a large sample series. On the contrary, other fluorescent compounds are formed by interactions, e.g., fluorescence (excitation maximum at 355 nm, emission maximum at 440 nm) was observed in oxidizing mixtures of soybean oil with soy proteins (Liang 1999). Differences in fluorescence during oxidation are very pronounced, e.g., in yellowtail fish (*Seriola quinqueradiata*), it increased from 9.45 to 28.34 nmol/mg within 12 d while in storage at 0°C. The fluorescence correlated with the malonaldehyde content. The method could be recommended for the assay of oxidized lipid-protein products.

The determination of available lysine is also a good marker. Interaction products of linoleic acid hydroperoxides with lysine were fractionated on silica columns. Lysine also reacts with aldehydes and epoxides so that its losses are a good indicator of oxidized lipid-protein interactions (Lederer 1996). The determination of available lysine has the advantage that the changes occurring during storage are quite pronounced, e.g., up to 70% of available lysine was lost during frozen storage of mackerel at –20°C for 11–33 wk (Zotos *et al.* 1995). The decomposition of other amino acids is less pronounced, but also noteworthy, e.g., tryptophan and histidine.

Among more sophisticated methods, the determination of protein solubility and protein hydrophobicity should be mentioned because they are relatively simple. Protein polymers are also used, but the polymerization may be due to protein oxidation by air oxygen, not only by lipid oxidation products. The digestibility of proteins is evident on the basis of the lower reactivity of interaction products with chymotrypsin, trypsin, and other proteases (Zamora and Hidalgo 2001). Changes in the utilization of proteins in nutrition are important markers of lipid-protein interactions. In fish protein concentrate mixed with oxidized oil, the digestibility decreased from 93.8 to 64.3%, and the utilization of protein nitrogen from 86.7 to 50.7% (Devadasan *et al.* 1985).

Some markers indicate the content of products of oxidized lipid-protein interactions. The browning reaction is frequently mentioned. Another, more complicated method is the determination of lysinoalanine or 6-*N*-hexyllysine. The FAST index determined by fluorometry showed good correlation with losses of available lysine in heated foods, such as extruded cereals or fried salmon (Birlouez-Aragon *et al.* 2001).

A very simple method is the determination of fluorescence, but only certain imine compounds show fluorescence. Dihydropyridine derivatives produced by the interaction of malonaldehyde with amino derivatives show very intensive fluorescence, unlike acyclic products, which possess no fluorescence activity (Kikugawa *et al.* 1984). The excitation and emission maxima differ for different proteins, amino acids bound to lipids, or their oxidation products. Two maxima were obtained in most cases. The pyrrolization, accompanied by fluorescence changes, is another interesting procedure.

The interactions destroy the allergenic groups in some proteins or produce substances with allergenic activities. Antibodies reacted strongly with hexanal-modified lysine, even when they did not react with the precursors, lysine and hexanal (Smith *et al.* 1999). Therefore, the immunochemical methods are sometimes used, such as the enzyme-linked immunosorbent assay (ELISA) (Goodridge *et al.* 2003). Proteins formed adducts with malonaldehyde, 4-hydroxynonenal, and other aldehydic lipid oxidation products. These adducts could be detected by indirect ELISA (Lynch *et al.* 2001). In a hexanal-bovine serum albumin model system, the results of the ELISA analysis were correlated with headspace gas chromatography and TBARS methods (Zielinski *et al.* 2001). The use of the best markers can never be equivalent to direct determination of covalently bound lipids, but it may give satisfactory information in most cases.

Browning Reactions Resulting from Oxidized Lipid-Protein Interactions

The browning due to interactions of oxidized lipids and proteins was observed on stored fish white muscle, where the staining is a serious quality defect. It is caused by polymerization of unsaturated imines formed by interactions of primary or secondary amine groups of protein with lipid oxidation products (Pokorný and Sakurai 2002). The discoloration is more intensive in the case of more polyunsaturated fatty acids, bound in lipids, e.g., linolenic or linoleic acids, compared with oleic acid (Zamora and Hidalgo 2003). Therefore, the browning is particularly evident in fish. The degree of browning also depends on the protein content in the system or on free amino acids present in the system (Hutapea *et al.* 2004). For example, cod fillets caused more intensive darkening of frying oil than potato slices (Tynek *et al.* 2001). Aldolization catalyzed by amine groups is very important in the early stages of browning so that macromolecular brown pigments have a very low nitrogen content, e.g., 0.02% (Fujimoto and Kaneda 1973).

The browning intensity and character of discoloration are measured either simply at a single wavelength, e.g., at 430 or 460 nm, or the trichromatic method is used, and the L*, a* and b* values are determined. The luminance L* of the chloroform-methanol extract decreased in stored fish with the increasing content of bound lipids and the increasing amount of blocked lysine groups (Pokorný *et al.* 1974). The intensity can be measured by reflectance on the surface, or soluble colored compounds are extracted, and the color intensity of the extract is measured. The solubilization may be achieved by the use of trichloroacetic acid. The macromolecular brown pigments are, naturally, only partially soluble either in water or in organic solvents (Chengchu *et al.* 2000). It is possible to separate pigments in the lipophilic and the hydrophilic fractions, and to determine the coloration separately in each fraction (Wakao and Pazos 1984). The majority of brown pigments are lipophilic. The evaluation of the degree of discoloration can be combined with fluorometric measurements.

Reactions of Lipids and Oxidized Lipids with Carbohydrates

Interaction of Native (Fresh) Lipids with Carbohydrates

Reactions of lipids with sugars depend on many factors. The presence or absence of water is very important. In the absence of water, various environmental factors, such as thermal or mechanical energy, are necessary to increase the mobility of lipids, thereby causing better contact between the phases and a higher reaction rate of the original medium (Lechtinen 2003). The presence of water in the medium enables easier interactions, and increases the rate of reactions between lipids and sugars. Sugars are very soluble in water, forming highly polar structures that do not react directly with lipids because lipids are hydrophobic. Polymeric carbohydrates, e.g., amylose, amylopectin, odextrins, can form helical structures in which the inner side is hydrophobic. Therefore, they can receive hydrophobic substances (Tomasik 2000, Tomasik and Schillingh 1998).

Inclusion compounds of amylose and lipids are the best known and widely occurring examples. Amylose is commonly present in starch. It can form complexes with native lipids, but also lipidic additives, such as monoacylglycerols and free fatty acids. The complex formation depends on several factors, such as the presence of water in the system (Wasserman and LeMaste 2000), the degree of hydrolysis (Shamekh *et al.* 1998; Goffman and Bergman 2003) or lipid oxidation (Castello *et al.* 1999, Bienkiewicz and Kolakowska 2003), the length of the fatty acid chain and the degree of its unsaturation (Schroeder and Hoseney 1978, Godet *et al.* 1995, Bienkiewicz and Kolakowska 2003), but also the source of starch and degree of its depolymerization (Gelders *et al.* 2004). Other macromolecular carbohydrates, such as maltodextrin, pullulan or α-cyclodextran can also entrap lipids.

Isolation and Analysis of Oxidized Lipids Bound with Carbohydrates

The degree of lipid degradation has a great effect on the reactivity of lipids with carbohydrates. Lipolytic enzymes and lipoxygenases contribute to the interactions. During the lipolysis of triacylglycerols, monoacylglycerols and free fatty acids are produced; these react more easily with hydrophobic groups of starch helices than triacylglycerols. Free unsaturated fatty acids in particular are less resistant than saturated fatty acids to the influence of external factors and to the transformations catalyzed by lipoxygenases. A decrease in the content of linoleic acid during dough kneading in the presence of lipoxygenases was due not only to oxidation, but also to the interaction with starch (Graveland 1970, Lechtinen *et al.* 2003). Bienkiewicz and Kolakowska (2004) observed that the degree of oxidation of lipids has an effect on their interaction with starch. Oxidized lipids formed complexes with starch less easily than fresh lipids in model systems containing amylose and fish lipids. However, during subsequent technological operations, such as heating or freezing of the products, oxidized lipids were extracted from dough less easily than fresh lipids. Different formation of fish lipid complexes with starch was observed in the case of mixtures with amylopectin. During processing of the mixture, oxidized lipids were bound more strongly and to a greater extent than fresh lipids (Bienkiewicz and Kolakowska 2003). Differences in complex formation between lipids with amylose or amylopectin were reported by others (Huang and White 1993, Villwock *et al.* 1999).

The degree of starch depolymerization limits the possibility of the formation of an inclusion complex between starch and lipids. The length of fatty acid chain has great influence on the formation of inclusion complexes. Gelders *et al.* (2004) observed that the degree of depolymerization affects the inclusion of C_{22} fatty acids. Complex formation is easiest in the case of fatty acids containing 12–20 carbon atoms in the chain, and *trans* fatty acids form complexes more easily than *cis* fatty acids (Stauffer 2001). Binding of this type causes more intensive polymerization of carbohydrates because of cross-linking (Seidel *et al.* 2001).

Concluding Remarks

Oxidized lipids bound into complexes with amino acids, proteins, and carbohydrates are retained more strongly by the nonlipid components than fresh lipids. Their isolation requires special procedures and knowledge of the composition and the history of the material. Oxidized lipids may be changed in the course of isolation processes so that the degree of oxidation is determined only in the extractable lipid fraction. Analysis of markers of oxidation can be obtained by simpler procedures; therefore, they are used to evaluate the degree of oxidation of the total lipids.

References

Alaiz, M., Hidalgo, F.J., and Zamora, R. (1997) Antioxidative Activity of Nonenzymatically Browned Proteins Produced in Oxidized Lipid/Protein Reactions, *J. Agric. Food Chem. 45*, 1365–1369.

Bienkiewicz, G., and Kolakowska, A. (2003) Effects of Lipid Oxidation on Fish Lipids—Amylopectin Interaction, *Eur. J. Lipid Sci. Technol. 105*, 410–418.

Bienkiewicz, G., and Kolakowska, A. (2004) Effects of Thermal Treatment on Fish Lipids—Amylose Interaction, *Eur. J. Lipid Sci. Technol. 106*, 376–381.

Birlouez-Aragon, I., Leclerc, J., Quedraogo, C.I., Birlouez, E., and Grognet, J.F. (2001) The FAST Method, a Rapid Approach of the Nutritional Quality of Heat-Treated Foods, *Nahrung 45*, 201–205.

Castello, P., Potus, J., Baret, L.J., and Nicolas, J. (1999) Effects of Mixing Conditions and Wheat Flour Dough Composition on Lipid Hydrolysis and Oxidation Levels in the Presence of Exogenous Lipase, *Cereal Chem. 76*, 476–482.

Chengchu, L., Morioka, K., Itoh, Y., and Obatake, A. (2000) Contribution of Lipid Oxidation to Bitterness and Loss of Free Amino Acids in the Autocatalytic Extract from Fish Wastes: Effective Utilization of Fish Wastes, *Fish. Sci. 66*, 343–348.

Coni, E., Di Benedetto, R., Di Pasquale, M., Masella, R., Modesti, D., Mattei, R., and Carlini, E.A. (2000) Protective Effect of Oleuropein on LDL Oxidizability in Rabbits, *Lipids 35*, 45–54.

Devadasan, K., Viswanathan Nair, P.G., and Anthony, P.D. (1985) Effect of Oxidation of Dietary Fish Lipids on the Quality of Proteins in Diet, *Fish. Technol. 22*, 70–73.

El-Zeany, B.A., Pokorný, J., Šmidrkalov, E., and Davidek, J. (1975) Reaktionen oxydierter Lipide mit Eiweisstoffen, 9. Bildung von Proteinoligomeren, *Nahrung 19*, 327–334.

Esterbauer, H. (1991) Effect of Antioxidants on Oxidative Modifications of LDL, *Ann. Med. 23*, 573–581.

Esterbauer, H., Jurgens, G., Quehenberger, O., and Koller, E. (1987) Autoxidation of Human LDL: Loss of Polyunsaturated Fatty Acids and Vitamin E and Generation of Aldehydes, *J. Lipid Res. 28*, 495–509.

Fujimoto, K., and Kaneda, T. (1973) Nitrogen Content of the Browning Substances, *Nippon Suisan Gakkaishi 39*, 179–183.

Gelders, G.G., Vandrestukken, T.C., Goesaert, H., and Delcour, J.A. (2004) Amylose-Lipid Complexation: A New Fractionation Method, *Carbohydr. Polym. 56*, 447–458.

Godet, M.C., Tran, V., Colonna, P., and Buleon, P. (1995) Inclusion/Exclusion of Fatty Acids in Amylose Complexes as a Function of the Fatty Acid Chain Length, *Int. J. Biol. Macromol. 17*, 405–408.

Goffman, F.D., and Bergman, C. (2003) Hydrolytic Degradation of Triacylglycerols and Changes in Fatty Acid Composition in Rice Bran During Storage, *Cereal Chem. 80*, 459–463.

Goodridge, C.F., Beaudry, R.M., Pestka, J.J., and Smith, D.M. (2003) ELISA for Monitoring Lipid Oxidation in Chicken Myofibrils Through Quantification of Hexanal-Protein Adducts, *J. Agric. Food Chem. 51*, 7533–7539.

Graveland, A. (1970) Enzymatic Oxidation of Linoleic Acid and Glycerol 1-Monolinoleate in Doughs and Flour-Water Suspension, *J. Am. Oil Chem. Soc. 47*, 352–361.

Guo, M.R., Hendricks, G.M., and Kinstedt, P.S. (1998) Component Distribution and Interaction in Powdered Infant Formula, *Int. Dairy J. 8*, 333–339.

Guo, S.T., Ono, T., and Mikami, M. (1997) Interaction Between Protein and Lipid in Soybean Milk at Elevated Temperature, *J. Agric. Food Chem. 45*, 4601–4605.

Hidalgo, F.J., and Zamora, R. (2002) Methyl Linoleate Oxidation in the Presence of Bovine Serum Albumin, *J. Agric. Food Chem. 50*, 5463–5467.

Huang, J.J., and White, P.J. (1993) Waxy Corn Starch: Monoglyceride Interaction in Model System, *Cereal Chem. 70*, 42–47.

Hutapea, E.B., Parkányiová, L., Parkányiová, J., Miyahara, M., Sakurai, H., and Pokorný, J. (2004) Browning Reactions Between Oxidized Vegetable Oils and Amino Acids, *Czech J. Food Sci. 22*, 99–107.

Janitz, W. (1987) Interaktionen der Fette und Eiweisse des Fleisches, I. Dynamik der Fettoxydation und quantitative Bestimmung der Eiweiss-Fett Komplexe, *Z. Ernaehrwiss. 26*, 62–72.

Kikugawa, K., Ido, Y., and Mikami, A. (1984) Fluorescent Products Derived from the Reaction of Primary Amines, Malonaldehyde, and Monofunctional Aldehydes, *J. Am. Oil Chem. Soc. 61*, 1574–1581.

Kolakowska, A., and Kwiatkowska-Zienkowicz, L. (1991) Wplyw sposobu izolowania lipidσ[3]w rybnych na ich utlenienie, *Zesz. Nauk. Akad. Roln. Szczecin 143*, 102–110.

Lam, H.S., Proctor, A., Nyalala, J., Morris, M.D., and Smith, W.G. (2004) Quantitative Determination of Low Density Lipoprotein Oxidation by FTIR and Chemometric Analysis, *Lipids 39*, 687–692.

Lechtinen, P. (2003) *Reactivity of Lipids During Cereal Processing*, Ph.D. thesis, University of Helsinki, Helsinki, Finland.

Lechtinen, P., Kiiliaien, K., Lechtomaki, I., and Laakso, S. (2003) Effects of Heat Treatment on Lipid Stability in Processed Oats, *J. Cereal Sci. 37*, 215–221.

Lederer, M.O. (1996) Reactivity of Lysine Moieties Toward γ-Hydroxy-α,β-Unsaturated Epoxides: A Model Study on Protein-Lipid Oxidation Product Interaction, *J. Agric. Food Chem. 44*, 2531–2537.

Lederer, O., Schuler, A., and Ohmenhäuser, M. (1999) Reactivity of Lysine Moieties Toward an Epoxyhydroxylinoleic Acid Derivative: Aminolysis Versus Hydrolysis, *J. Agric. Food Chem. 47*, 4611–4620.

Liang, J.H. (1999) Fluorescence Due to Interactions of Oxidizing Soybean Oil and Soy Proteins, *Food Chem. 66*, 103–108.

Ludwig, I., Ludwig, E., and Pingel, B. (1989) Eine Mikromethode zur Bestimmung der Fettbindekapazität, *Nahrung 33*, 99–101.

Lynch, M.P., Faustman, C., Silbart, L.K., Rood, D., and Furr, H.C. (2001) Detection of Lipid-Derived Aldehydes and Aldehyde:Protein Adducts *In Vitro* and in Beef, *J. Food Sci. 66*, 1093–1099.

Michajlik, A., and Bartnikowska, E. (1999) *Lipidy i Lipoproteiny Osocza*, PZWL, Warszawa.

Nielssen, H. (1981) Reaction of Individual Phospholipids with Proteins, *Lipids 16*, 215–222.

Pokorný, J. (1963) Üeber die Bildung von Komplexverbindungen bei der Reaktion oxydiert-er Lipide mit Eiweisstoffen, *Fette Seifen Anstrichm. 65*, 278–284.

Pokorný, J., and Kolakowska, A. (2002) Lipid-Protein and Lipid-Saccharide Interactions, in *Chemical and Functional Properties of Food Lipids* (Sikorski, Z.E., and Kolakowska, A., eds.), pp. 345–362, CRC Press, Boca Raton, FL.

Pokorný, J., and Sakurai, H. (2002) Role of Oxidized Lipids in Nonenzymic Browning Reactions, in *The Maillard Reaction in Food Chemistry and Medical Science*, pp. 373–374, Elsevier, Amsterdam.

Pokorný, J., El-Zeany, B.A., and Janiček, G. (1974) Browning Reactions of Oxidized Fish Lipids with Proteins, *Proceedings of the IVth International Congress on Food Science Technology 1*, 217–223.

Pokorný, J., Janiček, G., and Davidek, J. (1975) Determination of Interaction Products of Proteins with Lipids, *Zesz. Probl. Postepow. Nauk. Roln. 167*, 135–170.

Pokorný, J., Davidek, J., Novotná, E., Valentová, H., Janitz, W., Kminek, M. (1986) Effect of Lipid Hydroperoxides on Animal Proteins Under Conditions of Storage and Food Preparation, *Nahrung 30*, 416–418.

Pokorný, J., Davidek, J., Chi, T.H., Valentová, H., Matejiček, J., Dlasková, Z. (1988) Mechanism of Lipoprotein Formation from Interactions of Oxidized Ethyl Linoleate with Egg Albumin, *Nahrung 32*, 343–350.

Pokorný, J., Davidek, J., Chocholatá, V., Pánek, J., Bulantová, H., Janitz, W., Valentová, H., and Vierecklová, M. (1990) Interactions of Oxidized Ethyl Linoleate with Collagen, *Nahrung 34*, 159–169.

Pokorný, J., Reblová, Z., Kouřimská, L., Pudil, F., and Kwiatkowska, A. (1992) Effect of Interactions with Oxidized Lipids on Structure Change and Properties of Food Proteins, *4th Symposium on Food Proteins*, pp. 232–235, VCH, Weinheim.

Sakai, T., Nasu, A., and Habiro, A. (1998) Changes in Protein Carbonyls and Malonaldehyde Contents in Stored Fish Meat, *Fish. Sci. 64*, 495–496.

Sarma, J., Reddy, V.S., and Srikar, L.N. (2000) Effect of Frozen Storage on Lipids and Functional Properties of Proteins in Dressed Indian Oil Sardine, *Food Res. Int. 33*, 815–820.

Schroeder, L.F., and Hoseney, R.C. (1978) Mixograph Studies, II. Effect of Activated Double Bond Compounds on Dough-Mixing Properties, *Cereal Chem. 55*, 348.

Seidel, C., Kulicke, W., Hess, C., Hartman, B., Lechner, D., and Lazik, W. (2001) Influence of the Cross-Linking Agent on the Gel Structure of Starch Derivatives, *Starch 53*, No. 7, Abstracts.

Shamekh, S., Mustranta, A., Poutanen, K., and Forssell, P. (1998) Enzymatic Hydrolysis of Barley Starch Lipids, *Cereal Chem. 75*, 624–628.

Shenouda, S.Y.K., and Pigott, G.M. (1977) Fish Myofibril Protein and Lipid Interaction in Aqueous Media as Detected by Isotope Labeling, Sucrose Gradient Centrifugation, Polyacrylamide Electrophoresis and Electron Paramagnetic Resonance, Report Work, *Nat. Sea Grant Prog.*, 1–30.

Smith, S.A., Pestka, J.J., Gray, J.I., and Smith, D.M. (1999) Production and Specificity of Polyclonal Antibodies to Hexanal-Lysine Adducts, *J. Agric. Food Chem. 47*, 1389–1395.

St. Angelo, A.J., and Ory, R.L. (1975) Effects of Lipoperoxides on Proteins in Raw and Processed Peanuts, *J. Agric. Food Chem. 23*, 141–150.

Stapelfeldt, H., and Skibsted, L.H. (1994) Modification of β-Lactoglobulin by Aliphatic Aldehydes in Aqueous Solution, *J. Dairy Res. 61*, 209–219.

Stauffer, C.E. (2001) *Emulgatory*, WNT, Warszawa.

Suhur, S., and Howell, N.K. (2002) Effect of Lipid Oxidation and Frozen Storage on Muscle Proteins of Atlantic Mackerel, *J. Sci. Food Agric. 82*, 579–586.

Suomela, J.-P., Ahotupa, M., Sjävall, O., Kurvinen, J.-P., and Kallio, P. (2004a) Diet and Lipoprotein Oxidation: Analysis of Oxidized Triacylglycerols in Pig Lipoproteins, *Lipids 39*, 639–647.

Suomela, J.-P., Ahotupa, M., Sjävall, O., Kurvinen, J.-P., and Kallio, H. (2004b) New Approach to the Analysis of Oxidized Triacylglycerols in Lipoproteins, *Lipids 39*, 507–512.

Tomasik, O., and Schilling, H.C. (1998) Complexes of Starch with Inorganic Guests, *Carbohydrate Chem. Biochem. 53*, 263–339.

Tomasik, P. (2000) Sacharydy, in *Chemia Zywności* (Sikorski, Z.E., ed.), WNT Waeszawa, OOM od.

Tsutsui, T., Li-Chan, E., and Nakai, S. (1986) A Simple Fluorometric Method for Fat-Binding Capacity as an Index of Hydrophobicity of Proteins, *J. Food Sci. 51*, 1268–1272.

Tynek, M., Hazuka, Z., Pawlowicz, R., and Dudek, M. (2001) Frying Medium During Deep-Frying of Food Rich in Proteins and Carbohydrates, *J. Food Lipids 8*, 251–261.

Viljanen, K., Kivikari, R., and Heinonen, M. (2004) Protein-Lipid Interactions During Liposome Oxidation with Added Anthocyanin and Other Phenolic Compounds, *J. Agric. Food Chem. 52*, 1104–1111.

Villwock, V., Eliasson, A.-C., Silverio, J., and BeMiller, J.N. (1999) Starch-Lipid Interaction in Common Waxy ae du and ae su2 Maize Starches Examined by Differential Scanning Calorimetry, *Cereal Chem. 76*, 292–298.

Wakao, A., and Pazos, M. (1984) Browning Reaction in Canned Sardines: The Role of Lipids and Amines, *Bol. Investig. Tecnol. Pesquado 2*, 67–71.

Wasserman, L.A., and LeMaste, M. (2000) Influence of Water on Potato Starch-Lipid Interactions. An Electron Spin Resonance Probe Study, *J. Sci. Food Agric. 80*, 1608–1616.

Xiong, Y.L., and Decker, E.A. (1995) Alterations of Muscle Protein Functionality by Oxidative and Antioxidative Processes, *J. Muscle Foods 6*, 139–160.

Zamora, R., and Hidalgo, F.J. (2001) Inhibition of Proteolysis in Oxidized Lipid-Damaged Proteins, *J. Agric. Food Chem. 49*, 6006–6011.

Zamora, R., and Hidalgo, F.J. (2003) Comparative Methyl Linoleate and Methyl Linolenate Oxidation in the Presence of Bovine Serum Albumin at Several Lipid/Protein Ratios, *J. Agric. Food Chem. 51*, 4661–4667.

Zielinski, T.L., Smith, S.A., Pestka, J.J., Gray, J.I., and Smith, D.M. (2001) ELISA to Quantify Hexanal-Protein Adducts in a Meat Model System, *J. Agric. Food Chem. 49*, 3017–3023.

Zotos, A., Hole, M., and Smith, G. (1995) The Effect of Frozen Storage of Mackerel on Its Quality When Hot Smoked, *J. Sci. Food Agric. 67*, 43–48.

Zotos, A., Petridis, D., Siskos, I., and Gougoulias, C. (2001) Production and Quality Assessment of a Smoked Tuna Product, *J. Food Sci. 66*, 1184–1190.

Index

L

M

P